Efficient Desalination by Reverse Osmosis

Efficient Desalination by Reverse Osmosis

A guide to RO practice

Stewart Burn and Stephen Gray

IWA PUBLISHING

Published by **IWA Publishing**
Alliance House
12 Caxton Street
London SW1H 0QS, UK
Telephone: +44 (0)20 7654 5500
Fax: +44 (0)20 7654 5555
Email: publications@iwap.co.uk
Web: www.iwapublishing.com

First published 2016

British Library Cataloguing in Publication Data
A CIP catalogue record for this book is available from the British Library

ISBN: 9781780405056 (Paperback)
ISBN: 9781780405049 (eBook)

Contents

Acknowledgements

Victoria University is acknowledge for supporting and facilities for Prof Stephen Gray to contribute to this book.

CSIRO Manufacturing can be acknowledged for their support in providing facilities for Prof Stewart Burn to undertake the development of this book.

Chapter 1

Introduction: Desalination by reverse osmosis

Stewart Burn[1] *and Stephen Gray*[2]

[1]*Visiting Scientist, CSIRO Manufacturing. Adjunct Professor, Institute of Sustainability and Innovation, Victoria University. Adjunct Professor, Department of Civil, Environmental and Chemical Engineering, RMIT University.*

[2]*Director, Institute of Sustainability and Innovation, Victoria University*

Climate change in many nations is predicted to reduce the long-term yield of dams and groundwater systems, which when linked to increased urbanisation and population growth will see increasing water shortages in many cities. Higher temperatures and reduced precipitation could also increase urban water demand because many cities use about 30–40% of residential water for irrigating domestic gardens and public parks. More water tends to be used when the weather is warm and dry as water for irrigation and evaporative cooling. In many cities, new sources of water will be required to meet the increasing demand for water, giving an opportunity to undertake new solutions. Whilst options such as stormwater capture and reuse (potable or non potable) are possible for some cities, for many the combination of these with desalination or desalination alone is a reliable and attractive option, as exemplified by the desalination plant built at Kwinana, Western Australia and shown in Figure 1.1.

Water desalination has a long history. Initial impetus was the need for potable water from the sea, or from brackish groundwater in the case of arid and remote communities. Early applications of desalination were small-scale plants deploying a range of technologies. However with the technological developments in Reverse Osmosis, most new plants use this technology because it has a proven history of use and low energy and capital costs compared with other available desalination technologies. This has led to the recent trend for larger seawater desalination plants in an effort to further reduce costs, and 1,000 MLD seawater desalination plants are projected by 2020.

Figure 1.1 Desalination plant at Kwinana, Western Australia. Image: Western Australian Water Corporation.

Reverse osmosis (Figure 1.2) uses a membrane to filter and remove salt ions, large molecules, bacteria, and disease-causing pathogens from sea water by applying pressure to the water on the feed side of a semi-permeable membrane. The salt is retained on the concentrated side of the membrane and pure water passes to the other.

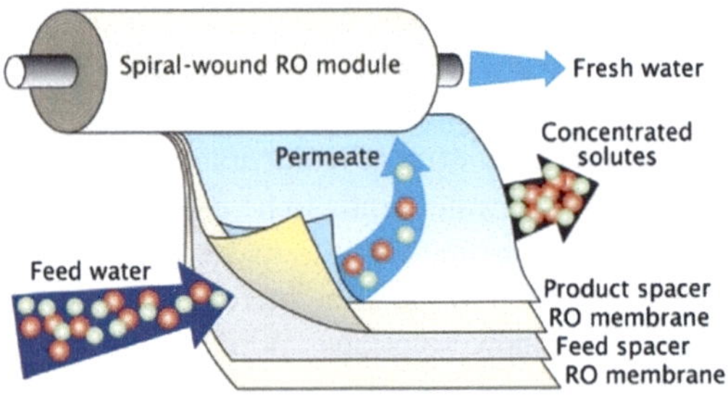

Figure 1.2 Schematic of the reverse osmosis (RO) process (Adapted from Burn, 2011).

Reverse osmosis has a number of shortcomings. Although the membrane is impervious to salt, it can let through small neutrally charged compounds such as N-nitrosodimethylamine (NDMA) and boron, requiring further or enhanced treatment before the desired water quality is met. Reverse osmosis removes all the naturally occurring salts to give un-buffered water that is deficient in calcium and other essential minerals, making it corrosive to distribution systems and inappropriate to drink. Minerals are, therefore, added back into the water to stabilise its corrosive nature and make it palatable for potable use. Brine streams are inevitable part of desalination, and management of concentrated brine requires careful attention, particularly in environmentally sensitive regions.

It is expected that the existing trend to use reverse osmosis for urban and industrial water desalination will continue and research is examining ways to make the process more efficient and reduce the amount of energy needed. For high salinity feed waters such as seawater, energy costs are a significant proportion of the operating cost and improvements in consumption stands to provide significant financial benefit. A range of emerging technologies increase efficiency by either pre-treating water, reducing membrane fouling, improving the throughput of water and rejection of pollutants, reducing the pressure at which the systems operate or recovering energy from the brine stream.

Pre-treatment is essential for all Reverse Osmosis plants and is needed to prevent fouling of reverse osmosis membranes. Inorganic salts, colloidal and particulate matter, organic compounds and microorganisms present in the feed water reduce membrane efficiency and lifespan. The main pre-treatment used is coagulation. However, coagulation only removes some pollutants and can produce small flocculants that fouling downstream membranes. New coagulants formulated for a number of water sources aim to greatly improve flocculent size, capture more pollutants, reduce membrane fouling, and can be easily washed from membranes. Technologies are also being developed to allow membrane surfaces to be treated with natural poly saccharides that have excellent anti-fouling properties.

Several emerging technologies have the potential to improve the efficiency of reverse osmosis membrane. For example, improved polymer membranes based on controlling the pore shape as shown in Figure 1.3 and nancomposite materials such as those based on carbon nanotechnology that could produce membranes composed of forests of microscopic tubes. However, it may be decades before some of these are mature enough for commercial application.

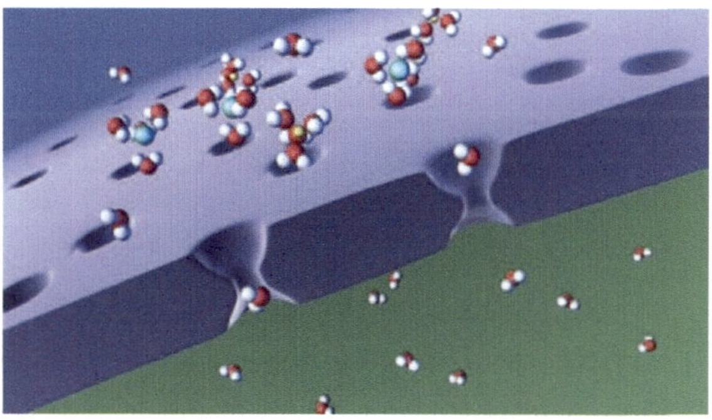

Figure 1.3 Schematic of a new polymer membrane which uses microscopically small hour- glass pores to allow water molecules to pass through the membrane while larger salt ions and other molecules are unable to pass through The polymer mimics the shape of micropores found in nature (*Source*: CSIRO).

This book recognises that desalination by reverse osmosis has progressed significantly over the last decades, and provides an up to date review of the state of the art for the reverse osmosis process, issues that arise from desalination operations, environmental issues and ideas for research that will bring further improvements in this technology.

REFERENCES

Burn S. (2011). Future urban water supplies. In: Water Science and Solutions for Australia, I. Prosser (ed.), CSIRO Publishing, Victoria, Australia.

Chapter 2

The process of reverse osmosis

Sergio G. Salinas-Rodriguez[1], Jan C. Schippers[1] and Maria D. Kennedy[1,2]*

[1]UNESCO-IHE Institute for Water Education, Environmental Engineering and Water Technology Department, Westvest 7, 2611 AX, Delft, The Netherlands

[2]Delft University of Technology, Faculty of Civil Engineering, Delft, The Netherlands

**Corresponding author: s.salinas@unesco-ihe.org*

2.1 INTRODUCTION

Reverse osmosis (RO) systems are capable of separating dissolved ions from a feed stream (Figure 2.1). In RO systems, feed water is split into two streams: one has no (low) salinity and the other one has high salinity. The low salinity stream is known as 'permeate or product water' while the high salinity stream is known as 'concentrate, brine, or reject'.

The quantity of water (Q_w) flowing through a membrane is proportional to the differential pressure feed-permeate (ΔP), membrane surface area (A) and permeability of the membrane (K_w). This relationship is expressed with the following equation:

$$Q_w = \frac{dV}{dt} = (\Delta P - \Delta \pi) \cdot K_w \cdot A \tag{2.1}$$

where:

Q_w = permeate flow (m^3/h)

V = total filtered volume water (permeate) (L or m^3)

t = time (h, min, s)

ΔP = differential pressure (pressure feed – pressure permeate) (bar)

$\Delta\pi$ = osmotic pressure difference (bar)
(osmotic pressure feed – osmotic pressure permeate)
K_w = permeability constant for water ($m^3/m^2/s/bar$)
A = surface area of the membrane(s) (m^2)
$(\Delta P - \Delta\pi)$ = net driving pressure (NDP) (bar)

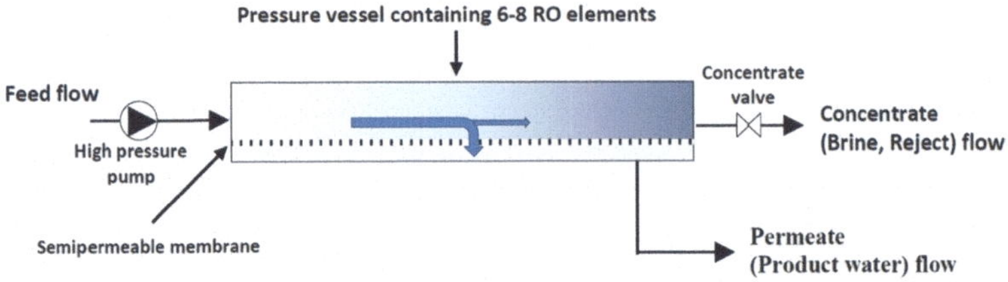

Figure 2.1 Schematic of a reverse osmosis system (Adapted from DOW Manual).

In membrane technology, flux is defined as the ratio of the permeate flow and surface area of the membrane. It is expressed as:

$$J = \frac{Q_w}{A} = \frac{1}{A} \cdot \frac{dV}{dt} \tag{2.2}$$

Flux (J) is the permeate flow through a membrane surface area (Q_w/A) ($m^3/m^2 \cdot h$ or $L/m^2 \cdot h$). Then,

$$J = \frac{1}{A} \cdot \frac{dV}{dt} = (\Delta P - \Delta\pi) \cdot K_w \tag{2.3}$$

EXAMPLE 2.1: Flux and permeate flow of spiral wound membrane elements

Assuming: Pressure is 10 bar; permeability of membrane is: *a)* 1 $L/m^2/h/bar$ and *b)* 5 $L/m^2/h/bar$; membrane surface area is 35 m^2; and no osmotic pressure.

Question: Calculate the flux and permeate flow under the given conditions.

Answer:

a) 10 bar × 1 $L/m^2/h/bar$ = 10 $L/m^2 \cdot h$ then 10 $L/m^2/h$ × 35 m^2 = 350 L/h
b) 10 bar × 5 $L/m^2/h/bar$ = 50 $L/m^2 \cdot h$ then 50 $L/m^2/h$ × 35 m^2 = 1,750 L/h

2.2 OSMOTIC PRESSURE

For Reverse osmosis to work, a pressure greater than the osmotic pressure needs to be applied to the seawater side of the membrane. Osmotic pressure is the pressure which needs to be applied to a solution to prevent the inward flow of water across a semipermeable membrane (Voet *et al.* 2001). Osmotic pressure is also defined as the minimum pressure needed to cancel out osmosis. Figure 2.2 illustrate the process of reverse osmosis in which the applied pressure needs to overcome the osmotic pressure head to force water to pass through the semipermeable membrane.

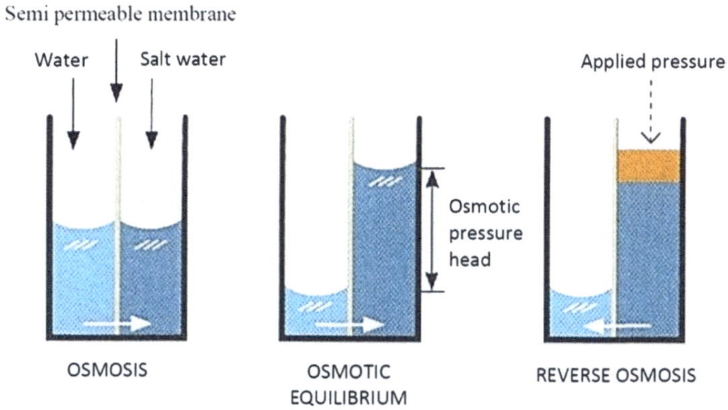

Figure 2.2 Illustration of reverse osmosis principle.

The osmotic pressure reduces the effect of hydraulic pressure, as a consequence, the effective pressure or **net driving pressure** is equal to the hydraulic pressure minus the osmotic pressure.

$$NDP = \Delta P - \Delta \pi \quad (2.4)$$

where:

ΔP = differential hydraulic pressure (pressure feed – pressure permeate) (bar)
$\Delta \pi$ = difference osmotic pressure (osmotic pressure feed – osmotic pressure permeate) (bar)

In membrane filtration, the osmotic pressure hinders the water flow as illustrated in Figure 2.3.

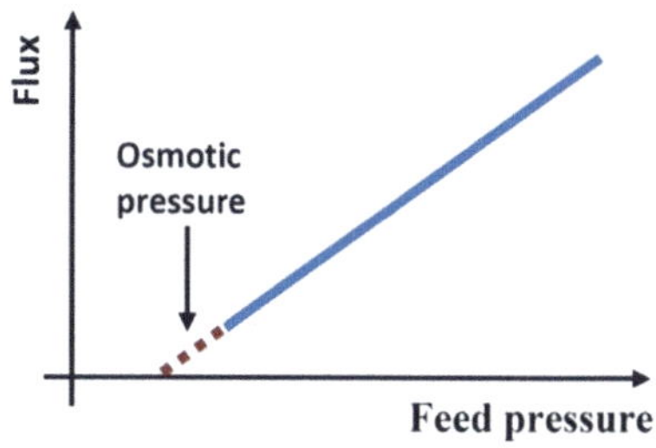

Figure 2.3 Schematic of osmotic pressure hindering flux.

In reverse osmosis systems the recovery at which the RO system operates is governed by the osmotic pressure of the feed water.

2.2.1 Calculation of osmotic pressure

In practice, feed water can be classified according to the amount of salts it contains as follows: brackish water with total dissolved solids-salts (TDS) between 1,000–5,000 mg/L, highly brackish 5,000–15,000 mg/L, saline 15,000–30,000 mg/L and seawater with TDS content higher than 30,000 mg/L. In Table 2.1 the osmotic pressure is presented for low salinity brackish water and for seawater.

Table 2.1 Classification of brackish water and seawater.

Water Type	Salinity, mg/L	Osmotic Pressure, bar
Brackish water	1,000	~0.7
Sea water	35,000	~25

In practice, salinity about 1,300 to 1,400 mg/L equals to 1 bar of osmotic pressure. From this figure, the osmotic pressure (in bar) can be estimated with a rule of thumb, as follows:

$$\pi \approx 0.7 \times 10^{-3} \times C \tag{2.5}$$

Where the salt concentration (C) is in milligrams per liter (mg/L). Hydranautics suggests that an approximation for osmotic pressure may be made by assuming that 1,000 mg/L of Total Dissolved Solids (TDS) equals about 11 psi (0.76 bar) of osmotic pressure (Hydranautics, 2001).

On the other hand, the *feed* osmotic pressure can be calculated more accurately by using the equation provided by ASTM (2000) based on the van't Hoff equation:

$$\pi_f = 8.308 \times \varphi \times (T_f + 273.15) \times \sum m_i \tag{2.6}$$

where:

π_f = osmotic pressure (kPa)
φ = Osmotic coefficient.
= Osmotic coefficients estimates for brackish and seawater, 0.93 and 0.90, respectively.
T_f = temperature of feed stream (°C)
Σm_i = summation of molalities of all ionic and non-ionic constituents in the water

Membrane manufacturers make use of similar formulas, for instance DOW in its Filmtec technical manual (DOW, 2005) recommends the following equation for the feed water osmotic pressure:

$$\pi_f = 1.12 \times (T + 273) \times \sum m_i \tag{2.7}$$

where:

π_f = osmotic pressure (psi). 1 psi = 6.8948 kPa
T = temperature of water (°C)
Σm_i = summation of molalities of all ionic and non-ionic constituents in the water

2.3 WATER FLOW

Theory suggests that the chemical nature of the membrane is such that it will absorb and pass water preferentially to dissolved salts at the solid/liquid interface. This may occur by weak chemical bonding of the water to the membrane surface or by dissolution of the water within the membrane structure (Solution Diffusion Theory). The chemical and physical nature of the membrane (e.g., surface charge and pore size) determines its ability to allow the preferential transport of water over salt ions.

In constant flux systems, the flux in a RO system can be expressed as:

$$J = (\text{Net Driving Pressure}) \cdot K_w = (\Delta P - \Delta\pi) \cdot K_w \tag{2.8}$$

with:

$$\Delta P = P_f - P_p \tag{2.9}$$

and

$$\Delta\pi = \pi_f - \pi_p \tag{2.10}$$

where:

J_w = water flux (L/m²/h)
ΔP = hydraulic differential pressure (pressure feed – pressure permeate) (bar)
$\Delta\pi$ = osmotic pressure difference between the feed water and product water (permeate) (bar)
K_w = permeability constant for water ($m^3/m^2/s/bar$) ($L/m^2/h/bar$)

EXAMPLE 2.2: Net driving pressure

In a seawater RO plant the feed pressure equals 25 bar. Assuming that the salinity of sea water is 35,000 mg/L and the RO recovery is 50% under normal conditions.

Question: What is the production at 25 bar?

Answer:
Feed osmotic pressure = 35 × 0.7 = 24.5 bar
NDP = 25 – 24.5 = 0.5 bar ≈ 0.
Concentrate osmotic pressure is much higher than feed osmotic pressure. Therefore, no production is expected.

EXAMPLE 2.3: Net driving pressure, flux and permeate production

A seawater RO element operates at a pressure of 60 bar.

Questions: What is the flux and the permeate production?

Assume: Salinity = 35,000 mg/L; membrane permeability $K_w = 0.8$ L/m²/h/bar; membrane surface area per element = 35 m² and Recovery very low (less than 5%).

Answers:
Flux: (60–25) bar × 0.8 L/m²/h/bar = 28 L/m²/h
Production: 28 L/m²/h × 35 m² = 980 L/h

2.3.1 Salt rejection

The salt rejection (*SR*) is by definition the ratio of the salt concentration in the feed water minus the salt concentration in the product water over the salt concentration in the feed water and it is expressed as percentage, as follows:

$$SR = \frac{C_f - C_p}{C_f} \times 100\% \tag{2.11}$$

$$SR = \left(1 - \frac{C_p}{C_f}\right) \times 100\% \tag{2.12}$$

Where C_f is the salt concentration in the feed water and C_p is the salt concentration in the product water.

2.3.2 Salt passage

The salt passage (*SP*) is by definition the ratio of the salt concentration in the product water over the salt concentration in the feed water expressed as percentage, as follows:

$$SP = \frac{C_p}{C_f} \times 100\% \tag{2.13}$$

Salt passage is the opposite of salt rejection.

$$SP = 100\% - SR \tag{2.14}$$

EXAMPLE 2.4: Salt rejection and salt passage

A sea water RO plant is processing water with a salinity of 45,000 mg/L. The product water has a salinity of 500 mg/L.

Question: Calculate the Salt Rejection of the plant and calculate the Salt Passage of the plant.

Answers:
$SP = (C_p/C_f) \times 100\% = (500/45{,}000) \times 100\% = 1.1\%$
$SR = 100\% - SP = 98.9\%$

2.4 SALT FLOW

Water can pass a reverse osmosis membrane; salts as well, however, at a much lower rate. The transport of salts through RO membranes is due to diffusion. Diffusion is a result of the motion of ions in water and the tendency of salts to move from high concentration to low concentration. Diffusion is a slow process, but cannot be neglected.

The salinity of the product water (C_p) depends on the relative rates of water and salt transport through a membrane. This relationship is expressed by the following equation:

$$C_p = \frac{Q_s}{Q_w} \tag{2.15}$$

Where Q_s is defined by the following equation:

$$Q_s = \Delta C \cdot K_s \cdot A \tag{2.16}$$

With

$$\Delta C = C_f - C_p \tag{2.17}$$

where:

Q_s = Flow rate of salt through membrane (kg/s)
ΔC = Salt concentration differential across membrane (kg/m^3) = $C_f - C_p$
K_s = Membrane permeability coefficient for salt (m^3/m$^2 \cdot$ s)
A = Membrane area (m^2)

Replacing the formula of Q_s in the formula of C_p, we have:

$$C_p = \frac{\Delta C \cdot K_s \cdot A}{(\Delta P - \Delta \pi) \cdot K_w \cdot A} \tag{2.18}$$

Replacing terms,

$$C_p = \frac{(C_f - C_p) \cdot K_s}{(P_f - P_p - \Delta \pi) \cdot K_w} \tag{2.19}$$

Dividing the whole equation by C_f, and rearranging the equation, we have the salt passage:

$$\frac{C_p}{C_f} = \frac{\left(C_f - \dfrac{C_p}{C_f}\right) \cdot K_s}{(P_f - P_p - \Delta\pi) \cdot K_w} \tag{2.20}$$

Then,

$$\text{Salt passage } (SP) = \frac{C_p}{C_f} \times 100\% \tag{2.21}$$

Looking at the right side of the equation, since C_p is small compared to C_f, the ratio C_p/C_f is much smaller than 1, therefore, the salt transport (Q_s) is constant at a certain C_p and is independent of the pressure. As a consequence, the salt passage ($SP = C_p/C_f$) is lower at high pressure (P_f) and vice versa. This is because the same quantity of salt (Q_s) will be diluted by a larger volume of (product) water and vice versa.

The salt flux (J_s) is defined as the salt transport per membrane area per hour (mg/m²/h) (mg per m² per hour). The salt flux is proportional with the concentration difference between membrane surface at: feed water side (C_f) and product water side (C_p), and the permeability coefficient of membrane for salts (ions) K_s.

$$J_s = (C_f - C_p) \times K_s \tag{2.22}$$

Remark: The larger the pores, the larger the permeability for salt and water. In general C_p is low in comparison with C_f. So, C_p can be neglected in this formula.

$$J_s \approx C_f \times K_s \tag{2.23}$$

where:

J_s = salt flux (mg/m²/h)
C_f = concentration at feed side membrane (mg/L)
C_p = concentration at product side membrane (mg/L)
K_s = Permeability for salt [(mg/m²/h) / (mg/L)]

EXAMPLE 2.5: Salt passage and salt concentration permeate

The flux in a RO membrane is 10 L/m²/h. Salt concentration in feed water is 41,000 mg/L and K_s value is 0.08 [(mg/m²/h)/(mg/L)] = 0.08 L/m²/h.

Questions: What is the salt concentration in the permeate stream? What is the salt passage?

Answer:
In 1 hour 10 L water is produced by 1 m² membrane surface water. In 1 hour the following amount of salts will pass 1 m² membrane surface area.

$$C_{\text{perm}} = \frac{J_s}{J_w} = \frac{(C_f - C_p) \times K_s}{J_w} \approx \frac{C_f \times K_s}{J_w}$$

$J_s = \text{mg/m}^2\text{/h} \approx C_f \times K_s = 41{,}000 \text{ mg/L} \times 0.08 \text{ L/m}^2\text{/h} = 3{,}280 \text{ mg/m}^2\text{/h}$
This amount of salt namely; 3,280 mg arrives in 10 L.
Resulting in (3,280 mg salt/10 L water) = 328 mg/L salt in the product water/permeate.
The formula applied is:

$$SP = \frac{C_p}{C_f} \times 100\% = \frac{328}{41{,}000} \times 100\% = 0.8\%$$

or

$$SR = 100\% - SP = 100\% - 0.8\% = 99.2\%$$

EXAMPLE 2.6: Salt Passage and salt concentration permeate

Considering the previous example, the water flux is increased from 10 L/m^2·h to 20 L/m^2·h.

Questions: What will be salt concentration in the permeate stream, salt passage and salt rejection?

Answer:
Salt transport is the same (since the concentration of the feed water does not change) and water transport (flux) is the double, so C_p will be two times lower. Consequently, *SP* will be two times lower.

$$C_p = \frac{J_s}{J_w} = \frac{C_f \times K_s}{J_w}$$

C_f and K_s are constant. J_w is 20/10 = 2 times higher.
As a consequence C_p will be 2 times lower, so (328 mg/L/2) = 164 mg/L
Salt passage will be: $SP = (164/41{,}000) \times 100\% = 0.4\%$ instead of 0.8%
Salt rejection will be: $SR = 100\% - 0.4\% = 99.6\%$ instead of 99.2%.

2.5 RECOVERY

Recovery (*R*) is also known as conversion. Recovery is by definition the part of feed water that is converted in product water and is expressed as percentage.

$$R = \frac{Q_p}{Q_f} \times 100\% \tag{2.24}$$

where:

Q_p = Product water flow rate (m³/h)
Q_f = Feed water flow rate (m³/h)

Recovery affects salt passage and product flow. As recovery increases, the salt concentration on the feed-concentrate side of the membrane increases, which increases the salt transport and the permeate salinity (C_p).

High salt concentration in the feed-concentrate solution increases the osmotic pressure, which consequently reduces the net driving pressure (*NDP*). As a result, the product water flow rate is reduced and the permeate salinity (C_p) increased.

A mass balance can be performed in a RO system to define the concentration factor as illustrated in

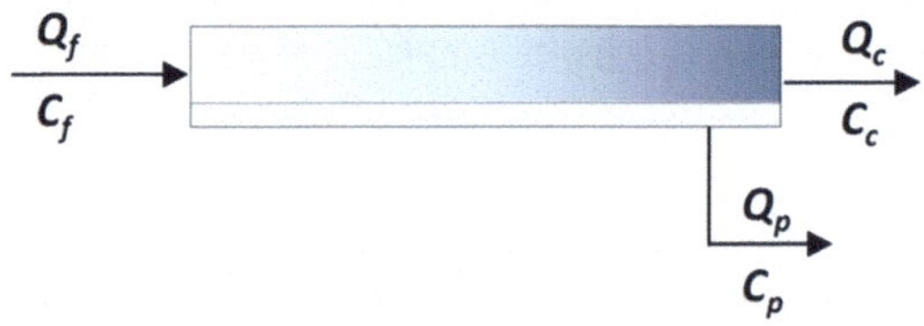

Figure 2.4 Mass balance in a RO system.

$$Q_f = Q_p + Q_c \tag{2.25}$$

Then multiplying by the concentration of salts in each stream, we have:

$$Q_f \cdot C_f = Q_c \cdot C_c + Q_p \cdot C_p \tag{2.26}$$

Re-arranging the equation for the feed concentration:

$$C_f = \frac{Q_p}{Q_f} \cdot C_p + \frac{Q_c}{Q_f} \cdot C_c \tag{2.27}$$

From the definition of recovery we can modify the previous equation.

$$\frac{Q_c}{Q_f} = \frac{Q_f - Q_p}{Q_f} = \frac{Q_f}{Q_f} - \frac{Q_p}{Q_f} = 1 - R \tag{2.28}$$

Substituting Eq. (2.28) into Eq. (2.27), we have:

$$C_f = R \cdot C_p + (1 - R) \cdot C_c \tag{2.29}$$

From the definition of recovery:

$$R = \frac{Q_p}{Q_f} = \frac{Q_f - Q_c}{Q_f} = 1 - \frac{Q_c}{Q_f} \tag{2.30}$$

The **concentration factor in a RO system** is by definition:

$$CF = \frac{C_c}{C_f} \tag{2.31}$$

From equation (2.21) we have:

$$C_p = C_f \cdot (1 - SR) \tag{2.32}$$

Replacing Eq. (2.32) in Eq. (2.29) we have

$$C_f = R \cdot C_f \cdot (1 - SR) + (1 - R) \cdot C_c \tag{2.33}$$

And

$$\frac{C_f}{C_f} = \frac{R \cdot C_f \cdot (1 - SR)}{C_f} + \frac{(1 - R) \cdot C_c}{C_f} \tag{2.34}$$

$$1 = R \cdot (1 - SR) + (1 - R) \cdot \frac{C_c}{C_f} \tag{2.35}$$

Then

$$(1 - R) \cdot \frac{C_c}{C_f} = 1 - R \cdot (1 - SR) \tag{2.36}$$

Then

$$C_c = \frac{C_f \cdot [1 - R \cdot (1 - SR)]}{1 - R} \tag{2.37}$$

Or

$$CF = \frac{[1 - R \cdot (1 - SR)]}{1 - R} \tag{2.38}$$

Since the salt passage, is usually low, we may assume that the total amount of salt entering the plant ($Q_{\text{feed}} \times C_{\text{feed}}$) will concentrate in the concentrate (brine). As a result:

$$CF = \frac{1}{1 - R} \tag{2.39}$$

And

$$C_c = \frac{C_f}{1 - R} \tag{2.40}$$

In the concentrate stream, concentrations of salts are increased, including sparingly soluble compounds. In seawater mainly: calcium carbonate while in brackish/fresh for example, calcium carbonate, calcium sulfate, strontium sulfate, barium sulfate and silica (SiO_2).

As soon as the solubility is exceeded, precipitation/scaling might occur, resulting in lower membrane permeability (K_w). Precipitation of calcium carbonate can be avoided by acid dosing and/or antiscalants. Super saturation to a certain extend is allowable by antiscalant dosing.

EXAMPLE 2.7: Conversion/Recovery/Concentration

A reverse osmosis plant is treating seawater with salinity of 40,000 mg/L. *Having:* feed flow of 500 m^3/h and permeate flow of 200 m^3/h.

Questions: What is the recovery of the plant? What is the concentrate flow? What is the salinity in the concentrate? What is the concentration factor?

Answers:
Recovery: $Q_p/Q_f = (200/500) \times 100\% = 40\%$
Concentrate flow: $Q_f - Q_p = 500 - 200 = 500$ m^3/h
Salinityconcentrate: $CF \times C_f = 1/(1-R) \times C_f = 1/(1-0.4) \times 40{,}000 = 66{,}667$ mg/L

Assuming a salt rejection of 100%, the concentration factor for various recoveries is detailed in Table 2.2:

Table 2.2 Concentration factor versus recovery.

Recovery, %	Concentration Factor
30	1.4
40	1.7
50	2.0
75	4.0
90	10.0

In practice, brackish water plants operate at 75%, some up to 90% recovery. In brackish water reverse osmosis systems the maximum recovery is mainly governed by scaling potential of feed water. Seawater reverse osmosis systems normally operate at 30–50%, and the maximum recovery is governed by the high osmotic pressure.

EXAMPLE 2.8: Feed flow

Total capacity of an RO plant: 1,000 m^3/h
Number of units: 3
Recovery/conversion: 45%

Question: What is the feed flow per unit?

Answer:
$R = Q_p/Q_f$ or $Q_f = Q_p/R$
$Q_f = (1{,}000/3)/0.45 = 740$ m^3/h per unit.
Total feed flow is $3 \times 740 = 2{,}220$ m^3/h

EXAMPLE 2.9: Recovery, brine concentration, osmotic pressure

A sea water RO plant is producing 200 m^3/h permeate.
Feed flow is 500 m^3/h; Feed water has a salinity of 47,000 mg/L

Questions: What is the recovery of the plant? What is salinity of the concentrate? What is the osmotic pressure in the feed water and in the concentrate? What should be the feed pressure at least? How many arrays are in the system?

Answers:
Recovery is usually 40%.
Concentration factor: $CF = 1/(1 - 0.4) = 1.667$
Salinity concentrate: $C_c = 47{,}000 \times \mathrm{CF} = 78{,}300$ mg/L
Feed osmotic pressure: $\pi_f = 47 \times 0.7 = 32.9$ bar
Concentrate osmotic pressure: $\pi_c = 78.3 \times 0.7 = 54.8$ bar
Pressure higher than 32.9 bar is required to exceed the osmotic pressure.
One stage.

2.6 CONCENTRATION POLARIZATION

As water flows through a membrane and salts are rejected by the membrane, the retained salts can accumulate at the membrane surface where their concentration will gradually increase. The concentration build-up at the membrane will generate a diffusive flow back to the bulk of the feed, but after a given period of time steady state conditions will be established. Steady-state conditions are reached when the convective salt flow to the membrane surface is balanced by the salt flux through the membrane plus the diffusive flow from the membrane to the bulk.

Under steady-state conditions, the concentration at the membrane surface (C_m) is constant. The cross flow along the membrane surface enhances back diffusion of salts to the bulk. This increase in salt concentration at the membrane surface is called concentration polarization.

Concentration polarization is the accumulation of salts (ions) at the membrane surface. As a result the concentration at the membrane surface (C_m) is higher than in the bulk (or feed water) (C_b).

This phenomenon results from: water flow through a membrane; salts (ions) are rejected; retained salts (ions) accumulate at the membrane surface.

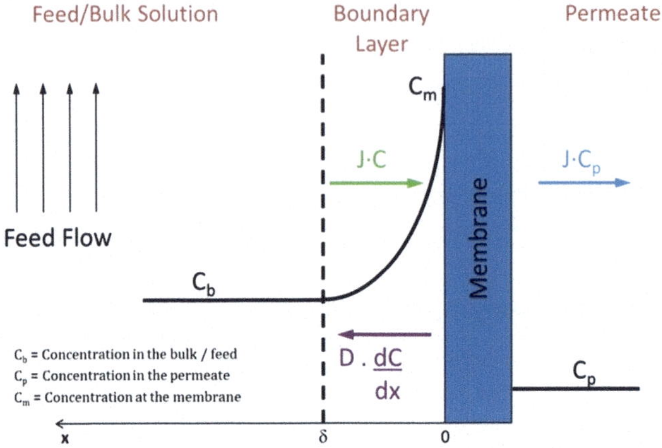

Figure 2.5. Concentration polarization – Concentration profile under steady state conditions (Mulder, 2003).

At steady-state the convective transport of salt to the membrane is equal to the sum of the permeate salt flow plus the diffusive back transport of salt:

$$J \cdot C_p = D \cdot \frac{dC}{dx} + J \cdot C \tag{2.41}$$

With the following boundary conditions:

- When $x = 0$ then $C = C_m$
- When $x = \delta$ then $C = C_b$

After integration:

$$ln\left(\frac{C_m - C_p}{C_b - C_p}\right) = \frac{J \cdot \delta}{D} \tag{2.42}$$

Or after re-arranging the previous formula we have

$$\frac{C_m - C_p}{C_b - C_p} = e^{\left(\frac{J \cdot \delta}{D}\right)} \tag{2.43}$$

The ratio of the diffusion coefficient (D) and the thickness of the boundary layer (δ) is called the mass transfer coefficient (k).

$$k = \frac{D}{\delta} \tag{2.44}$$

The intrinsic retention (R_{int}) of the membrane is the salt concentration at the membrane minus the concentration of salt in permeate over the salt concentration at the membrane.

$$R_{int} = 1 - \frac{C_p}{C_m} \tag{2.45}$$

Then, C_m/C_b becomes:

$$\frac{C_m}{C_b} = \frac{e^{\left(\frac{J}{k}\right)}}{R_{int} + (1 - R_{int}) \cdot e^{\left(\frac{J}{k}\right)}} \tag{2.46}$$

The ratio C_m/C_b is called the *concentration polarization factor*. This ratio increases (i.e., the concentration at the membrane (C_m) surfaces increases) with increasing flux (J), with increasing retention (R_{int}) and with decreasing mass transfer coefficient (k).

When the salt is completely retained by the membrane ($R_{int} = 1$ and $C_p = 0$), then

$$\frac{C_m}{C_b} = e^{\left(\frac{J}{k}\right)} \tag{2.47}$$

This is the basic equation for concentration polarization which demonstrates the two factors (the flux 'J' and the mass transfer coefficient 'k') and their origin (membrane part 'J', hydrodynamics 'k') responsible for concentration polarization.

The pure water flux (specific permeability) is determined by the membrane used and this parameter is not subject to further change once the membrane has been selected. On the other hand, the mass transfer coefficient depends strongly on the hydrodynamics of the system and can therefore be varied and optimized.

2.6.1 Control of concentration polarization

The following actions can be considered to control concentration polarization in RO membranes:

- Decreasing the flux (J). The higher the permeate flow (Q_p) in an element, the higher the transport of salts (ions) to the membrane surface. As a result accumulation will be higher and concentration polarization will be higher.
- Increasing the mass transfer coefficient (k). k is mainly determined by the diffusion coefficient and the flow velocity. Because the diffusivity of solutes cannot be increased (only by changing the temperature), k can only be increased by increasing the feed velocity along the membrane or by changing the module shape and dimensions (decreasing module length or increasing the hydraulic diameter). The higher the cross flow along the

membrane surface the higher the back diffusion. As a result accumulation will be reduced and concentration polarization will be lower.

2.6.2 Effects of concentration polarization

Concentration polarization has several negative effects on the performance of reverse osmosis systems. These effects are described below.

- A higher osmotic pressure at membrane surface, than in bulk feed water, resulting in a lower Net Driving Pressure. Consequently higher feed pressure is required to maintain the same flux (capacity).
- Increased salt transport (Q_s) due to higher salt concentration at the membrane surface. As a result a lower salt rejection (higher salt passage).
- Due to higher concentration of sparingly soluble salts (e.g., calcium carbonate, calcium sulfate) at the membrane surface, the possibility of precipitation (scaling) will increase.
- Reduced water transport through the membrane (Q_w).
- Higher rates of fouling due to suspended and colloidal matter, organic polymers, etc. accumulating at the membrane surface.

2.6.3 Concentration polarization factor

The concentration polarization factor (β) can be calculated with the following formula:

$$\beta = \frac{C_m}{C_b} = e^{\left(\frac{J}{k}\right)} \tag{2.48}$$

In practice, the formula is simplified to:

$$\beta = K_p \cdot e^{\left(\frac{Q_p}{Q_{f\,avg}}\right)} \tag{2.49}$$

where:

K_p is a proportionality constant depending on the module geometry.

This simplification is justified by the fact that: i) Q_p is proportional to J, and ii) $Q_{f\text{avg}}$ is the average feed flow and is proportional to k and k is almost proportional to the cross flow velocity (v).

Using the arithmetic average of feed and concentrate flow as average feed flow, the concentration polarization factor can be expressed as a function of the permeate recovery rate of a membrane element R_i.

$$\beta = K_p \cdot e^{\left(\frac{2 \cdot R_i}{2 - R_i}\right)} \tag{2.50}$$

where:

β = CPF = concentration polarization factor
K_p = constant depending on type (manufacturer) membrane (usually 0.99)
Q_p = permeate flow of an element (m^3/h)
Q_c = concentrate flow in an element (m^3/h)
R_i = recovery of a membrane element

The value of the concentration polarization factor of 1.2, which is the recommended Hydranautics limit, corresponds to 18% permeate recovery for a 40 inches long membrane element (Hydranautics, 2001). Formulas 6.9 and 6.10 and frequently applied by Koch and Hydranautics membranes manufacturers (Hydranautics, 2001).

EXAMPLE 2.10: Concentration polarization

In a reverse osmosis element the feed flow equals: 8 m^3/h; product flow equals 1 m^3/h; the average TDS (total dissolved salts) in this element equals 43,000 mg/L.

Question: What is the average concentration polarization factor in this element? What is the average salt concentration (*TDS*) at the membrane surface?
Assume: $K_p = 0.99$

Answers:
$\beta = 0.99 \times \exp^{(1/8)} = 1.12$
$1.12 \times 43{,}000 = 48{,}160$ mg/L

Dow Filmtec (Dow, 2005) applies for their elements the formula:

$$\beta = e^{(0.7 \cdot R)} \tag{2.51}$$

where:

β = Concentration Polarization Factor
R = Recovery

The recommended recovery (by Dow) per RO element varies with the quality of the feed water for example:

- Seawater 10–12%
- Filtered treated Domestic waste water 10–12%
- Pre-treated surface water 15–18%
- Softened well water 19–25%

2.7 MASS TRANSFER COEFFICIENT

The mass transfer coefficient (k), is related to the Sherwood number (Sh).

$$Sh = \frac{k \cdot d_h}{D} = a \cdot Re^b \cdot Sc^c \tag{2.52}$$

where:

a, b and c are constants.

With

$$Re = \frac{\rho \cdot v \cdot d_h}{\eta} \tag{2.53}$$

and

$$Sc = \frac{\eta}{\rho \cdot D} \tag{2.54}$$

where:

k = Mass transfer coefficient (m/s)
Re = Reynolds number, dimensionless
Sc = Schmidt number, dimensionless
d_h = Hydraulic diameter (m)
v = Flow velocity (m/s)
ρ = Water density (kg/m^3)
η = Dynamic viscosity (kg/m/s)
D = Diffusion coefficient (m^2/s)

Correlations for mass transfer coefficients depend on physical characteristics of the system and the flow conditions (e.g., laminar or turbulent) as shown in Table 2.3. The mass transfer coefficient (k) is mainly a function of the feed flow velocity (v), the diffusion coefficient of the solute (D), the density and the module shape and dimensions. Of these parameters, flow velocity and diffusion coefficient are the most important.

Table 2.3 Mass transfer coefficients in various flow regimes.

	Laminar	Turbulent
Tube	$Sh = \frac{k \cdot d_h}{D} = 1.62 \cdot \left(\frac{Re \cdot Sc \cdot d_h}{L}\right)^{0.33}$ (2.55)	$Sh = 0.04 \cdot Re^{0.75} \cdot Sc^{0.33}$ (2.56)
Channel	$Sh = 1.85 \cdot \left(\frac{Re \cdot Sc \cdot d_h}{L}\right)^{0.33}$ (2.57)	$Sh = 0.04 \cdot Re^{0.75} \cdot Sc^{0.33}$ (2.58)

In the channel of the feed spacer of a spiral-wound RO element, Schock and Miquel (1987) found that the concentration polarization mass transfer coefficient could be predicted by the following equation, when calculations for the velocity in the channel and the hydraulic diameter took the presence of the spacer into account:

$$k = 0.023 \cdot \frac{D}{d_h} \cdot Re^{0.875} \cdot Sc^{0.25} \tag{2.59}$$

2.8 WATER TEMPERATURE

Temperature has an effect on membrane permeability (K_w). The higher the water temperature the higher the permeability will be. The change in permeability is about 3% per °C. K_w is also linked with the viscosity of water and requires the use of a temperature correction factor (TCF).

$$TCF = 1.03^{(t-25)} \tag{2.60}$$

When dealing with the membrane permeability, the correction will be as follows:

$$K_{wt} = K_{25°C} \times 1.03^{(t-25)} \tag{2.61}$$

where:

t = temperature in °C.
K_{wt} = membrane permeability at temperature 't'
K_{w25} = membrane permeability at 25 °C.

As a result of the temperature effect of viscosity and therefore on membrane permeability, the required pressure to achieve or keep a certain flux (capacity) will be lower at higher temperatures.

EXAMPLE 2.11: Effect of temperature on required feed pressure

A brackish water element operates at 15°C (t_1) and a feed pressure of 20 bar (P_1). In summer season the water temperature increases to 30°C (P_2).

Question: What will happen with the required feed pressure in summer time (P_2), if we want to keep the product flow constant? Assume that osmotic pressure can be neglected.

Answer: At higher temperature the viscosity is lower.
As a consequence the permeability will be higher. So the required pressure will be . . . bar.
How much lower will be the pressure?
To calculate, we apply the Temperature Correction Factor (TFC).
$TCF = (1.03)^{(t_1 - t_2)}$
Combined with: $P_2 = P_1 \times TFC$
$P_2 = 20 \times (1.03)^{(15-30)}$ bar {$P_1 = 20$ bar, $t_1 = 15$°C and $t_2 = 30$°C}
$P_2 = 20 \times 0.63$ bar
$P_2 = 12.6$ bar, which is (20 – 12.6)/20 = 36.7% lower.
With rule of thumb of '3% per °C' we get: 3 × 15% = 45% lower.

EXAMPLE 2.12: Where does the TCF come from?

$P_2 = P_1 \times TCF$
$TCF = (1.03)^{(t_1 - t_2)}$
Since: $J = P \times K_w$
follows: $P_2 = J/K_{w2}$ and $P_1 = J/K_{w1}$ so, $P_2/P_1 = K_{w1}/K_{w2}$
and from: $K_{w1} = K_{w25oC} \times (1.03)^{(t_1 - 25)}$ and $K_{w2} = K_{w25oC} \times (1.03)^{(t_2 - 25)}$
follows: $K_{w1}/K_{w2} = (1.03)^{(t_1 - t_2)} = TCF$ or $P_2 = P_1 \times TCF$

The membrane permeability and the salt passage tend to increase with temperature. Salt permeability is connected with the diffusion of salt ions through the membrane. The **diffusion coefficient** is defined with the following formula:

$$D = \frac{k_B \cdot T}{6 \cdot \pi \cdot \eta \cdot r} \tag{2.62}$$

where:
k_B = Boltzmann constant
T = absolute temperature 273 + t °C
η = viscosity of water
r = radius of ion

In the diffusion coefficient equation, the viscosity is a dominant factor. Viscosity will decrease with temperature.

A frequently applied formula for normalizing the salt permeability is:

$$K_{st} = K_{s25^\circ C} \times 1.03^{(t-25)} \tag{2.63}$$

where:
K_{st} = Salt permeability at temperature 't'
$K_{s25^\circ C}$ = Salt permeability at 25°C
t = Temperature in °C

The consequence of the previous equation is that the higher the temperature the higher the salt passage will be. Similar to the effect of temperature on required pressure, the effect on salt passage can be derived as follows:

$$C_{p\,t_2} = \frac{C_{p\,t_1}}{TCF} = \frac{C_{p\,t_1}}{1.03^{(t_1 - t_2)}} \tag{2.64}$$

EXAMPLE 13: Normalization of permeate salinity

A RO spiral wound element is treating river water at 5°C at a flux of 25 L/m^2/h. The salinity (sodium chloride) of the permeate is 5 mg/L.

Question: What will be the salinity in the summer period when the temperature of the water increases to 25°C. The flux is kept at the same level.

Answer:
The salt permeability will increase with approximately 3% per °C.
Since the flux (J) and C_f are constant, C_p will increase with approximately (25−5) = 20°C times 3% per °C or about 60%.
So, C_p will increase approximately with 60%, from 5 mg/L to 8 mg/L.

Using the *TFC* we get the more accurate answer:
$C_{p\,t_2} = C_{p\,t_1}/TCF = C_{p\,t_1}/1.03^{(t_1-t_2)}$; $\{C_{p\,t_1} = 5$ mg/L and $t = 5$°C and $t = 25$°C$\}$
$C_{p\,t_2} = 5$ mg/L/$TCF = C_{p\,t_1}/1.03^{(5-25)}$
$C_{p\,t_2} = 5$ mg/L/0.55 = 9 mg/L

2.9 SUMMARY

In this chapter the basic principles governing reverse osmosis have been defined. Principles such as: osmotic pressure, water flow through a membrane, net driving pressure, salt passage, salt flow, recovery, concentration polarization, mass transfer coefficient, water temperature, and membrane permeability have been explained.

2.10 REFERENCES

ASTM 2000. Standard Practice for Standardizing Reverse Osmosis Performance Data. *Designation: D4516 – 00* West Conshohocken, ASTM.

DOW 2005. *FILMTEC™ Reverse Osmosis Membranes – Technical Manual* (Vol. Form No. 609-00071-0705). http://www.dowwaterandprocess.com/en/resources/reverse-osmosis-technical-manual (accessed 23 June 2015).

Hydranautics 2001. Terms and equations of reverse osmosis. http://www.membranes.com/docs/trc/termsequ.pdf (accessed 23 June 2015).

Mulder M. 2003. Basic Principles of Membrane Technology. Dordrecht / Boston / London, Kluwer Academic.

Schock G. and Miquel A. 1987. Mass transfer and pressure loss in spiral-wound modules. *Desalination*, **64**, 339.

Voet D., Voet J. G. and Pratt C. W. 2001. Fundamentals of Biochemistry (Rev. ed.). New York.

Chapter 3

Filtration technologies for pretreatment of seawater desalination based on reverse osmosis

Seung-Hyun Kim[1], Sangho Lee[2], Leonard D. Tijing[3] and Ho Kyong Shon[3]*

[1]Department of Civil Engineering, Kyungnam University, Wolyoung-dong, Changwon, 631–701, Republic of Korea

[2]School of Civil and Environmental Engineering, Kookmin University, Seongbuk-gu, Seoul, 136–702, Republic of Korea

[3]Centre for Technology in Water and Wastewater (CTWW), School of Civil and Environmental Engineering, University of Technology, Sydney (UTS), P.O. Box 123, Broadway, NSW 2007, Australia

**Corresponding author: shkim@kyungnam.ac.kr*

3.1 INTRODUCTION

Exponential population growth and rapid industrialization has led to an increased demand for fresh water sources. Desalination of seawater using reverse osmosis (RO) is considered to be one of the best options for the production of fresh water. The use of seawater RO (SWRO) has increased over the past 10 years due to the availability of new materials and reduction in costs (Valavala *et al.* 2011). However, one big hurdle in SWRO is the decline of productivity of RO membranes with time (Kumar *et al.* 2006). This productivity decline is mainly attributed to membrane scaling and fouling, which is considered a major challenge in the SWRO process (Shon *et al.* 2009a). Foulants such as inorganic/organic compounds, colloidal particles, biological matters, or their combination could form layers of deposit on the membrane surface that can reduce water productivity (Shon *et al.* 2009b). Seawater sources are considered to cause greater fouling compared to other sources such as surface water and groundwater (Greenlee *et al.* 2009), thus it is necessary for seawater to be pretreated prior to the RO process, to minimize the occurrence of membrane fouling and to maintain high feed water quality.

Pretreatment serves as the first line of defense to membrane fouling as it initially filters out or reduces to acceptable levels, the undesirable fouling materials before it reaches the RO membrane. Poor feed water quality can lead to higher chances of membrane fouling, reduced membrane lifetime, higher maintenance cost, short operation period, and even shutdown of operation. Depending on the pretreatment method, pretreatment can alter the physico-chemical and/or biological properties of the feed water resulting in lower fouling and improved SWRO performance. One important advantage of pretreatment is that it can simply be added to the existing SWRO installation, thus it is the first choice in SWRO plants to combat fouling. There are many different conventional and advanced pretreatment methods available for use in SWRO desalination (see Fig. 3.1 and Table 3.1). Among the conventional methods are coagulation/flocculation, pH adjustment, scale inhibition, lime softening, and media filtration. Selecting the appropriate pretreatment technique to use should carefully consider the following factors: feed water quality, membrane material, module type, recovery ratio, and final water quality. In recent years, increasing research has been focused on the use of membrane-based advanced pretreatment method mostly on microfiltration (MF) and ultrafiltration (UF) and a few on nanofiltration (NF) for SWRO (Ebrahim *et al.* 1997; Wolf *et al.* 2005; Ang *et al.* 2014).

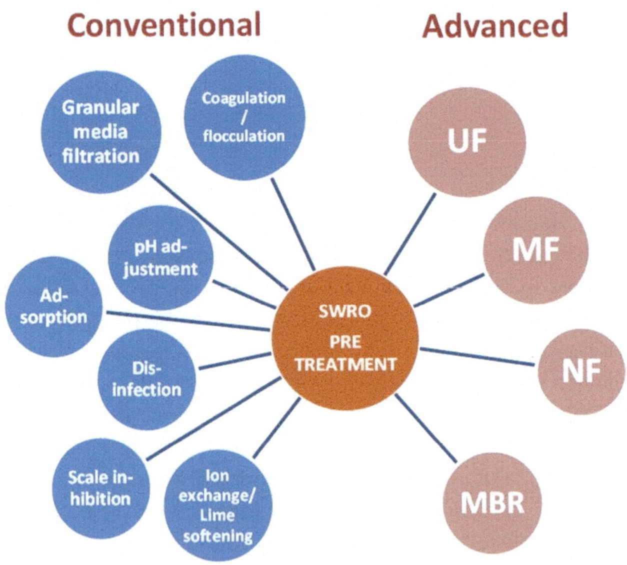

Fig. 3.1 Different pretreatment methods for seawater reverse osmosis desalination.

Table 3.1 Classification of pretreatment methods.

Classification	Pretreatment
Physical pretreatment	Adsorption, media filtration, cartridge filtration, MF, UF, NF, UV pretreatment
Chemical pretreatment	pH adjustment, scale inhibitor, ion exchange, lime softening, flocculation, chlorination, dechlorination,
Biological pretreatment	MBR, media biofiltration, conventional biological treatment
Electrical pretreatment	Electromagnetic treatment

Pretreatment can generally be classified into four groups according to its treatment mechanism, that is, physical, chemical, biological, and electrical pretreatment (see Table 3.1). Pretreatment can remove soluble salts (hardness), colloids (silt, Fe, Al and silica), solids (suspended and particulate organics), biological material, oxidizing agents and dissolved organic matter.

Physical pretreatment are those processes predominantly involving applied physical forces. These include adsorption, media filtration, cartridge filtration, membrane-based separation technologies such as MF, UF and NF, and use of ultraviolet (UV) light. This method mostly removes colloidal particles and suspended solids. Additionally, UF using activated carbon can partially remove dissolve organic matter. Chemical pretreatment are those methods that involve the addition of chemicals or use chemical reactions. Examples include flocculation, pH adjustment, lime softening, scale inhibition, ion exchange, chlorination and dechlorination. Pretreatment processes that utilize microbial activity are referred to as biological pretreatment. The process breaks down biodegradable organic compounds and some nutrients into carbon dioxide and sludge. This includes the conventional biological treatment, media biofiltration, and the advanced membrane bioreactor (MBR) technique. Electrical pretreatment uses electrical charging and electromagnetic force, which affect the orientation of charges in water, and results to the formation of microscopic nuclei which causes calcium carbonate to precipitate within the water. Electric and electromagnetic pretreatment removes sparingly soluble salts, colloidal particles and suspended solids. However, this pretreatment is not often used in membrane filtration.

Conventional pretreatments are widely applied in many SWRO desalination plants, however, in recent years, there is a surge in the application of UF/MF aspretreatment techniques due to their ability to provide silt density index (SDI) values less than 2, leading to less downtime (Vedavyasan, 2007). In many cases, utilizing one conventional pretreatment technique such as filtration is not sufficient to reduce the fouling tendency in SWRO. Thus, an integration of several pretreatment methods is usually considered combining the benefits of separate pretreatment processes to improve the effectiveness of the overall pretreatment process (Valavala *et al.* 2011).

Among the conventional methods, granular membrane filtration is reported to be the most widely used pretreatment in many large and medium size desalination plants around the world (Voutchkov, 2009). However, in the past decade, UF/MF pretreatment has also seen increasing use in SWRO. Recent studies have reported on the use of NF and MBR as well. Both conventional and advanced methods present different advantages and disadvantages when used as pretreatment, which depends on the source water quality and origin. In this chapter, the different advantages and challenges of both granular media filtration and membrane-based filtration (i.e., focusing on UF/MF) as SWRO pretreatment are presented, and the different factors to be considered for the selection of pretreatment techniques are also discussed.

3.2 GRANULAR MEDIA FILTRATION

3.2.1 Background

The most common form of pretreatment for RO seawater desalination plants is granular media filtration (GMF). A schematic representation of the GMF method is shown in Fig. 3.2. The GMF has been successfully used as the polishing step in water treatment for more than a century. The GMF removes fine suspended particles remaining in water after sedimentation. Small particles, which pass through sedimentation, are retained in the granular media during filtration process. Similarly, the GMF removes suspended particles in seawater in order to protect RO membranes when it is employed as pretreatment.

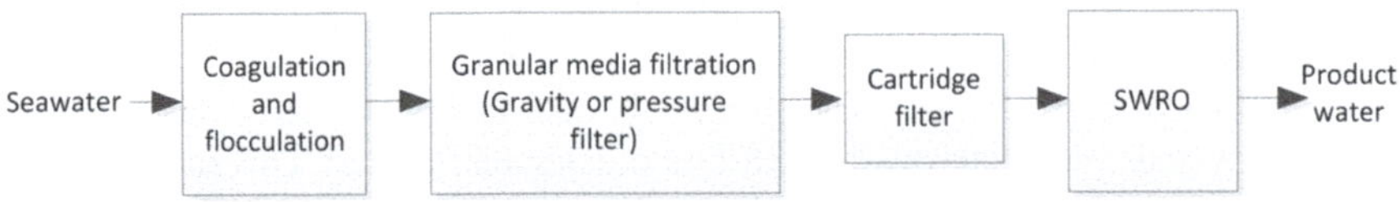

Fig. 3.2 Schematic of the granular media filtration as pretreatment to SWRO.

GMF systems can be classified as either gravity or pressure filtration, with incoming seawater passes through the filter media under gravity in gravity filtration, and under pressure in pressure filtration. The GMF can be also classified as conventional filtration, direct filtration, or in-line filtration depending on the upstream process. Seawater passes through processes such as rapid mixing, flocculation, and sedimentation before filtration in conventional filtration. Coagulant addition into seawater occurs during the rapid mixing phase, flocs are promoted to grow during flocculation, and the subsequent large particles settle under gravity during sedimentation. The sedimentation process is omitted in direct filtration, while the flocculation step can be omitted in in-line filtration, wherethe water is directly sent to a filter after coagulant addition. Obviously, either direct or in-line filtration has an advantage over conventional filtration due to less capital and operating costs, since a sedimentation basin is not required, thus reducing capital costs. The operating cost reduction is related to the lower levels of sludge produced. On the other hand, there is less response time to a change in source water quality either in direct or in-line filtration. Therefore, source water of good and stable quality is generally required for direct or in-line filtration.

According to the American Water Works Association (AWWA) committee report (Report, 1980), water meeting the following criteria can be a good candidate for direct filtration.

- Color < 40 CU
- Turbidity < 5 NTU
- Algae < 2,000 asu/mL
- Iron < 0.3 mg/L
- Manganese < 0.05 mg/L

Direct or in-line filtration is most commonly employed in seawater desalination probably due to the low turbidity of seawater. The AWWA criteria specify that the turbidity of the source water for direct filtration should be less than 5 NTU. The seawater turbidity is generally lower than 5 NTU, according to the literature. Table 3.2 presents some of the turbidity values of seawater from different parts of the world. The data shown in Table 2 clearly indicate that the turbidity of most seawater remains well below the direct filtration criteria of 5 NTU, except for occasional turbidity peaks. Turbidity peaks can occur during heavy rain if the intake is located near a river and ship movements can also cause turbidity peaks. However, not all the seawater around the world is of low turbidity. The Venezuelan seawater turbidity is reported to be 5 ~ 20 NTU (van Hoof *et al.* 2001) and in these cases where the seawater turbidity is high conventional filtration is preferred over direct or in-line filtration.

Table 3.2 Some of the seawater turbidity reported in literature.

Seawater Location	Turbidity (NTU)	Reference
Straits of Gibraltar	0.7	Brehant *et al.* (2002)
Red Sea	0.2 ~ 1.1	Glueckstern *et al.* (2002)
Gulf of Oman	0.12 ~ 0.56	Bonnelye *et al.* (2004)
Persian Gulf	0.45 ~ 1.13	Bonnelye *et al.* (2004)
Masan Bay (Korea)	0.6 ~ 7.7 (average = 2.3)	Kim *et al.* (2009a)
Raw seawater at Perth desalination plant	<3	Vollprecht (2013)

A number of factors affect the efficiency of the filtration process, especially the filtrate quality and the head loss. Yao and his research group (Yao, 1968; Yao &

M'Melia, 1968; Yao *et al.* 1971) developed equation (3.1) to predict the filtrate quality assuming that sedimentation is dominant over the Brownian motion for the filtration efficiency, whilst equation (3.2) or the Carman-Kozeny equation describes the head loss of a fixed bed.

$$\ln\frac{N_{\text{out}}}{N_{\text{in}}} = -\frac{3}{2}\cdot(1-\varepsilon)\alpha\cdot\frac{(\rho_p-\rho_w)gd_p^2}{18\mu V_o}\cdot\frac{L}{d_c} \tag{3.1}$$

$$\frac{\Delta h}{L} = k\cdot\frac{\mu}{\rho_w g}\cdot\frac{(1-\varepsilon)^2}{\varepsilon^3}\cdot S_o^2\cdot V_o \tag{3.2}$$

where:

N_{in} = number concentration of influent particles
N_{out} = number concentration of effluent particles
ε = porosity of filter bed
α = attachment efficiency factor
ρ_p = particle density
ρ_w = water density
g = acceleration of gravity
d_p = particle diameter
d_c = grain diameter
μ = viscosity
L = filter bed depth
V_0 = filtration velocity
Δh = head loss
k = Carman-Kozeny coefficient
S_o = Surface area per unit volume of grain

These equations show that the following factors are important in determining the filter performance with regards to filtrate quality and head loss.

- Particle; diameter (d_p), density (ρ_p)
- Viscosity (μ)
- Filtration rate (V_0)
- Filter bed; depth (L), porosity (ε)
- Grain diameter (d_c)

Among these, particle diameter and density are closely related to raw seawater characteristics. Raw seawater may contain large and dense particles, or small and light particles, or any combination. Although large and dense particles can yield the better filtrate quality than small and light particles, these parameters cannot be controlled in terms of operation unless raw seawater is coagulated, which can modify the particle characteristics. The viscosity is related to the seawater temperature and the viscosity of seawater becomes high at low temperature, and vice versa. Unfortunately, it is practically impossible to control the seawater viscosity in filter operation.

Unlike those parameters related to seawater characteristics, filtration rate or velocity (V_0), filter bed depth (L), porosity (ε), and grain diameter (d_c) can be engineered. When filtration takes place in a deep bed with small grains and low porosity, better filtrate quality can be obtained. These parameters are also related to the head loss of a filter bed. Use of small grains and low porosity increases the head loss. The head loss increases in proportion to the filtration rate and bed depth according to equation (3.2).

These engineering-related parameters (V_0, μ, ε, d_c) are present in both equations (3.1) and (3.2). According to these equations, high filtration rate produces more volume of filtrate, but results in poor quality of filtrate and high head loss. On the other hand, bed porosity and grain size have conflicting effects on the the filtrate quality and the head loss. Use of a filter bed with large porosity yields poor quality of filtrate, but keeps the head loss low. Use of small grains yields better quality of filtrate at the sacrifice of high head loss.

3.2.2 Coagulation

The importance of proper coagulation for the successful operation of filtration has been highlighted by many researchers (Fuller, 1898; O'Melia, 1985a, 1985b; Cleasby *et al.* 1989, 1992). Coagulation modifies the particle characteristics so that the particles can easily attach to the filter media, resulting in better removal of particles during filtration. Among various characteristics, coagulation increases the particle size, which can lead to better filtrate quality. Since the particle size gets bigger with increasing coagulant dosage, the higher dosage results in better filtrate quality as long as the raw water maintains its buffer capacity. Coagulant dosage is also related to the head loss development and head loss is related to the porosity of the filter bed. The lower the porosity, the higher the head loss. As the coagulation dosage increased, more particles are captured by the filter media, which decreases the porosity of filter bed, leading to fast head loss development. Therefore, the filtrate quality and the head loss should be considered for the determination of the optimum dosage.

Sedimentation influences the floc characteristics. The formation of large flocs is preferred in conventional filtration, because sedimentation is included upstream of filtration. Easily settleable large flocs are removed during sedimentation, and the remaining small flocs are then removed during filtration using conventional filtration. A different approach is required in direct or in-line filtration, in which a sedimentation stage is not included. Since large flocs cause rapid clogging of filters, coagulation should promote the formation of pinpoint flocs rather than large flocs. In other words, small and filterable flocs are preferred for either in direct or in-line filtration, while large and settleable flocs are preferredfor conventional filtration. Because the floc size is closely related to the coagulant dosage, large flocs form at high coagulant dosages, and small flocs at low dosages. Consequently, higher coagulant dosage is generally required in conventional filtration compared to either in direct or in-line filtration.

Coagulant dosage is also closely related to filtration performance. High coagulant dosage results in better filtrate quality and high head loss, as indicated in equations (3.1) and (3.2). Consequently, the optimum dosage should be carefully selected consideringboth the filtrate quality and head loss. An example of selecting the optimum coagulant dosage is presented here for the pore controllable filter (PCF, SSeng, Republic of Korea), which is a pressure fiber filter. The PCF was operated as a pretreatment to seawater desalination of reverse osmosis system. Ferric chloride, which was used as a coagulant, was added before seawater entering the PCF at a dose rate of 0 ~ 8 mg/L.

The filtrate quality was represented by turbidity and SDI. The variation of the filtrate turbidity and SDI at different coagulant dosages and filtration rates is shown in Table 3.3. According to Table 3.3, the filtrate quality improved as the coagulant dosage increased. Average filtrate turbidity was 0.50 NTU when the PCF was operated at 5 m/h without coagulant addition. It gradually lowered to 0.43 NTU at 2 mg/L, 0.24 NTU at 4 mg/L, and 0.18 NTU at 8 mg/L as the coagulant dosage increased up to 8 mg/L. Similarly, average filtrate SDI value decreased from 4.3 without coagulant addition to 3.8 at 2 mg/L, 3.3 at 4 mg/L, and 2.2 at 8 mg/L. This result indicates that higher coagulant dosage is beneficial for good filtrate quality. Another advantage of higher coagulant dosage was the stabilization of the filtrate quality. As shown in Fig. 3.3, the filtrate turbidity was unstable when the PCF influent was coagulated at low dosages (0 and 2 mg/L), but it stabilized at high dosages (4 and 8 mg/L). Figure 3.3 shows the monitoring results of the filtrate turbidity when the PCF was operated at 5 m/h with various coagulant dosages (0 ~ 8 mg/L).

Table 3.3 Filtrate qualities of PCF at different coagulant dosages and filtration rates.

	5 m/h		7.5 m/h		10 m/h	
	Turbidity NTU	SDI	Turbidity NTU	SDI	Turbidity NTU	SDI
0	0.50 ± 0.16	4.3	0.64 ± 0.05	4.7 ± 0.15	0.67 ± 0.37	4.6 ± 0.15
2 mg/L	0.43 ± 0.07	3.8 ± 0.76	0.44 ± 0.04	4.1 ± 0.25	0.45 ± 0.05	3.8 ± 0.15
4 mg/L	0.24 ± 0.05	3.3 ± 0.42	0.13 ± 0.06	3.5 ± 0.06	0.24 ± 0.07	3.5 ± 0.13
8 mg/L	0.18 ± 0.07	2.2 ± 0.57	0.13 ± 0.06	3.4 ± 0.20	0.16 ± 0.02	2.6 ± 0.25

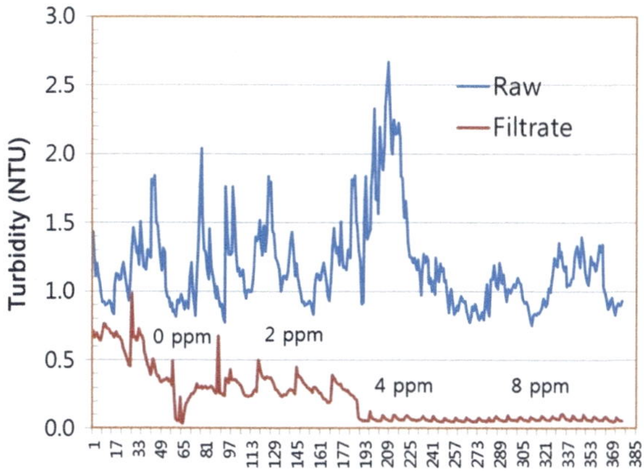

Fig. 3.3 Filtrate turbidity variation at different coagulant doses (filtration rate of 5 m/h).

However, the addition of high coagulant dosage resulted in fast head loss development as shown in Fig. 3.4, which clearly shows that the pressure linearly increased with filtration time and the extent of the increase was different depending upon the coagulant dosages. The linear increase indicates the in-depth filtration, meaning that particle removal occurred over an entire filter bed, which is a typical behavior of GMF. Steeper pressure increase was observed at higher coagulant dosages. The PCF backwash was initiated when the operating pressure reached the value set to 0.7 bar or the trans-PCF pressure of 0.5 bar. When the PCF influent was not coagulated, the pressure increase was minimal and the operating pressure even after 1,000 min of operation was far less than 0.7 bar. As the coagulant dosage increased, the pressure increase became steeper. When the PCF influent was coagulated at 2 mg/L, the operating pressure reached 0.7 bar after 500 min of filtration. The duration shortened to 165 min at 4 mg/L and 90 min at 8 mg/L. It seems that coagulation affects the porosity of a filter bed. As the coagulation dosages increase, larger flocs are produced, decreasing the porosity of a filter bed. Since the porosity is inversely related with the head loss, faster pressure increase was observed at higher coagulant dosages. This result clearly indicates that lower coagulant dosages are beneficial if lower head loss development is desired. After considering the filtrate quality and the head loss development, 4 mg/L of ferric chloride was selected as the optimum dosage of the PCF operation in this case because the acceptable filtrate quality ($SDI < 4$) and the reasonable filtration duration (165 min) could be obtained at this dosage.

3.2.3 Coagulant type

In pretreatment of seawater desalination, ferric chloride is most commonly used as a coagulant because of the low solubility of ferric salts and the thus the low potential biological toxicity of the residual metal. There are some studies using different coagulants for pretreatment. Jeong *et al.* (Jeong *et al.* 2011) used poly ferric silicate (PFS) as coagulant and compared its performance with ferric chloride. According to his study, PFS was more effective than ferric chloride in turbidity removal as well as dissolved organic carbon (DOC) removal at low doses. The addition of 3 mg/L (as Fe) of ferric chloride resulted in 57% DOC removal, while that of 1.2 mg/L (as Fe) for PFS resulted in 68% DOC removal. Ferric chloride was more effective in removal of hydrophobic fractions, while PFS was more effective in removal of hydrophilic fractions. However, use of PFS resulted in faster fouling of the MF compared with ferric chloride. They attributed the increased fouling to the smaller but much larger amount of flocs produced by PFS. Jeong and colleagues (Jeong *et al.* 2011) reported similar results for titanium chloride ($TiCl_4$) used as coagulant for pretreatment of seawater desalination. At the same doses (5 mg/L as Fe), ferric chloride removed 60% of incoming DOC and $TiCl_4$ 68%. Titanium chloride was more effective in the removal of hydrophilic fractions than ferric chloride.

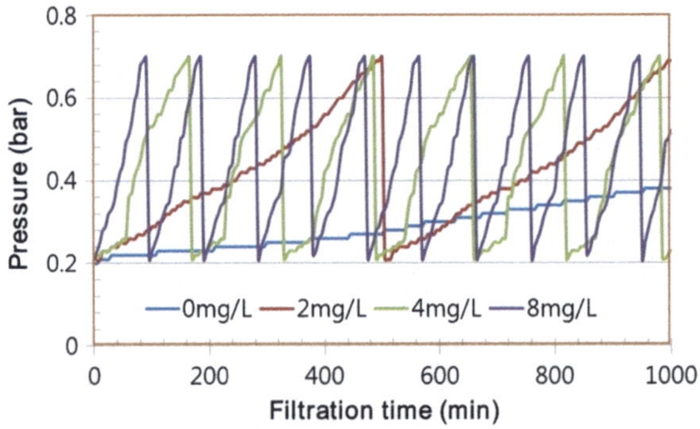

Fig. 3.4 Pressure increase with time at different coagulant doses (filtration rate of 5 m/h).

There is a study comparing floc characteristics such as size and shape and the kinetics of floc formation for different types of coagulant (Kim *et al.* 2009a). According to their study, ferric chloride and titanium chloride produced bigger flocs than aluminum salts. Flocs (Al floc and Ti floc) formed resulting from use of ferric chloride and titanium salts were more spherical than Al floc, leading to faster settling. In terms of kinetics, the use of ferric chloride resulted in faster floc formation than aluminum salts.

3.2.4 Filtration rates

Filtration rates are closely related to filtration performance. Equations (3.1) and (3.2) predict that the filtrate quality deteriorates and the head loss of a fixed bed increases as filtration rates increase. Experimental results examining the effect of filtration rates on the filtration performance are shown in Table 3.3 and Fig. 3.5. According to Table 3.3, the effect of filtration rates on the filtrate turbidity was different depending on coagulation dosages. When the PCF influent was not coagulated, the filtrate quality deteriorated as the filtration rates increased. Average filtrate turbidity increased from 0.50 NTU to 0.67 NTU as the filtration rates increased from 5 m/h to 10 m/h. This effect was minimized when coagulation pretreatment was provided. Average filtrate turbidity was 0.43 NTU at 5 m/h, 0.44 NTU at 7.5 m/h, and 0.45 NTU at 10 m/h when the PCF influent was coagulated at 2 mg/L. Similar results were observed at different dosages. Unlike the filtrate turbidity, the filtrate SDI values increased as the filtration rates increased, but the extent of the change was insignificant. The filtrate SDI deteriorated from 4.3 to 4.6 as the filtration rates increased from 5 m/h to 10 m/h without coagulation pretreatment. Similarly, the filtrate SDI increased from 3.3 at 5 m/h to 3.5 at 10 m/h when the PCF influent was coagulated at 4 mg/L.

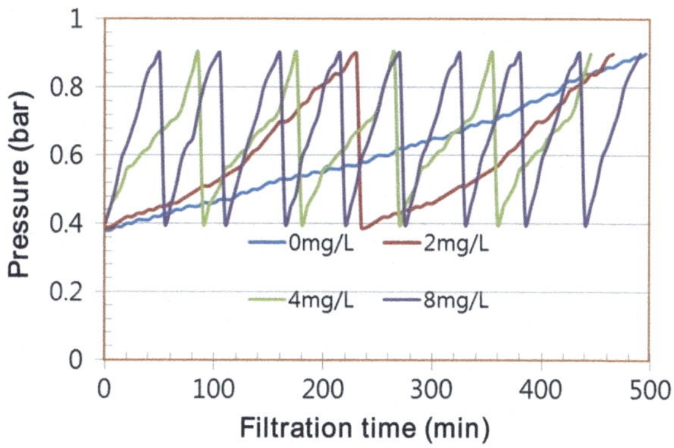

Fig. 3.5 Pressure increase with time at different coagulant doses (filtration rate of 10 m/h).

According to equation (3.2) which describes the head loss of the fixed bed, the head loss is related linearly to the filtration rate. If filtration rates are doubled, so does the head loss of the fixed bed. As predicted from equation (3.2), the initial operating pressure of 0.2 bar increased to 0.4 bar as the filtration rates were increased from 5 m/h to 10 m/h, as shown in Figures 3.4 and 3.5. The filtration rates also affected the head loss development during filtration, with faster pressure increases observed at higher filtration rates. Assuming the PCF is backwashed at the trans-PCF pressure of 0.5 bar (terminal pressure of 0.9 bar), the filtration duration of one cycle reduced from 500 min at 5 m/h to 230 min at 10 m/h when the PCF was coagulated at 2 mg/L. The duration reduced from 165 min to 85 min at 4 mg/L of coagulation.

The filtration rates are also an important factor for filtrate stability. When the PCF was operated at high filtration rates (10 m/h), the filtrate turbidity was unstable even at a high coagulant dosage (4 and 8 mg/L), as shown in Fig. 3.6. The filtrate stability was closely related to the coagulation pretreatment and the filtration rates. When the filter influent was poorly coagulated (up to 2 mg/L of coagulation), the filtrate turbidity was unstable even at lower rate (5 m/h), as shown in Fig. 3.3. As the coagulation dosage was increased to 4 mg/L, the filtrate turbidity became stable. However, the filtrate turbidity was not so stable at high filtration rate. After considering the filtrate quality and the pressure increase, the filtration rate of PCF has been determined as 5 m/h.

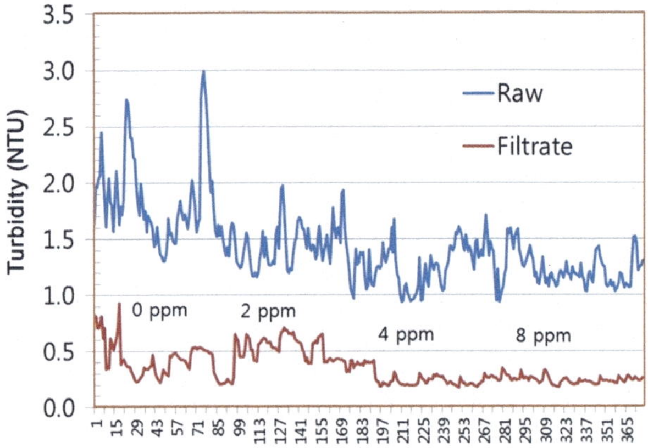

Fig. 3.6 Filtrate turbidity variation at different coagulant doses (filtration rate of 10 m/h).

3.2.5 Media

Filtration can be classified depending on the number of filter media as mono media filtration, dual media filtration, and triple media (multi media) filtration. Typically, sand is used in mono media filtration. Anthracite and sand are used in dual media filtration. Anthracite, sand, and garnet are used in triple media filtration. Anthracite is lighter than sand and is placed on top of sand in dual media filtration. Garnet is placed below the sand in triple media filtration. Dual media filtration is generally favored for pretreatment in seawater desalination.

Besides these media, there are several commercial products such as Filtralite®, AFM, and pumice. Filtralite is an expended clay filter media manufactured by Weber. AFM is an activated filter media, manufactured by Dryden Aqua, who claims that their products have a high negative zeta potential and catalytic self-sterilizing properties. Pumice, which is a volcanic rock, is also supplied as a filter media by several manufacturers. Filtralite can replace anthracite because of its low density (1.0 ~ 1.7 g/cm^3). As shown in equations (3.1) and (3.2), media properties such as size and porosity are important for the filtration performance. Use of small media with low porosity leads to better filtrate quality, but higher head loss of the fixed bed. Therefore, filter media should be selected in consideration of media properties, regardless of the media type. For example, low head loss development might occur due to its high porosity when Filtralite is used as filter media. According to the manufacturer, the Filtralite porosity is in the range of 0.40 ~ 0.73.

3.2.6 Backwash

Backwash is important for successful filtration. The purpose of backwash recovers the impaired filter performance back to its original level during filtration. A dirty filter is usually backwashed by water. However, backwash by wash water alone is generally insufficient to clean a dirty filter due to limited media contact (Amirtharajah, 1978). Subsequently, air scouring is often adopted in order to increase the media contact during backwash. There is a study (Kim *et al.* 2011) which attempted to find the optimum backwash method of dual media filter used for pretreatment in seawater desalination. They compared the effectiveness of backwash methods such as water wash alone at fluidization velocity, air scouring followed by fluidized water wash, and simultaneous air scouring and sub-fluidized water wash. The backwash effectiveness was evaluated by backwash waste turbidity, mass of dislodged particles during backwash, head loss development, and initial breakthrough. According to their study results, simultaneous air scouring and sub-fluidized water wash was the most effective backwash method. This result was supported by the hydrodynamic evaluation using computational fluid dynamics (CFD) technique. According to the CFD analysis, the highest shear rate occurred around the filter media when a dirty filter was backwashed by the simultaneous air scouring and sub-fluidized water wash.

3.3 MEMBRANE FILTRATION

3.3.1 Background

Compared with conventional granular filtration (GMF), application of Microfiltration (MF) and Ultrafiltration (UF) membrane filtration for seawater pretreatment is relatively new. Nevertheless, the number of seawater desalination plants with membrane pretreatment is increasing (Bush *et al.* 2010; Voutchkov, 2010). Figure 3.7 illustrates a schematic diagram of seawater desalination plant using membrane pretreatment. There are three main components for the membrane pretreatment: (1) screen; (2) pretreatment for MF/UF; and (3) MF/UF. In order to protect the membranes from physical damage, the source seawater has to be pre-screened. Depending on the characteristics of feed water, a pretreatment prior to MF/UF may be also required. The cartridge filter for this process may be placed between MF/UF and screen.

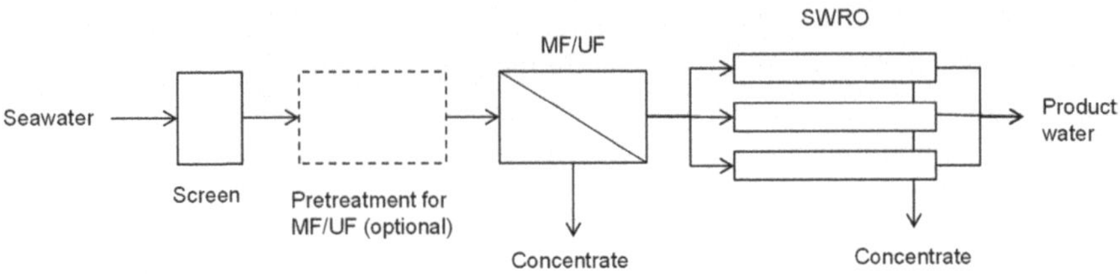

Fig. 3.7 Schematic of SWRO desalination process using membrane pretreatment.

MF and UF are physical filtration processes where a contaminated water flow passes through a porous membrane to separate particles from filtrate water. MF membranes are used for particle separation in the range size of 0.08 to about 2.0 μm. On the other hand, UF membranes are suitable for particle separation in the range of 0.002 to about 0.1 μm. In MF and UF processes, particle separation occurs according to their size. MF and UF membranes are efficient in removing turbidity and non-soluble and colloidal organics contained in the source seawater. Turbidity and the Silt Density Index (SDI) can be lowered below 0.1 NTU and 3 over 90% of the time, respectively. However, membrane pretreatment does not remove significant amounts of dissolved organics and marine microorganisms, which typically cause SWRO membrane biofouling. For comparison, granular media filters, depending on their configuration, loading rate and depth, could remove 20% to 60% of soluble organics from source seawater.

The removal efficiency of particulate matters by MF/UF is closely related to the pore size of the membrane. The rejection mechanism of MF/UF is described based on the pore flow model (Bungay *et al.* 1986). In this model, the rejection (R) of particles by the membrane is:

$$R = \left[1 - 2\left(1 - \frac{a}{r}\right)^2 + \left(1 - \frac{a}{r}\right)^4\right] \times 100\% \tag{3.3}$$

where a is the radius of the particle and r is the radius of the membrane pore. For example, the membrane pore size should be less than $0.929a$ to achieve over 99% rejection of particles with the radius of a.

The permeability of the membrane is also related to the pore size. Although the flow of water through membranes can be modeled using empirical equations, the Hagen-Poiseuille model can be used to demonstrate this flow of water (Baker, 2004). The Hagen-Poiseuille equation considers that the membrane has a number of parallel cylindrical pores, which are parallel or oblique to the membrane surface. Assuming that the capillaries are uniform and cylindrical, the water flux through a MF/UF membrane is:

$$J = \frac{\Delta P}{\eta R_m} = \frac{\Delta P}{\eta}\left(\frac{\varepsilon_m r^2}{8\tau x_m}\right) \tag{3.4}$$

where J is the flux(L/m^2-hr); ΔP is the pressure difference (Pa); x_m is the membrane thickness (m); η is the viscosity of liquid (Pa-sec), ε_{m} is the surface porosity of the membrane; and τ is the tortuosity of the membrane pore. It is evident from equations (3.3) and (3.4) that the particle rejection increases and membrane permeability decreases with a decrease in the membrane pore size.

Hollow fiber modules are the most commonly used type for MF/UF membranes. Hollow fiber membrane modules are formed in two basic geometries (Baker, 2004). The first is the shell-side feed design where a loop or a closed bundle of fibers is contained in a pressure vessel. The system is pressurized from the shell side; permeate passes through the fiber wall and exits through the open fiber ends. This design is easy to fabricate and allows very large membrane areas to be contained in an economical system. The second type of hollow fiber module is the bore-side feed type. The fibers in this type of unit are open at both ends, and the feed fluid is circulated through the bore of the fibers.

MF and UF membrane processes are divided into two categories depending on the type of the driving filtration force: pressurized and vacuum-driven membrane systems (Baker, 2004). Pressurized UF and MF systems consist of membrane elements installed in pressure vessels. Vacuum-driven pretreatment systems include a series of membrane modules immersed in filtration tanks (Di Profio *et al.* 2011). Usually the vacuum-driven systems operate in a range of 0.1 to 0.8 bar, whilst pressure-driven systems typically run at pressures of 0.1 to 2.5 bar depending on the types of the membrane modules.

MF/UF have a wider spectrum of particle removal capabilities than conventional GMF (Voutchkov, 2010). Single or dual granular-media filters usually have lesser removal efficiency for fine particles, silt and colloidal materials. The physical and chemical analyses of the filtrate from the MF system are superior to the conventional pretreatment (Brehant *et al.* 2002). An assessment study showed that the total unit water cost was lower for the MF system and this indicated that the membrane-based system was economically competitive (Ebrahim *et al.* 2001). MF/UF are also less prone to upsets caused by seasonal changes in source seawater temperature, pH, turbidity, color, and size and type of particles because their removal mechanisms for particles. However, membrane filtration as a pretreatment is not the solution to all seawater applications (Kumar *et al.* 2006). For instance, it has been reported that the use of MF pretreatment is less efficient to control SWRO fouling than GMF under algal bloom conditions (Voutchkov, 2010). In general, the use of membranes for pretreatment is recommended for large capacity plants with open intakes or horizontally sub-sea wells, either independently or in conjunction with other standard processes (Peñate *et al.* 2012).

3.3.2 Coagulation

The pretreatment efficiency of conventional media filtration is highly dependent on the effectiveness of the chemical coagulation and flocculation ahead of the filtration process. However, the efficiency of the MF/UF pretreatment systems is less sensitive to the upstream chemical coagulation and flocculation than conventional media filtration. This is because the removal efficiency of particles and colloidal materials are mainly determined by the radius ratio of particle and pore (a/r) as shown in equation (3.3). Although coagulation and flocculation aid to increase the size of particles, the pore size of MF/UF is small enough to reject most particles in source seawater. Therefore, the coagulation and flocculation are of lesser importance for achieving high quality of pretreated water.

On the other hand, MF/UF may require coagulation and flocculation as its pretreatment for the control of its fouling. Although MF/UF are used to control fouling of SWRO membrane, MF/UF itself can suffer from serious fouling depending on the characteristics of source seawater. For instance, if the turbidity and/or total organic carbon (TOC) in the feed water are high, particles and organics deposit onto a MF/UF membrane surface or into membrane pores in a way that degrades the membrane's performance. Membrane fouling can cause severe flux decline and increase the operating costs of a treatment plant. Accordingly, proper conditioning of feed water for MF/UF is necessary to retard membrane fouling.

Among various techniques of feed water conditioning for MF/UF, coagulation prior to a MF or UF membrane has been widely adopted (Kwon *et al.* 1997). The coagulation aims at the removal of dissolved NOM by adsorption onto iron or aluminum salts and increase of colloidal matters by aggregation. Because of coagulation, a more porous filter cake is formed and the adsorption of organics on the membrane is reduced. Although coagulation may be used prior to MF/UF pretreatment systems, their coagulant dose is generally smaller than that of conventional media filter systems (Voutchkov, 2010).

As an example, Fig. 3.8 illustrates the effect of coagulation on UF fouling in membrane pretreatment of SWRO desalination. The normalized fluxes are compared between UF system and UF system with coagulation. $FeCl_3$ (3 mg/L) was used for the coagulation of seawater prior to UF. The $FeCl_3$ dose was determined by a series of jar tests. An inside-out hollow fiber UF (0.02 μm) made of polyester with styrene and methyl methacrylate (PESM) was used. The initial permeate flux was 60 L/m^2-hr. Without coagulation, the permeate flux decreases by 10% after 500 min of UF operation. On the other hand, virtually no flux decline was observed in the UF system with coagulation.

In Fig. 3.9, the transmembrane pressure (TMP) was compared for the UF system and the UF system with coagulation. At 60 L/m^2-hr, the UF system without coagulation suffered from serious fouling. The TMP increases from 0.75 bar to 1.6 bar and thus chemical cleaning was applied during the operation. However, the UF system with coagulation showed stable operation during a few months operation. The TMP ranged from 0.05 to 0.7 bar at 60 L/m^2-hr. When the flux was increased from 60 L/m^2-hr to 90 L/m^2-hr, the effect of increased flux on the TMP was negligible. This suggests that coagulation prior to UF allows effective control of fouling and operation under higher flux.

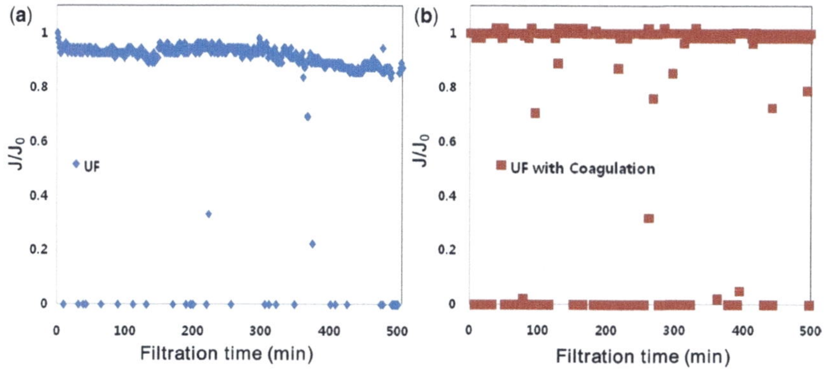

Fig. 3.8 Dependence of normalized UF flux on filtration time in UF pretreatment of seawater for SWRO desalination (a) UF (b) UF with upstream coagulation.

Although coagulation is effective to increase the performance of UF, it did not affect the quality of treated seawater. The turbidity and SDI for the treated seawater were almost constant for the UF and UF with coagulation. The fouling propensity of SWRO after UF pretreatment was not different even if the coagulation was applied. This is because UF alone could produce water that did not cause RO fouling. In fact, the average SDI was less than 3.

3.3.3 Flux and recovery

The two most important parameters to consider for the design of membrane pretreatment system are the design flux and feed water recovery. The design flux determines the amount of total membrane area and modules/elements needed to produce certain volume of seawater. The capital and equipment costs for the membrane system decrease as the design flux increases. However, an increase in the design flux leads to an increased risk of membrane fouling. Accordingly, this results in an increase in the operating cost.

In general, the fouling rate of MF/UF membranes are sensitive to the design flux. This is explained by the concept of critical flux. The critical flux for MF is that on start-up there exists a flux below which a decline of flux with time does not occur; above it fouling is observed (Field *et al.* 1995). This flux is the critical flux and its value depends on the hydrodynamics and probably other variables. In other words, the fouling rate is not linearly proportional to the design flux. Instead, it may exponentially increase with an increase in the design flux if the MF/UF operation is done above its critical flux condition.

Critical flux exists at the condition where the net force exerted on particles in the fluid is zero. In a crossflow MF/UF membrane, particle transport depends on two major actions: one action is moving the particles toward the membrane surface (negative direction) and the other involves shifting them away from the membrane surface (positive direction). The negative direction forces include van der Waals attraction (F_A), and permeation drag (F_d), while the positive direction actions include electrical double layer repulsion (F_R), Brownian diffusion (F_B), shear-induced diffusion (F_s), and lateral inertial lift (F_l). The gravity and buoyancy forces are equal to zero assuming that the density of particle is the same as that of the flowing liquid. The rate of momentum of a particle equals the sum of all the forces imposed on the particles in a fluid stream along a membrane channel. Thus the net force exerted on a particle along the membrane channel, F, is the sum of all forces noted above.

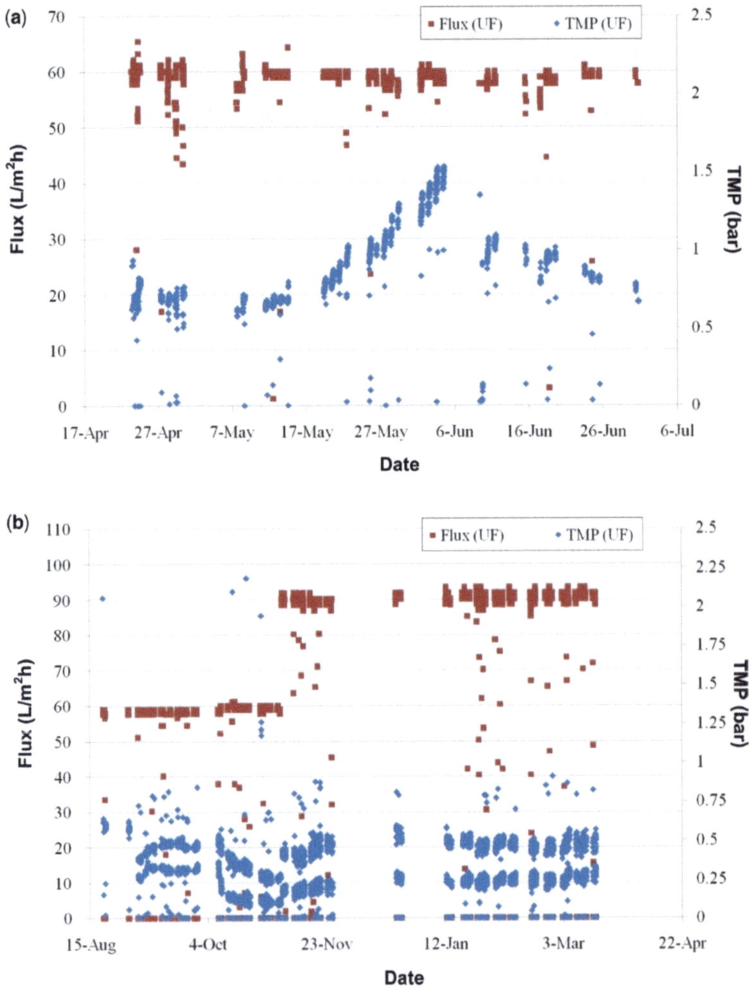

Fig. 3.9 Comparison of flux and transmembrane pressure in a UF pilot plant for seawater desalination (a) UF (b) UF with upstream coagulation.

Although critical flux can be estimated using a set of theoretical equation, it may be an ideal value and different from the 'real' critical flux in a pilot or full-scale MF/UF system. Accordingly, it should be experimentally determined for practical purposes. One of the techniques to determine critical flux is the flux step method (Di Profio *et al.* 2011). In this test, the slope of the TMP profile (d(TMP)/dt) at each flux step is determined. If there is a significant change above a certain flux, this is the point where there is a transition between sub-critical and critical flux operation. Rigorously speaking, the maximum flux at which the d(TMP)/dt remained low can be considered a 'threshold' flux instead of a critical flux as the threshold flux marks the transition from (very) low fouling to significant fouling rather than the boundary between no fouling and fouling. However, it may be used instead of the critical flux in practical applications because the purpose of critical flux determination is to identify the conditions for low fouling operation.

An example of the critical flux determination for UF membrane in a membrane pretreatment system for a SWRO desalination pilot plant is shown in Fig. 3.10. In this case, a rapid increase in TMP was observed at 100 L/m^2-hr, indicating that the critical flux exists between 90 L/m^2-hr and 100 L/m^2-hr.

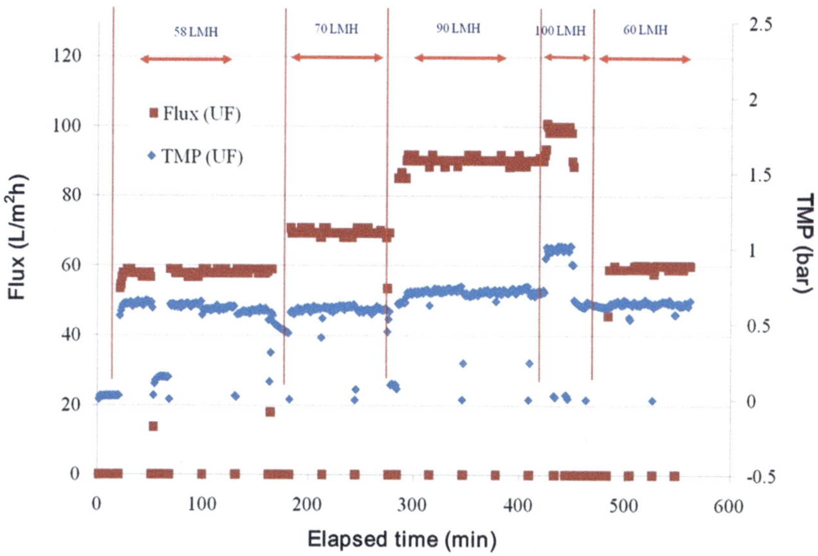

Fig. 3.10 Critical flux determination for UF pretreatment system using flux step method.

Feed water recovery indicates the fraction of the source seawater that is converted into filtrate suitable for seawater desalination. The recovery ratio of membrane pretreatment systems is usually in a range of 88% to 94% and is typically lower than the recovery rate of granular media pretreatment filters for the same source water (95% to 98%). To increase the recovery ratio, a multi-stage configuration for membrane pretreatment system may be used. Figure 3.11 shows an example of such a multi-stage configuration to achieve 99% of the recovery ratio.

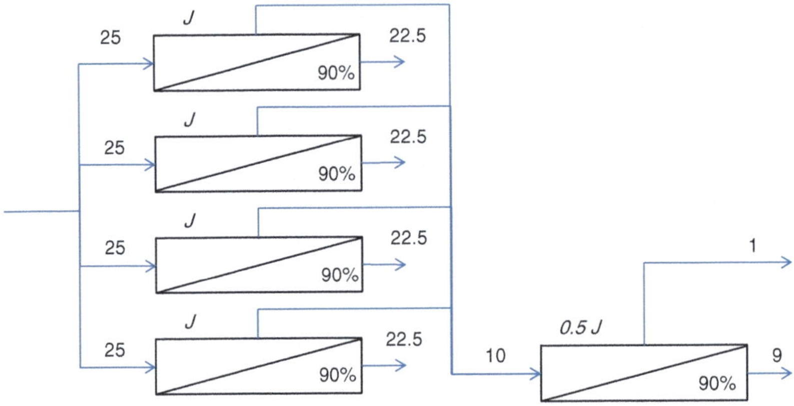

Fig. 3.11 Two stage configuration of membrane pretreatment designed for 99% recovery ratio.

3.3.4 Membranes

Various membranes have been applied for pretreatment in SWRO desalination. To date, UF membranes have found wider application for seawater pretreatment than MF membranes (Voutchkov, 2010). This is attributed to the ability of UF to provide a better removal efficiency of suspended organics, silt and colloids from the source seawater. Results from a comparative MF/UF study indicate that "tight" UF membranes (i.e., membranes that can reject compounds of size of 20 kilo-Dalton or kDa or smaller) can produce filtrate with lower SWRO membrane fouling potential than 0.1 μm MF membranes (Kumar *et al.* 2006). As shown in Fig. 3.12, the flux of RO membrane increases with decreasing the pore size of the MF/UF membranes (Lee *et al.* 2006).

If there are large amounts of particles whith a size comparable to the pores of the MF membrane, rapid and serious fouling may occur due to the blockage of the pores. Silt particles that are produced by naval ship traffic or ocean bottom dredging near the area of the intake may cause such a problem. In this case, UF is preferred to MF because UF membrane pores are an-order-of magnitude smaller than these of the MF membranes and thus not easily affected by such particles (Voutchkov, 2010). Nevertheless, the selection of MF and UF membranes should be done based on the results of a pilot testing.

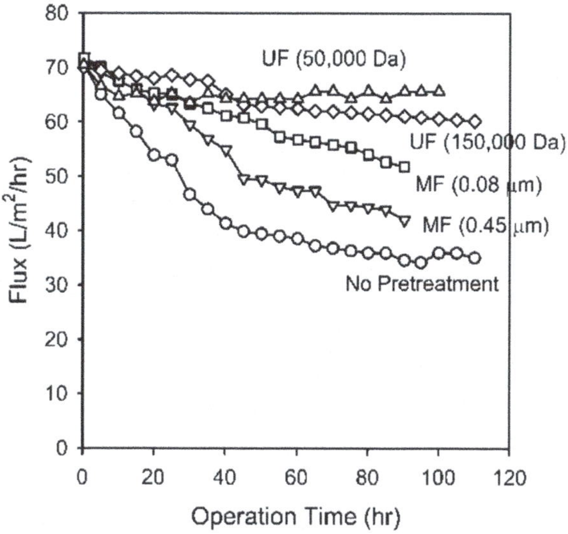

Fig. 3.12 Comparison of flux variation during RO among different pretreatments. (Adapted from Lee *et al.* 2006).

Recently, ceramic membranes have been introduced for the pretreatment of seawater desalination. These membranes have exhibited superior performance compared to conventional treatment membranes (Cui *et al.* 2013). Cui *et al.* (2011) showed that ceramic membranes are available to pre-treat seawater for RO, even under condition with very low temperature (Cui *et al.* 2011). Most fouling can be removed from ceramic membranes by employing the cleaning in place (CIP) methodology (Cui *et al.* 2013).

3.3.5 Cleaning

Fouling and cleaning are at the heart of the membrane application issues. For water applications, the CIP cleaning frequency ranged from 0.2 to 50 per year with a median of 4 per year (Regula *et al.* 2014). However, it is highly specific to the feed water characteristics, membrane types and operating conditions. Cleaning and its optimization (cleaning time, cleaning parameters, cleaning costs.) has been studied more deeply since the 1990s with the aim of reducing operative costs and maintain membrane performance. Nevertheless, relatively little information is available on the cleaning techniques of MF/UF membranes used for seawater pretreatment.

In MF/UF membranes, there are three methods to evaluate the effectiveness of cleaning (Regula *et al.* 2014):

- Physicochemical cleanliness is based on the detection of molecules on the membrane surface.

- Microbiological cleanliness is based on the detection of microorganisms on the membrane surface. The methods used are destructive, that is why this cleanliness parameter is generally determined by analyzing the microorganisms in the effluent itself.
- Hydraulic cleanliness is based on the measurement of permeability after cleaning/disinfection compared to the one before cleaning/disinfection. Although this method has drawbacks, it is widely applied in practice because of its simplicity.

In addition, there are three kinds of cleaning techniques for pilot-scale and full-scale membrane systems.

- Backwash (BW) refers to pumping water backwards through the membrane, sometimes including intermittent use of compressed air during the process. It is usually applied several times a day.
- Chemically enhanced backwash (CEBW) is a special backwash operation conducted once or twice a week with chemicals such as sodium hypochlorite, caustic soda or acid, added to the supply
- CIP uses more concentrated chemical solutions to clean the membrane and are typically conducted monthly or yearly.

These cleaning techniques should be applied in a proper way to retard the fouling of MF/UF membranes, otherwise, the performance of the membrane rapidly deteriorates. Fig. 3.13 shows how CEBW affects the overall efficiency of MF membrane used for seawater pretreatment. With CEBW, the permeability did not significantly change, however, the permeability started to decrease when CEBW was not applied.

Cleaning agents commonly refer to the chemical products which can efficiently remove matter accumulated onto and within the membrane during the production cycle. The cleaning efficiency will depend on the type of foulants, adequacy of cleaning agents, and the cleaning/disinfection conditions. However, in most cases the choices of cleaning/disinfection conditions are not deliberated; rather, they are the result of a trial/error process. The choice of cleaning/disinfection agents also takes into account other parameters such as (Wang *et al.* 2014):

- Compliance of the products with the Food and Drink Industries standards (FDI)
- Compliance of the products with membrane materials and other components: glue, seal, spacers, and so on
- Treatability of cleaning effluents
- Required volumes of water
- Stability of the cleaning/disinfection agent
- Rinsibility, foaming
- Price.

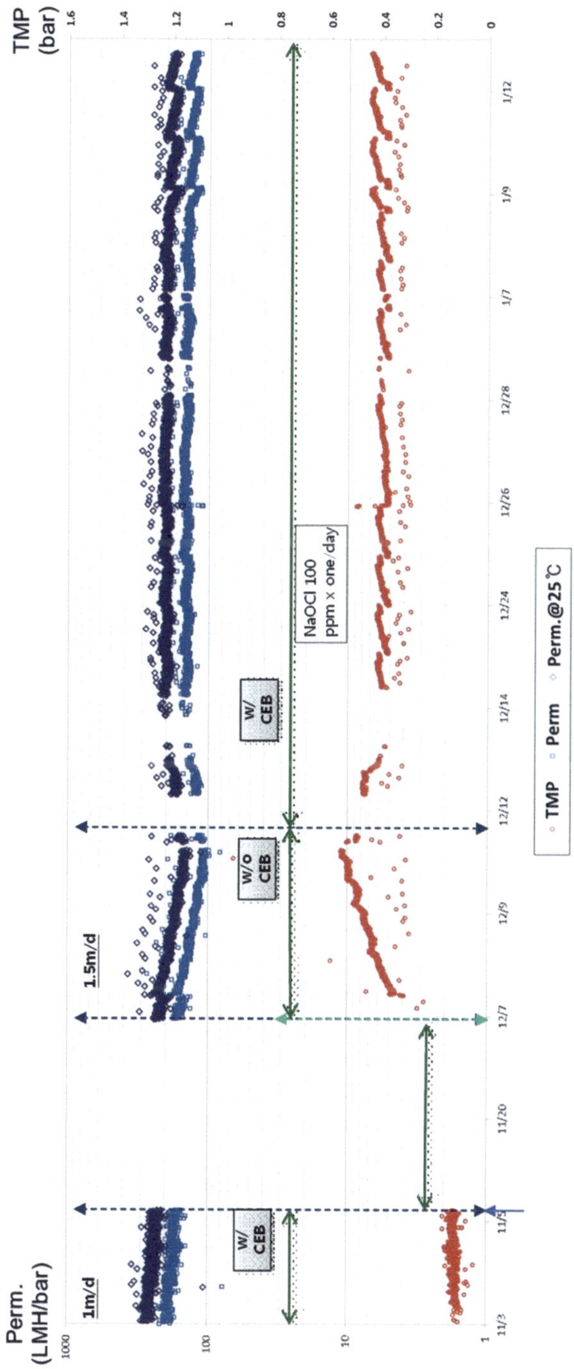

Fig. 3.13 Effect of CEBW on the performance of MF pretreatment system for SWRO desalination.

Table 3.4 shows an example of UF cleaning using different cleaning agents. First, the pure water flux of fouled membrane was examined to determine the total TMP of fouled membrane. The fouled membrane was then subjected to different cleaning procedures using chemical cleaning. When the TMP reaches to about maximum pressure (over 1.0), chemical cleaning is performed. Membrane cleaning by NaOH and citric acid significantly improved TMP recovery to 92% whereas, only NaOH cleaning did not yield that much TMP recovery. However, cleaning with only NaOH was affective in removing particulate fouling on the membrane surface. These results suggested that chemical cleaning of UF membrane filtration helped in reducing the deposition of strongly bonded particulate and organic foulants on the membrane surface thus improving the cleaning effectiveness. In addition, acid cleaning was found to be effective, suggesting that inorganic fouling on the membrane and TMP recovery were also significant.

Table 3.4 Comparison of cleaning efficiences under different cleaning conditions.

Cleaning Condition	TMP (Bar)	Recovery (%)
Initial	0.55	100
NaOH	0.89	62
H_2SO_4	0.67	82
NaOH + Citric Acid	0.60	92

3.4 SELECTION OF SEAWATER FILTRATION PRETREATMENT

The selection of a suitable pretreatment technique or a series of pretreatment methods will mainly depend on the quality of the feed water and the cost associated with the overall pretreatment process. In an ideal situation, the pretreatment should be able to filter out all other constituents in water except for the dissolved minerals, and with proper operating conditions ensuring no precipitation on membrane surface, thus allowing the SWRO process to be operated for a long time (Voutchkov, 2009). However, this is not the case for real SWRO process as pretreatment can remove most of the suspended solids but not all of them. Selecting the appropriate pretreatment methods should consider the life-cycle cost-benefit analysis, and it is recommended to have a side-by-side pilot testing of the conventional and membrane filtration systems to obtain performance information.

3.4.1 Water quality

The type and configuration of which pretreatment process to be used highly depends on the seawater quality as presented in Table 3.5. The threshold level of significance for each quality parameter based from full-scale desalination experience is also given in the Table. Generally, membrane filtration (UF/MF) is known to have a wider degree of removal capability for different particles (such as suspended raw water organics, disinfection by-product precursors, fine particles, silt and pathogens) in seawater compared to GMF. If the source water is exposed to wide variations in quality due to seasonal changes, membrane filtration seem to be the better choice as it is less prone to upsets by sudden change in seawater quality. This is attributed to its primary treatment mechanism which is by mechanical particle removal through membranes with small pore sizes. However, if the source water is of consistent quality (i.e., not varying much in quality even at different seasons), GMF seems to give a cost-effective choice.

Table 3.5 Characteristics of seawater that have impact in the selection of pretreatment method.

Source Seawater Quality Parameter	Pretreatment Issues and Considerations
Turbidity (NTU)	Levels above 0.1 mg/L are indicative of high potential for fouling. Spikes above 50 NTU for more than 1 h would require sedimentation or dissolved air flotation (DAF) treatment prior to filtration.
Total organic carbon (mg/L)	If below 0.5 mg/L – biofouling is unlikely. Above 2 mg/L – biofouling is very likely.
Silt density index (SDI)	Source seawater levels consistently below 2 year-around typically indicate that no additional filtration pretreatment is needed. SDI > 4 – pretreatment is necessary.
Total suspended solids (mg/L)	This parameter is needed to assess the amount of residuals generated during pretreatment. Does not correlate well with turbidity beyond 5 NTU.
Iron (mg/L)	If iron is in reduced form, SWRO membranes can tolerate up to 2 mg/L. If iron is in oxidized form, concentration > 0.05 mg/L would cause accelerated fouling.
Manganese (mg/L)	If manganese is in reduced form, SWRO membranes can tolerate up to 0.1 mg/L. If manganese is in oxidized form, concentration > 0.02 mg/L would cause accelerated fouling.
Silica (mg/L)	Concentrations higher than 20 mg/L may cause accelerated fouling. Analyze for colloidal silica if concentration > 20 mg/L.
Chlorine (mg/L)	Concentrations higher than 0.01 mg/L would cause RO membrane damage.

(Contineud)

Table 3.5 Characteristics of seawater that have impact in the selection of pretreatment method (*Continued*).

Source Seawater Quality Parameter	Pretreatment Issues and Considerations
Temperature (°C)	$T \leq 12$°C would cause significant increase in unit energy use. $T \geq 35$°C may cause accelerated mineral scaling and biofouling. $T > 45$°C may cause irreversible RO membrane damage.
Oil and grease (mg/L)	Concentrations higher than 0.02 mg/L would cause accelerated organic fouling.
pH (units)	Typical pH of seawater is 7.6 to 8.3. Long term exposure to pH < 4 and pH > 11 may cause membrane damage.

Source: Adapted from Voutchkov (2010).

Field studies in an SWRO plant in Jeddah observed that the changes in seawater quality due to climatic conditions such as dust storms and wind have resulted in partial or complete closure of conventional filtration due to its inability to produce acceptable SDI values (Al-Sheikh, 1997). Similar experience was also observed in the Addur SWRO desalination plant (Hussain *et al.* 1998). The Doha research plant reported unstable SDI values when using conventional pretreatment, with frequent backwashing, large usage of chemicals for coagulation and pH adjustment, and clogging problems in media filters (Ebrahim *et al.* 2001). The fluctuation in raw seawater quality can also lead to underdosing or overdosing of coagualants because it is quite difficult to control/predict the optimum dose since the dosage is based on the quality of the feed water. When appropriate SDI values are not achieved due to failure of GMF, this could lead to biofouling problems as the feed water could still contain nutrients that make up the food for the microorganisms to survive. One study (Glueckstern *et al.* 2002) compared the efficiency of conventional and UF membrane pretreatments. Both pretreatment methods showed similar performances at normal weather conditions, but when storms came (i.e., affecting the seawater quality), the coagulation-UF pretreatment obtained better overall performance (producing consistent good water quality) compared to the conventional pretreatment. Another study also investigated the performance of UF in seawater with algal blooms. It reported good feed water quality to RO with a consistent SDI of 2.2, which was much better than that obtained by conventional pretreatment, which fluctuated significantly with respect to turbidity (Pearce *et al.* 2004).

However, aside from the mentioned disadvantages, the use of sand media filtration could also offer some benefits when treating seawater with varying qualities, but not in all situations. For example, when seawater is exposed to sudden variation in pH such as during chemical spills, or large oil and grease spills; or exposure of seawater to very high temperatures or the existence of some contaminants/materials (such as sharp objects, shell particles, barnacles, etc.) that might cause irreversible damage to the UF/MF membranes, GMF would give a cost-effective choice as the cost of replacing the media is much cheaper compared to replacing all membrane elements for a same size plant (Voutchkov, 2009). To protect the membrane filter from physical damage due to sharp objects, microscreening is usually applied.

When selecting for granular filter media for SWRO pretreatment, the filter media type, uniformity, size and depth are of key importance. Deep dual media filters are recommended when the goal of pretreatment is to remove soluble organics from seawater through biofiltration, or to handle high content of pico-plankton. Tri-media filters are used for seawater containing large amount of silt or seawater with algal blooms. Granular media gravity filtration is reported to be more advantageous than granular media pressure filter when the source seawater has frequent exposure to heavy algal blooms. This is because the pressure filter, due to the application of pressure, could result to the breakdown of algal cells, which release cellular cytoplasm in seawater that serves as food for microorganisms leading to increased biofouling problem (Voutchkov, 2010). In terms of space and performance in small desalination plants, granular media pressure filters are very cost-competitive, with lesser space requirement, and are easier and faster to install and operate.

The temperature of seawater is an important parameter for consideration in the selection of the pretreatment process. When seawater temperature is equal or above 35°C, it is prone to mineral scaling and biofouling, and when $T > 45$°C, irreversible RO membrane damage could happen.

Between the two kinds of membrane filters (i.e., vacuum-driven and pressure-driven), the vacuum-driven membrane filtration is considered more advantageous when dealing with variable seawater quality especially in terms of spikes in turbidity (20 NTU or more). This is because pressure-driven membranes have lower capability of retaining solids due to its tight membrane vessel configuration with low retention volume. The vacuum-driven membrane filtration has higher volume and capacity to handle highly turbid water, and since it operates at lower trans-membrane pressure, it has lower tendency for membrane fouling and has more stable operation at variable load conditions. However, vacuum-driven membrane systems are more sensitive to the temperature/viscosity of source water. For example, the use of vacuum driven membrane pretreatment system at temperatures lower than 12°C will result in a less cost-effective process because the flux will be drastically lowered due to the huge increase in unit weight of seawater at low temperature (Voutchkov, 2009). In most cases, the pressure-driven membrane systems are often the membrane pretreatment choice for desalination plants with deep off-shore open ocean intakes, having an average temperature of 15°C, or having low turbidity and low organic and algal content (Voutchkov, 2010).

3.4.2 Cost

Aside from the pretreatment performance, the cost of installation and operation is an important factor for the selection of proper pretreatment strategy. Table 3.6 shows the cost analysis and differences in requirements between conventional granular media filtration pretreatment and advanced membrane (UF/MF) pretreatment.

Table 3.6 Analysis of the benefits between conventional and MF/UF pretreatments.

	Conventional pretreatment	MF/UF pretreatment	Benefits
Capital costs	Cost competitive with MF/UF	Slightly higher than conventional pretreatment. Costs continue to decline as developments are made	Capital costs of MF/UF could be 0–25% higher, whereas life cycle costs using either of the treatment schemes are comparable
Foot print	Calls for larger footprint	Significantly smaller footprint	Foot print of MF/UF could be 30–60% of conventional filters.
Energy requirements	Less than MF/UF as it could be gravity flow	Higher than conventional	MF/UF requires pumping of water through the membranes. This can vary depending on the type of membrane and water quality
Chemical costs	High due to coagulant and process chemicals needed for optimization	Chemical use is low, dependent on raw water quality	Less chemicals
RO capital cost	Higher than MF/UF since RO operates at lower flux	Higher flux is logically possible resulting in lower capital cost	Due to lower SDI values, RO can be operated at 20% higher flux if feasible, reducing RO capital costs
RO operating costs	Higher costs as fouling potential of RO feed water is high resulting in higher operating pressure. One experiences frequent cleaning of RO membranes.	Lower RO operating costs are expected due to less fouling potential and longer membrane life	The NDP (net driving pressure) is likely to be lower if the feed water is pretreated by MF/UF. Membrane cleaning frequency is reduced by 10–100%, reducing system downtime and prolonged element life.

Source: Adapted from Vedavyasan (2007).

The capital cost for membrane pretreatment technology is slightly higher than that of conventional pretreatment, however, the latter has relatively higher operating cost than the former. This is due to the excess usage of chemicals, filtrate and energy for conventional pretreatment. The usual life of UF/MF membranes is only around 5–10 years, whereas for the conventional pretreatment, the equipment and media last much longer, with low maintenance requirement. According to worldwide surveys, the main reason for early membrane replacement is not due to loss of productivity, but is due to loss of membrane integrity (Voutchkov, 2009). The shorter life of UF/MF membrane and subsequent membrane replacement is balanced by the lower expected expenses in RO replacement, and lesser cleaning frequency, as the membrane pretreatment produces good quality feed, that is, low SDI value. Recent advances in membrane technology also make membranes more cost competitive, however, the diversity of membrane elements and configurations could also pose a disadvantage in its future use.

In terms of space requirement, membrane technologies (UF/MF) have about a 30–60% smaller footprint compared to GMF (Vedavyasan, 2007; Kim *et al.* 2009a). The importance of lower space requirement becomes much more significant when an upgrade of an existing desalination plant is needed with limited space, or when the cost of the target site is high. The advantage of a lower footprint for membrane pretreatment could be of greater importance when high turbidity water is treated. For similar filter effluent, the membrane pretreatment could save 60% less land area than conventional pretreatment. One study conducted an economic comparison between UF and conventional pretreatment (Wolf *et al.* 2005). It reported 40% higher specific investment cost for UF compared to conventional pretreatment, but the cost is offset by the lower desalination system investment cost for UF, leading to more competitive specific investment cost for UF pretreatment (Ang *et al.* 2014).

Membrane pretreatment requires approximately two times more power for its operation compared to conventional GMF. This difference is due to the typical need of pumping power of membranes to create flow-driving pressure, source seawater pumping, and for backwashing. Conventional pretreatment usually utilizes a gravity granular filtration with minimum power requirements (Voutchkov, 2009).

The evaluation between conventional and advanced pretreatment should consider the extent of cleaning, waste stream quantity and quality, backwashing and maintenance requirements. Conventional granular media pretreatment in most cases does not use chemicals for cleaning the media (but chlorine is used occasionally), but requires coagulation chemicals for proper separation of solids. On the other hand, membrane pretreatment needs cleaning chemicals to clean its membranes. The conditioning chemicals for conventional pretreatment and the cleaning chemicals for membranes are found to be comparable in the total annual cost. However, the difference in cleaning cost lies in the extent of cleaning of RO system, wherein pretreatment by membrane could be more advantageous due to the production of better feed water quality, thus lessening the tendency of fouling on RO membranes. In addition to cleaning, the amount and quality of waste stream could add up to the overall cost of the desalination plant. Conventional pretreatment typically generates only one waste stream, that is, coming from filter backwash. However, membrane pretreatment generates two types of streams: from waste membrane wash water, and from membrane cleaning solution. The amount of generated waste streams is related to the volume of plant intake source, wherein with membrane pretreatment, larger plant intake source volume is needed which leads to appropriate increase in plant size, construction cost, and maintenance cost (Voutchkov, 2009).

The right pretreatment strategy should consider the overall life-cycle cost analysis of the desalination system, which requires careful judgment. Desalinated water from membrane pretreatment is reported to cost 5–10% more than that from conventional pretreatment, but if space is limited and land price is high, and chemical and energy costs are relatively low, then membrane pretreatment could be more cost advantageous (Voutchkov, 2009). The availability of new materials and advancements in membrane technologies leading to compatible, standardized and commercialized membrane products could allow membrane pretreatment to be highly competitive in the future. These days, increasing number of newer desalination plants are incorporating membranes as their pretreatment strategy. However, conventional pretreatment at present still dominates as the pretreatment of choice in desalination.

3.5 SUMMARY AND CONCLUSION

Pretreatment is a necessary process for efficient and more productive seawater reverse osmosis (SWRO) desalination. Seawater contains different constituents and contaminants such as colloidal particulates, organic matters, mineral salts, oils and grease, hydrocarbons, and microorganisms, which require extensive pretreatment prior to RO to remove the possible foulants. Pretreatment can be divided into conventional and advanced processes. The most commonly used pretreatment in many desalination plants around the world is conventional granular media filtration (GMF), which has provided water quality required for RO lines. However, several obstacles are faced by the current conventional GMF. Among these are unstable SDI values, a need for frequent backwashing, high usage of conditioning chemicals, and a high rate of water consumption. In recent years, membrane technologies (advanced pretreatment) such as UF and MF have increasingly garnered attention as an efficient pretreatment process for SWRO as they can provide more stable and reliable feed water quality at variable feed water conditions. Additionally, membrane pretreatment has a reduced footprint, is easier to operate, has a lower frequency for RO membrane replacement, making it more attractive. However, depending on the feed water quality conditions, integration of different pretreatment technologies may offer beneficial results in RO desalination.

Selecting the most appropriate pretreatment technologies should consider the overall life-cycle cost analysis accounting for all the costs related to installation and operation between conventional and advanced pretreatment systems. With the rapid development and advancement of membrane technologies, coupled with the experiences of the desalination industry using membrane pretreatment, the use of membranes as pretreatment looks more competitive and promising in the years to come.

3.6 REFERENCES

Al-Sheikh A. H. H. (1997). Seawater reverse osmosis pretreatment with an emphasis on the Jeddah Plant operation experience. *Desalination*, **110**(1), 183–192.

Amirtharajah A. (1978). Optimum backwashing of sand filters. *J. Environ. Eng. Div. Am. Soc. Civ. Eng.*, **104**(10), 917–932.

Ang W. L., Mohammad A. W., Hilal N. and Leo C. P. (2014). A review on the applicability of integrated/hybrid membrane processes in water treatment and desalination plants. *Desalination*, **363**, 2–18.

Baker R. W. (2004). Membrane technology and applications. John Wiley & Sons, Ltd, London.

Bonnelye V., Sanz M. A., Durand J.-P., Plasse L., Gueguen F. and Mazounie P. (2004). Reverse osmosis on open intake seawater: pre-treatment strategy. *Desalination*, **167**, 191–200.

Brehant A., Bonnelye V. and Perez M. (2002). Comparison of MF/UF pretreatment with conventional filtration prior to RO membranes for surface seawater desalination. *Desalination*, **144**(1–3), 353–360.

Bungay P. M., Lonsdale H. K. and dePintio M. N. (1986). Porous Membranes, in Synthetic Membranes: Science Engineering and Applications. D. Reidel Publishing Company, Dordrecht.

Bush M., Chu R. and Rosenberg S. (2010). Novel trends in dual membrane systems for seawater desalination: minimum primary pretreatment and low environmental impact treatment schemes. *IDA Journal*, **2**(1), 56–71.

Cleasby J. L., Dharmarajah A. H., Sindt G. L. and Baumann E. R. (1989). Design and Operation Guidelines for Optimization of the High-Rate Filtration Process: Plant Survey Results. American Water Works Association Research Foundation, Co.

Cleasby J. L., Dharmarajah A. H., Sindt G. L. and Baumann E. R. (1992). Design and Operation Guidelines for Optimization of the High-Rate Filtration Process: Plant Demonstration Studies. American Water Works Association Research Foundation, Co.

Cui Z., Peng W., Fan Y., Xing W. and Xu N. (2013). Ceramic membrane filtration as seawater RO pre-treatment: influencing factors on the ceramic membrane flux and quality. *Desalination and Water Treatment*, **51**(13–15), 2575–2583.

Cui Z., Xing W., Fan Y. and Xu N. (2011). Pilot study on the ceramic membrane pre-treatment for seawater desalination with reverse osmosis in Tianjin Bohai Bay. *Desalination*, **279**, 190–194.

Di Profio G., Ji X., Curcio E. and Drioli E. (2011). Submerged hollow fiber ultrafiltration as seawater pretreatment in the logic of integrated membrane desalination systems. *Desalination*, **269**(1–3), 128–135.

Ebrahim S., Bou-Hamed S., Abdel-Jawad M. and Burney N. (1997). Microfiltration system as a pretreatment for RO units: Technical and economic assessment. *Desalination*, **109**(2), 165–175.

Ebrahim S., Abdel-Jawad M., Bou-Hamad S. and Safar M. (2001). Fifteen years of R&D program in seawater desalination at KISR Part I. Pretreatment technologies for RO systems. *Desalination*, **135**(1), 141–153.

Field R. W., Wu D., Howell J. A. and Gupta B. B. (1995). Critical flux concept for microfiltration fouling. *Journal of Membrane Science*, **100**(3), 259–272.

Fuller G. W. (1898). The Purification of the Ohio River Water at Louisville, Kentucky: made to the president and directors of the Louisville Water Company. D. Van Nostrand Co, New York.

Glueckstern P., Priel M. and Wilf M. (2002). Field evaluation of capillary UF technology as a pretreatment for large seawater RO systems. *Desalination*, **147**(1), 55–62.

Greenlee L. F., Lawler D. F., Freeman B. D., Marrot B. and Moulin P. (2009). Reverse osmosis desalination: Water sources, technology, and today's challenges. *Water Res.*, **43**(9), 2317–2348.

Hussain A. R. and Ahmed A. H. (1998). The Addur SWRO desalination plant towards a full plant production. *Desalination*, **120**(1), 143–152.

Jeong S., Nateghi F., Nguyen T. V., Vigneswaran S. and Tu A. (2011). Pretreatment for seawater desalination by flocculation: Performance of modified poly ferric silicate (PFSi-δ) and ferric chloride as flocculants. *Desalination*, **283**, 106–110.

Kim S. H., Yoon J. S. and Lee S. (2009a). Determination of the optimum filtration conditions in pretreatment of seawater desalination by reverse osmosis. *Journal of Korean Society of Water and Wastewater*, **23**(2), 207–214.

Kim Y. M., Kim S. J., Kim Y. S., Lee S., Kim I. S. and Kim J. H. (2009b). Overview of systems engineering approaches for a large-scale seawater desalination plant with a reverse osmosis network. *Desalination*, **238**(1–3), 312–332.

Kim S.-H., Lim H.-K., Jeong W.-C. and Park N.-S. (2011). Optimum backwash method for granular media filtration of seawater. *Desalin Water Treat*, **32**(1–3), 431–436.

Kumar S. A., Adam S. and Pearce W. (2006). Investigation of seawater reverse osmosis fouling and its relationship to pre-treatment type. *Environmental Science and Technology*, **40**, 2037–2044.

Kwon D. Y., Vigneswaran S., Ngo H. H. and Shin H. S. (1997). An enhancement of critical flux in crossflow microfiltration with a pretreatment of floating medium flocculator/prefilter. *Water Science and Technology*, **36**(12), 267–274.

Lee S. and Lee C. H. (2006). Microfiltration and Ultrafiltration As a Pretreatment For Nanofiltration of Surface Water. *Separation Science and Technology*, **41**(1), 1–23.

O'Melia C. R. (1985a). Practices, pretreatment, and filtration. *Journal of Engvironment Engineering Division, ASCE*, **111**(6), 874–884.

O'Melia C. R. (1985b). The influence of coagulation and sedimentation on the fate of particles, associated pollutants, and nutrients in lakes. Chemical processes in lakes. Wiley-Interscience, W. Stumm, pp. 207–224.

Pearce G., Talo S., Chida K., Basha A. and Gulamhusein A. (2004). Pretreatment options for large scale SWRO plants: case studies of OF trials at Kindasa, Saudi Arabia, and conventional pretreatment in Spain. *Desalination*, **167**, 175–189.

Peñate B. and García-Rodríguez L. (2012). Current trends and future prospects in the design of seawater reverse osmosis desalination technology. *Desalination*, **284**, 1–8.

Report A. C. (1980). *Jour. AWWA*. **72**(7), 405–411.

Regula C., Carretier E., Wyart Y., Gésan-Guiziou G., Vincent A., Boudot D., et al. (2014). Chemical cleaning/disinfection and ageing of organic UF membranes: A review. *Water Research*, **56**, 325–365.

Shon H., Vigneswaran S., Zareie M., Aim R. B., Lee E., Lee J., *et al.* (2009a). Physicochemical pretreatment to seawater reverse osmosis (SWRO): organic characterization and membrane autopsy. *Desalination*, **236**(1), 282–290.

Shon H. K., Kim S. H., Vigneswaran S., Ben Aim R., Lee S. and Cho J. (2009b). Physicochemical pretreatment of seawater: fouling reduction and membrane characterization. *Desalination*, **238**(1–3), 10–21.

Valavala R., Sohn J., Han J., Her N. and Yoon Y. (2011). Pretreatment in reverse osmosis seawater desalination: A short review. *Environmental ENgineering Research*, **16**, 205–212.

van Hoof S. C. J. M., Minnery J. G. and Mack B. (2001). Dead-end ultrafiltration as alternative pre-treatment to reverse osmosis in seawater desalination: a case study. *Desalination*, **139**(1–3), 161–168.

Vedavyasan C. V. (2007), Pretreatment trends — an overview. *Desalination*, **203**(1–3), 296–299.

Vollprecht R. (2013). Maximising the Potential of the Perth Seawater Desalination Plant through varying Process Optimisations. International FO/PRO Summit Seoul, Korea.

Voutchkov N. (2009). SWRO pre-treatment systems: Choosing between conventional and membrane filtration. *Filtration & Separation*, **46, Supplement 1**, 5–8.

Voutchkov N. (2010). Considerations for selection of seawater filtration pretreatment system. *Desalination*, **261**(3), 354–364.

Wang Z., Ma J., Tang C. Y., Kimura K., Wang Q. and Han X. (2014). Membrane cleaning in membrane bioreactors: A review. *Journal of Membrane Science*, **468**, 276–307.

Wolf P. H., Siverns S. and Monti S. (2005). UF membranes for RO desalination pretreatment. *Desalination*, **182**(1–3), 293–300.

Yao K. M. (1968). Influence of suspended particle size on the transport aspect of water filtration. Ph.D. dissertation, University of Nortch Carolina.

Yao K. M. and M'Melia C. R. (1968). Particle transport in aqueous flow through porous media. 16th Annual Conference of the Hydraulic Division of ASCE, MIT, Cambridge, Massachussets, Available as ESE publication no. 210, Dept. of Environmental Sciences and Engineering, University of North Carolina, Chapel Hill, N.C., 1968.

Yao K.-M., Habibian M. T. and O'Melia C. R. (1971). Water and waste water filtration. Concepts and applications. *Environ Sci Technol*, **5**(11), 1105–1112.

Chapter 4

Reverse osmosis process' design and applications

Nikolay Voutchkov

President, Water Globe Consulting, LLC

Corresponding author: nvoutchkov@water-g.com

4.1 OVERVIEW

Based on the source water salinity they process, membrane based desalination plants can be divided into three main groups: nanofiltration (softening) plants (NF); brackish water reverse osmosis plants (BWRO) and seawater reverse osmosis plants (SWRO). In addition, depending on the number of sequential reverse osmosis (RO) systems for treatment of permeate and concentrate, RO system configurations could be divided into two main categories: (1) single and multiple pass RO systems; (2) single and multiple stage RO systems. In all types of desalination plants multiple pass and multiple stage RO systems could also be combined into configurations that allow achieving target RO system recovery and product water quality at optimum life-cycle cost of water production. The various types of systems and their practical application are discussed in the following sections.

4.1.1 Single and multiple pass RO systems

A NF or RO system where the saline source water is desalinated only once (i.e., source water passes through one membrane) is referred to as a single-pass RO system (see Figure 4.1). RO systems that are designed to re-treat permeate from a single-pass (also referred to as first-pass) RO system by reverse osmosis multiple times are termed multiple-pass RO systems.

Figure 4.1 depicts a general schematic of single-pass, two pass and three pass RO systems. Since each RO pass provides additional treatment of permeate produced by the previous RO pass, the overall system permeate water quality improves with each pass. Therefore, multiple pass RO systems are applied when source water salinity is relatively high and the target product water quality cannot be achieved by treating the saline source water by reverse osmosis only once.

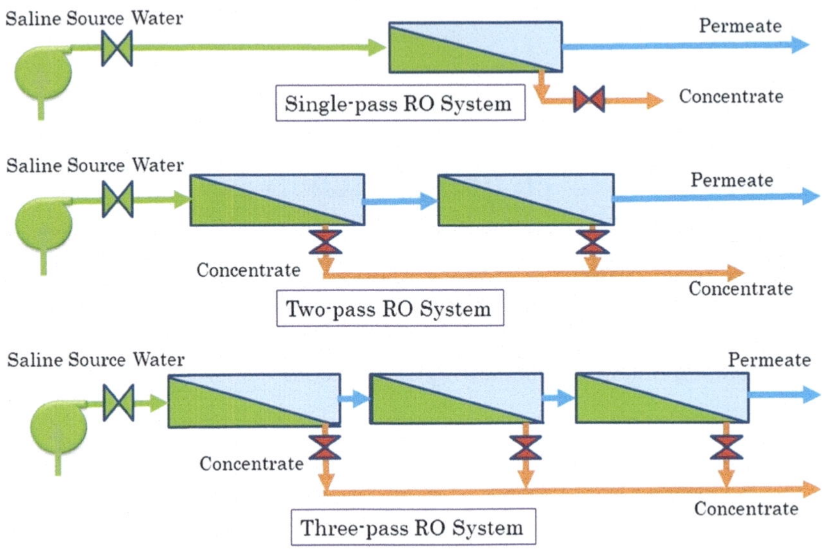

Figure 4.1 Single and multiple pass RO systems.

While multiple RO pass systems generate product water of very high quality, such systems are also more costly and produce less water than a single-pass RO system processing the same source water flow. Therefore, multiple-pass RO systems are typically applied when a single pass RO system cannot produce product water of the desired quality. Also, multiple-pass (usually two-pass) RO systems may be preferred when the source water has a very high fouling potential.

The rate of RO membrane fouling is exponentially proportional to the membrane flux, which in turn is proportional to the difference between the RO system feed and permeate pressures. By using a two-pass rather than a single-pass RO system, the feed pressure to the first pass can be reduced significantly, typically down to 65 to 75% of the total RO feed pressure. Such a reduction in pressure results in a measurable improvement of RO system operations; by reducing the frequency of RO membrane cleaning and lowering the overall differential/feed RO pressure. The use of a two-pass RO system, however, requires the installation of an inter-pass RO pump to deliver the remaining 25 to 35% of the total RO feed pressure needed for water production. The pump which boosts the permeate pressure from the first pass to the second pass is referred to as a booster or second pass RO feed pump.

A key challenge associated with the use of multiple-pass RO systems is that the overall recovery of such systems decreases with the number of installed RO passes, because a portion of the saline source water is converted into concentrate at each pass (see 7–2).

In order to reduce the total volume of concentrate and increase the overall water recovery to produce more fresh water from the same volume of source water, the concentrate generated by the individual RO passes can be treated by a separate RO system referred to as a 'stage'. Usually, the membrane vessels of the BWRO system stages are grouped in parallel and are referred to as 'arrays'. The stages are arranged in series such that adequate cross-flow and minimum concentrate flow is maintained in each stage.

Typically, the ratio between the number of vessels in the first stage and second stage of NF and BWRO systems is set at 2:1 and such configurations are referred to as a '2:1 array'. In three-stage NF and BWRO systems the typical array configuration is – 3:2:1 – that is, the number of vessels in the first stage is 3 times higher than the number in the third stage. The second stage has two-times higher number of vessels than the third stage.

Figure 4.2 depicts single-stage, two-stage and three-stage RO systems. Use of multiple stages allows improving the overall recovery of the entire RO system. However, the additional concentrate treatment steps also increase RO system costs. The optimum number of stages and passes for a given RO system depends on many factors, such as source water quality, target permeate water quality, fresh water production flow, and cost of equipment and energy. Therefore the optimum configuration should be determined based on the site-specific conditions of each project.

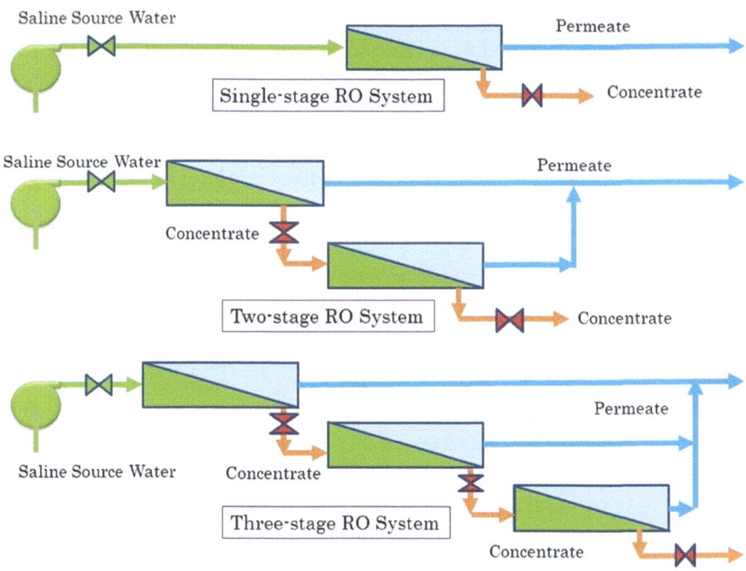

Figure 4.2 Single and multiple stage RO systems.

It is important to note that the single-pass and single-stage RO systems shown in Figures 4.1 and 4.2 are identical in configuration – that is a RO system in which the source water is treated only once can be referenced both as single-pass and single stage. The difference of nomenclature can be explained by the main difference in configuration of SWRO and BWRO systems. For SWRO desalination plants it is common that the source water undergoes RO treatment multiple times, while for NF and BWRO plants the concentrate rather than the permeate generated by the first RO treatment undergoes additional processing by reverse osmosis. As a result, a SWRO system, where the source water and permeate are processed through reverse osmosis in sequence is referred to as a two-pass SWRO system, while a BWRO system where the source water and the concentrate are processes through reverse osmosis in sequence is referred to as two-stage BWRO system.

4.2 NF SYSTEM CONFIGURATIONS

Nanofiltration plants usually treat relatively low salinity water (typically less than 1,000 mg/L) which has a high level of hardness, organic content, color, iron or other bi-valent ions such as nitrates. Typically, NF plants are equipped with two-stage RO systems and are designed to operate at 80 to 90% recovery. The most commonly used NF membrane configuration includes pressure vessels with seven elements per vessel and a 2:1 membrane array.

Figure 4.3 depicts the general process schematic of the largest operational NF plant in the US – the 151,000 m^3/day (40 MGD) Boca Raton plant is Florida. This plant processes well water of relatively low salinity (466 mg/L) but has high hardness (265 mg/L), alkalinity (265 mg/L); iron (0.3 mg/L) and NOM content (TOC = 12 mg/L). The plant is designed to operate for a total recovery of 85% and it has ten (10) two-stage NF trains with 2:1 arrays (72 vessels in the first stage and 36 vessels in the second stage). The plant uses Hydranuatics ESNA1-LF2 elements operating at an average flux of 20.7 Lmh (12.2 gfd) and producing an average of 20.1 m^3/day (5,300 gpd) per element. The plant permeate has total hardness of 50 to 80 mg/L; iron content below 0.04 mg/L and TOC of less than 1 mg/L (Bartels *et al.* 2007).

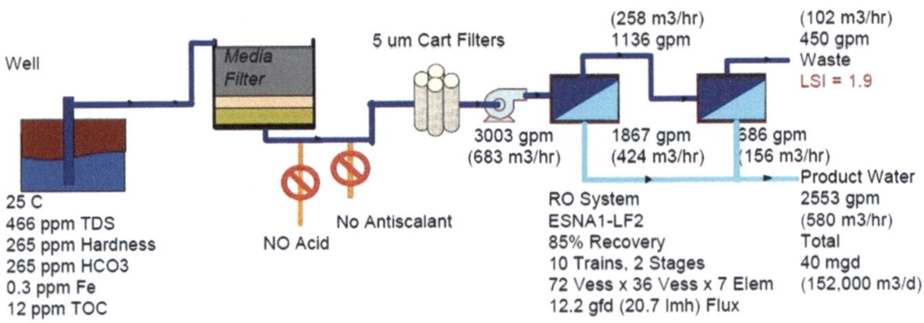

Figure 4.3 Schematic of Boca Raton NF plant, Florida.
Source: Hydranautics.

4.3 BWRO SYSTEM CONFIGURATIONS

Source water TDS concentration of BWRO plants typically ranges between 500 mg/L and 10,000 mg/L. Plants processing source water with salinity between 500 and 2,500 mg/L and in a range of 2,500 to 10,000 mg/L (or above) are referred to as low-salinity and high-salinity BWRO desalination facilities, respectively.

Figure 4.4 illustrates a typical process flow diagram of a low-salinity BWRO desalination plant. For such plants blending a portion (5% to 30%) of the source water flow (indicated on Figure 4.4 as 'Bypass Source Water') with RO permeate is very common practice for re-mineralization of the desalinated water. Low-salinity BWRO plants often process the source water through a single RO stage (pass) only. However, two-stage BWRO plants configured with 2:1 arrays are also very common.

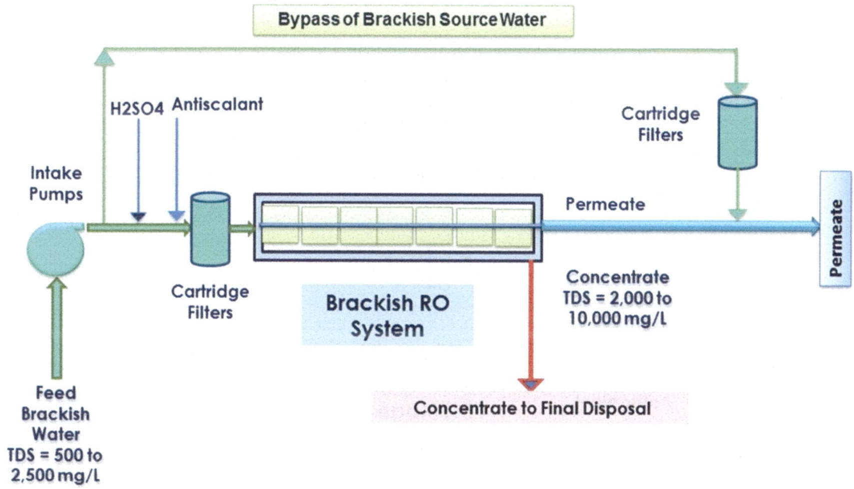

Figure 4.4 Schematic of typical low-salinity BWRO plant.

Table 4.1 provides an illustrative hypothetical example of the permeate water quality produced by low-salinity BWRO plant operating at blending ratio of 28.6% and permeate recovery of 85% (Wilf *et al.* 2007). In this specific example, the TDS of the source water and RO permeate are 647.3 mg/L and 215 mg/L, respectively.

Figure 4.5 depicts a typical process flow diagram for a high-salinity BWRO desalination plant. As indicated in this figure, brackish source water in this case is usually treated by a two-stage RO system where the concentrate generated by the first RO pass is processed through a second RO system after increasing (boosting) the concentrate pressure via an inter-stage booster pump. Under this configuration, the first stage permeate usually contributes 75% to 85% of the total permeate flow while the RO system which processes concentrate from the first RO pass (referred to as 'second stage') produces the remaining 15% to 25% of the total RO system permeate flow.

Table 4.1 Example of product water quality generated by low-salinity BWRO plant.

Water Quality Parameter	Source Water Quality	Blended Permeate Water Quality
Temperature, °C	25	25
pH	7.0	6.6
Ca^{2+}, mg/L	96	29
Mg^{2+}, mg/L	11.7	3.5
Na^{+}, mg/L	90	32.1
K^{+}, mg/L	6.5	2.4
HCO_3^-, mg/L	72.6	30.4
SO_4^{2-}, mg/L	158.4	47.2
Cl^-, mg/L	190.7	61
F^-, mg/L	0.2	0.1
SiO_2, mg/L	24.3	9.3
TDS, mg/L	647.3	215

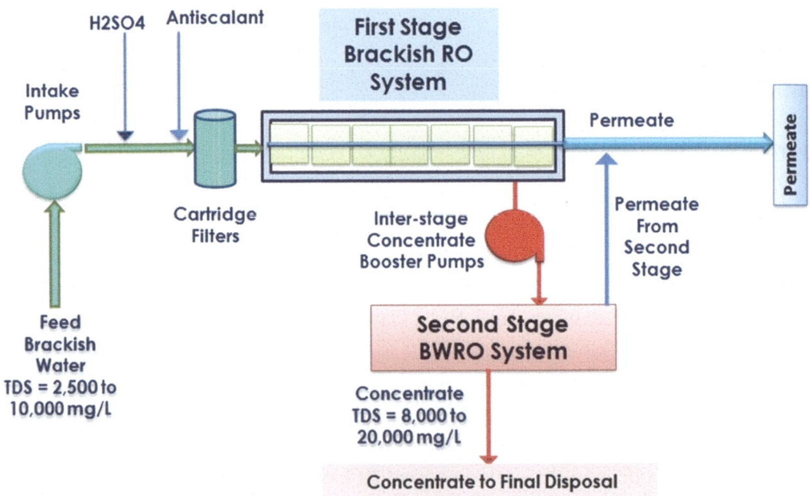

Figure 4.5 Typical schematic of high-salinity brackish water RO plant.

A hypothetical example of the permeate water quality produced by a high-salinity BWRO plant is presented in Table 4.2 (Wilf *et al.* 2007). In this example, the desalination plant uses a brackish well water source of salinity of 5,881 mg/L and processes this source through a two-stage BWRO system similar to that illustrated in Figure 4.5. The overall plant recovery is 80% (i.e., 80% of the source water is converted into low salinity permeate). The system is equipped with an inter-stage booster pump which directs the concentrate generated by the first BWRO stage to the second RO stage.

Table 4.2 Example of high-salinity BWRO plant permeate water quality.

Water Quality Parameter	Source Water Quality	Blended Permeate Water Quality
Temperature, °C	28	28
pH	7.0	6.0
Ca^{2+}, mg/L	105	1.0
Mg^{2+}, mg/L	130	1.3
Na^{+}, mg/L	1,837	84
K^{+}, mg/L	85	4.8
CO_3^{2-}, mg/L	0.3	0.0
HCO_3^{-}, mg/L	250	18.4
SO_4^{2-}, mg/L	479	5.1
Cl^{-}, mg/L	2,970	123
F^{-}, mg/L	1.4	0.1
NO_3^{-}, mg/L	5.0	1.4
SiO_2, mg/L	17.0	0.5
TDS, mg/L	5,881	240

The recovery of the first RO stage for the example presented in Table 4.2 is 64%. The second stage recovers approximately 44% of the concentrate produced by the first stage RO system representing 20% of the total system permeate flow.

4.4 SEAWATER SYSTEM CONFIGURATIONS

The SWRO system configurations that are most widely applied at present include: single-pass treatment, where the source water is processed by reverse osmosis only once (Figure 4.6); and two-pass RO treatment where the seawater is first processed through a SWRO system and then permeate produced by this system is re-processed by brackish RO membranes (see Figures 4.7 and 4.8).

4.4.1 Single-pass SWRO systems

Single-stage SWRO systems are designed to produce desalinated seawater (permeate) in one step using only a single set of RO trains operating in parallel. In general, between 800 and 900 SWRO membrane elements installed in 100 to 150 vessels are needed to produce 10,000 m^3/day (2.6 MGD) of permeate suitable for potable use in a single-stage SWRO system.

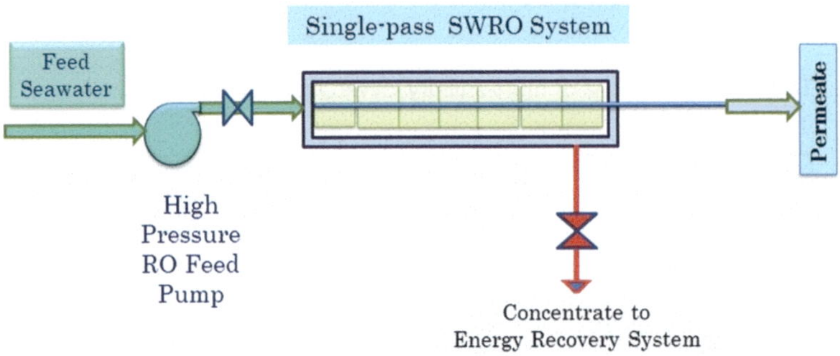

Figure 4.6 Single-pass SWRO system.

Under a typical single-stage SWRO system configuration, each RO train has a dedicated system of transfer pumps for pretreated seawater followed by a high-pressure RO feed pump. The high-pressure feed pump motor/operation is coupled with that of energy recovery equipment (see Figure 4.6).

Single-stage SWRO systems are widely used for production of drinking water. However, these systems have found limited industrial application mainly because of water quality limitations of the produced permeate. Even if using the highest-rejection RO membrane elements commercially available today (nominal minimum rejection of 99.85%), single-stage SWRO desalination systems typically cannot consistently yield permeate with TDS concentration lower than 200 mg/L, chloride level of less than 100 mg/L, and boron concentration lower than 0.5 mg/L, especially when source water temperatures exceed 18 to 20°C (64 to 68°F).

If enhanced boron removal is needed in such systems, high boron rejection membranes are used, and/or sodium hydroxide and antiscalant is added to the RO system feed water to increase pH to 8.8 or more, which in turn improves boron rejection.

4.4.2 Two-pass SWRO systems

Two-pass SWRO systems are typically used when either the source seawater salinity is relatively high (i.e., higher than 35,000 mg/L) and/or the product water quality requirements are very stringent. For example, if high salinity/high-temperature source water (such as Red Sea and Persian Gulf seawater for example) is used in combination with standard-rejection (99.6%) SWRO membranes, then single-stage

SWRO systems are not be able to produce permeate suitable for drinking water use. In this case, two-pass SWRO systems have proven to be a very efficient and cost-effective configuration for potable water production. RO systems with two or more passes are also widely used for production of high-purity industrial water.

The two-pass SWRO systems typically consist of a combination of a single-pass SWRO system and a single or multiple-pass brackish water RO (BWRO) system connected in series (Figures 4.7 and 4.8). Permeate from the SWRO system (i.e., first pass) is directed for further treatment to the BWRO system (i.e., second pass) to produce a high-quality TDS permeate. The concentrate form the pass-two BWRO system is returned to the feed of the pass-one SWRO system to maximize the overall desalination system production capacity and efficiency. Two-pass SWRO systems are classified in two main groups: conventional full two-pass systems and split-partial two-pass systems.

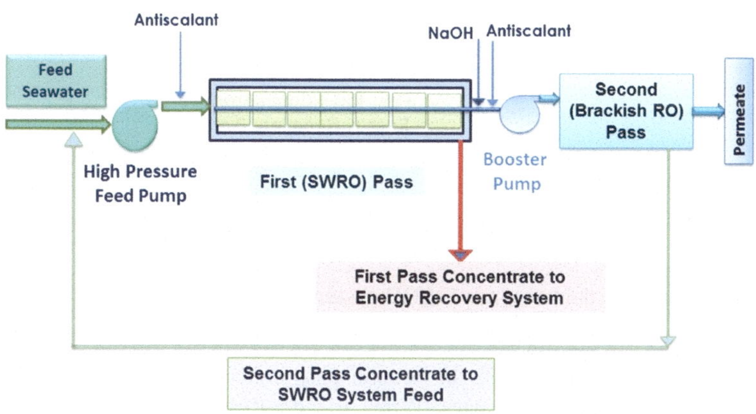

Figure 4.7 Conventional full two-pass SWRO system.

4.4.3 Conventional full-two pass SWRO systems

In conventional full two-pass SWRO membrane systems (see Figure 4.7), the source seawater is first treated by a set SWRO membrane trains (referred to as first RO pass) and then the entire volume of desalinated water from the first pass is processed through a second set of brackish water desalination membrane trains (Greenlee *et al.* 2009). If enhanced boron removal is needed, sodium hydroxide and antiscalant are added to the feed permeate of the second RO pass to increase pH and improve boron rejection.

4.4.4 Split-partial two-pass SWRO systems

In split-partial two-pass systems the second RO pass typically processes only a portion (50 to 75%) of permeate generated by the first pass. The rest of the low salinity permeate is produced by the front (feed) SWRO elements of the first pass. This low-salinity permeate is collected and without additional desalination it is directly blended with permeate produced by the second RO pass (see Figure 4.8).

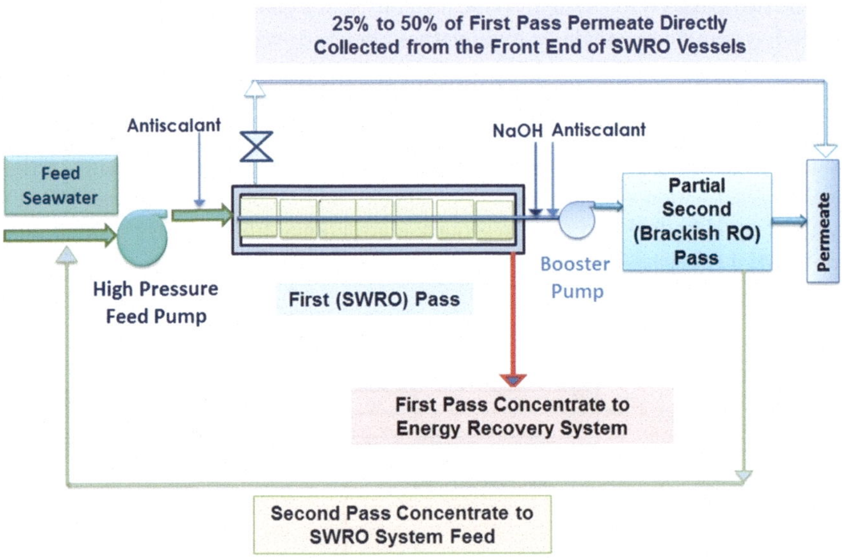

Figure 4.8 Split-partial two-pass SWRO system.

As indicated in Figure 4.8, the second pass concentrate is returned to the feed of the first RO system pass. When the desalination system is designed for enhanced boron removal, this concentrate will have pH of 9.5 to 11 and potentially could cause precipitation of calcium carbonate on the membranes. In order to avoid this challenge, typically antiscalant is added to the feed to the partial second pass (brackish RO) system. Long-term experience with such configuration indicates that this solution is very effective in preventing scaling of the first pass RO membranes by the recycled second pass concentrate. As the boron level in the front permeate stream is usually between 0.25 and 0.5 mg/L, no additional treatment of this stream is needed even if the plant is designed and operated for enhanced boron removal.

While the recycling of the second-pass concentrate returns a small portion of the source water salinity and therefore, it slightly increases the salinity of the seawater fed to the first RO pass, the energy use associated with this incremental salinity increase is significantly smaller than the energy savings of processing the entire volume of the first-pass permeate through the second pass. Under the split partial two pass configuration the volume of permeate pumped to the second RO pass and the size of this pass are typically 25 to 50% smaller than the volume pumped to the second RO pass under conventional once-through operation. Since pumping energy is directly proportional to flow, the energy costs for the second-pass feed pumps are reduced proportionally i.e., with 25 to 50%. For a SWRO system operating at 45% recovery, such savings will amount to 14 to 22% of the energy of the first pass RO pump.

The concentrate returned from the second pass carries only 1 to 2% of additional salinity to the first pass RO feed, which reduces the energy benefit from such recovery proportionally – that is, by 1 to 2% only. As a result, the overall energy savings of the use of split partial two pass RO system as compared to conventional two pass RO system are between 12 and 20%. Practical experience with large SWRO desalination plants indicates that the average total RO system life-cycle cost savings associated with applying such a SWRO system configuration are typically between 14 to 16%.

At present, most new SWRO desalination systems are designed with split partial second pass configuration because this configuration allows reducing the size of the second pass RO system and the overall fresh-water production costs. It should be noted that split-partial second RO pass systems can be configured in several alternatives, which may involve the use of the same or different type of membrane elements within the first pass SWRO system.

An example of a plant with partial-second pass configuration is the 95,000 m^3/day (25 MGD) Tampa Bay seawater desalination facility in Florida. The second pass at this facility is designed to treat up to 30% of the permeate produced by the first-pass SWRO system as needed in order to maintain the concentration of chlorides in the plant product water below 100 mg/L.

The partial second pass at the Tampa Bay water seawater desalination plant was installed to provide operational flexibility and to accommodate the wide fluctuations of source water salinity (16,000 to 32,000 mg/L) and temperature (18 to 40°C/64 to 104°F). Typically, the product water quality target chloride concentration of 100 mg/l at this plant is achieved by only operating the first-pass of the system. However, when source water TDS concentration exceeds 28,000 mg/L and/or the source water temperature exceeds 35°C (95°F), the second pass is activated to maintain adequate product water quality. The percent of first-pass permeate directed for additional treatment through the second pass, is a function of the actual combination of source water TDS and temperature and is adjusted based on the plant product water chloride level.

4.4.5 Product water quality of single and two-pass SWRO systems

Tables 4.3 through 4.7 present summaries of the range of permeate water quality produced by typical single-pass and partial two-pass seawater desalination systems processing five different seawater sources: Pacific Ocean water; Atlantic Ocean water; Mediterranean seawater; Persian Gulf seawater and Red Sea water.

Table 4.3 Reverse osmosis permeate water quality seawater source – Pacific Ocean.

Water Quality Parameter	Pacific Ocean Source Seawater Quality	Permeate Water Quality	
		Single Pass SWRO System	Split Partial Two Pass RO System
Temperature, °C	14–28	15–29	16–30
pH	8.0	6.3–7.2	7.6–7.8
Ca^{2+}, mg/L	358	0.6–1.1	0.2–0.5
Mg^{2+}, mg/L	1,720	1.8–2.8	0.07–0.10
Na^{+}, mg/L	9,900	78–134	9–20
K^{+}, mg/L	600	3.0–6.0	0.43–0.60
CO_3^{2-}, mg/L	2.0	0.0	0.0
HCO_3^{-}, mg/L	170	1.8–2.5	0.4–0.7
SO_4^{2-}, mg/L	2,570	2.6–5.3	0.7–1.3
Cl^{-}, mg/L	18,100	130–195	13–20
F^{-}, mg/L	2.1	0.9–1.2	0.7–0.9
NO_3^{-}, mg/L	0.00	0.00	0.00
B^{-}, mg/L	4.5	0.7–1.2	0.3–0.5
Br^{-}, mg/L	73	0.6–0.9	0.2–0.4
TDS, mg/L	33,500	220–350	25–45

Table 4.4 Reverse osmosis permeate water quality seawater source – Atlantic Oocean.

Water Quality Parameter	Atlantic Ocean Source Seawater Quality	Permeate Water Quality	
		Single Pass SWRO System	Split Partial Two Pass RO System
Temperature, °C	16–30	17–31	18–32
pH	8.0	6.3–7.2	7.7–8.0
Ca^{2+}, mg/L	410	0.7–1.4	0.3–0.5
Mg^{2+}, mg/L	1,302	1.6–2.4	0.35–0.8
Na^{+}, mg/L	10,506	83–160	11–25
K^{+}, mg/L	390	2.4–4.5	0.35–0.50
CO_3^{2-}, mg/L	2.0	0.0	0.0
HCO_3^{-}, mg/L	145	1.4–2.0	0.5–0.8
SO_4^{2-}, mg/L	2,720	2.4–5.8	1.1–1.2
Cl^{-}, mg/L	19,440	146–220	20–34
F^{-}, mg/L	2.5	1.0–1.6	0.8–1.2
NO_3^{-}, mg/L	0.00	0.00	0.00
B^{-}, mg/L	4.5	0.7–1.2	0.3–0.5
Br^{-}, mg/L	78	0.8–1.1	0.3–0.5
TDS, mg/L	35,000	240–400	35–65

Table 4.5 Reverse osmosis permeate water quality seawater source – Mediterranean Sea.

Water Quality Parameter	Mediterranean Source Seawater Quality	Permeate Water Quality	
		Single Pass SWRO System	Split Partial Two Pass RO System
Temperature, °C	16–28	17–29	18–30
pH	8.1	6.3–7.2	7.9–8.1
Ca^{2+}, mg/L	480	1.0–2.0	0.35–0.45
Mg^{2+}, mg/L	1,558	1.9–2.8	0.5–1.0
Na^{+}, mg/L	12,200	98–196	15–34
K^{+}, mg/L	480	3.0–5.5	0.8–1.8
CO_3^{2-}, mg/L	5.6	0.0	0.0
HCO_3^{-}, mg/L	160	1.7–2.4	0.5–0.8
SO_4^{2-}, mg/L	3,190	2.9–6.3	1.4–2.95
Cl^{-}, mg/L	22,340	169–260	25–52
F^{-}, mg/L	1.4	0.7–1.1	0.5–0.8
NO_3^{-}, mg/L	0.00	0.00	0.00
B^{-}, mg/L	5.0	0.9–1.5	0.4–0.6
Br^{-}, mg/L	80	0.9–1.3	0.35–0.6
TDS, mg/L	40,500	280–480	45–95

Table 4.6 Reverse osmosis permeate water quality seawater source – Red Sea.

Water Quality Parameter	Red Sea Source Seawater Quality	Permeate Water Quality	
		Single Pass SWRO System	Split Partial Two Pass RO System
Temperature, °C	22–33	23–34	24–35
pH	7.0–8.0	6.8–7.8	7.6–8.0
Ca^{2+}, mg/L	500	1.1–2.1	0.5–0.7
Mg^{2+}, mg/L	1,540	1.8–3.4	0.7–1.0
Na^{+}, mg/L	13,300	142–220	20–38
K^{+}, mg/L	489	3.2–6.5	1.2–1.8
CO_3^{2-}, mg/L	2.4	0.0	0.0
HCO_3^{-}, mg/L	142.4	1.4–2.0	0.5–1.0
SO_4^{2-}, mg/L	3,100	2.8–6.2	1.9–2.6
Cl^{-}, mg/L	22,840	195–276	29–58
F^{-}, mg/L	0.9	0.5–0.7	0.3–0.5
NO_3^{-}, mg/L	0.00	0.00	0.00
B^{-}, mg/L	5.3	1.2–1.7	0.45–0.80
Br^{-}, mg/L	80	1.0–1.4	0.45–0.60
TDS, mg/L	42,000	350–520	55–105

Table 4.7 Reverse osmosis permeate water quality seawater source – Persian Gulf.

Water Quality Parameter	Persian Gulf Source Seawater Quality	Permeate Water Quality	
		Single Pass SWRO System	Split Partial Two Pass RO System
Temperature, °C	18–35	19–36	20–37
pH	6.0 – 7.0	5.1–6.0	5.1–6.0
Ca^{2+}, mg/L	570	1.4–2.6	0.6–0.8
Mg^{2+}, mg/L	1,600	2.0–3.6	0.9–1.3
Na^{+}, mg/L	14,100	142–228	25–45
K^{+}, mg/L	530	4.3–6.8	1.5–2.2
CO_3^{2-}, mg/L	4.2	0.0	0.0
HCO_3^{-}, mg/L	155	1.8–2.3	0.6–0.9
SO_4^{2-}, mg/L	3,300	3.1–6.5	2.1–3.2
Cl^{-}, mg/L	24,650	222–305	37.5–64
F^{-}, mg/L	1.5	0.9–1.2	0.5–0.8
NO_3^{-}, mg/L	0.00	0.00	0.00
B^{-}, mg/L	6.3	1.3–2.5	0.7–1.0
Br^{-}, mg/L	83	1.2–1.5	0.60–0.80
TDS, mg/L	45,000	380–520	70–120

4.4.6 Four-stage SWRO systems

In addition to the single and two-pass SWRO systems described above, another reverse osmosis configuration that has found practical application in a number of large SWRO desalination plants that have to meet a very stringent boron limit of 0.4 mg/L or less (i.e., Ashkelon, Hadera, and Sorek SWRO plants in Israel) is the four-stage SWRO system depicted in Figure 4.9 (Gorenflo *et al.* 2007).

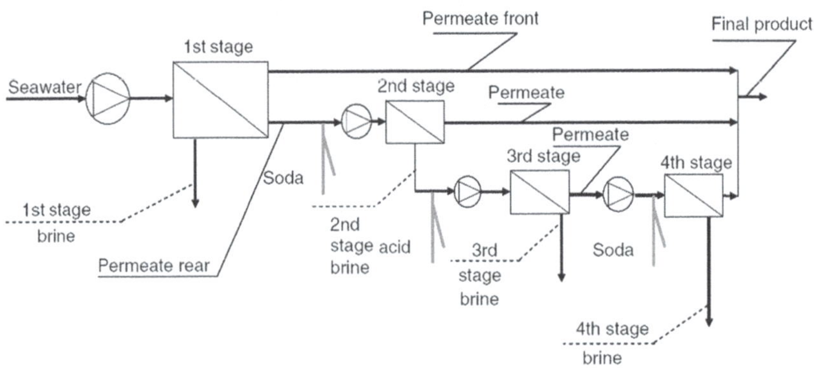

Figure 4.9 Ashkelon four-stage SWRO system (Gorenflo *et al.*, 2007).

Table 4.8 presents a summary of the source water quality and the design and actual product water quality of the Ashkelon SWRO plant in Israel, which was the first facility where such a reverse osmosis configuration was used and therefore, has the longest track record of successful performance (Dreizin *et al.* 2008).

Table 4.8 Reverse osmosis permeate water quality ashkelon seawater desalination plant.

Water Quality Parameter	Mediterranean Source Seawater Quality	Fresh Water Quality	
		Permeate	Finished Water
Temperature, °C	16–28	17–29	18–30
pH	8.1	7.5–8.5	8–8.5
Ca^{2+}, mg/L	483	0.2–0.5	90–110
Mg^{2+}, mg/L	1,557	0.5–0.8	0.5–0.8
Na^{+}, mg/L	12,200	6–30	30–39
K^{+}, mg/L	481	1.5–1.8	1.5–1.8
HCO_3^-, mg/L	162	0.6–0.8	45 – 50
SO_4^{2-}, mg/L	3,186	1.0–1.4	1.0–1.8
Cl^-, mg/L	22,599	10–15	10–15
F^-, mg/L	1.4	0.1–0.2	0.8–1.0
NO_3^-, mg/L	0.00	0.00	0.00
B^-, mg/L	5.3	0.2–0.3	0.2–0.3
Br^-, mg/L	80	0.2–0.3	0.2 – 0.3
TDS, mg/L	40,679	20–40	180–220

4.4.7 Two-stage SWRO systems

Two-stage SWRO membrane systems are mainly used to maximize the overall desalination plant recovery and reduce the volume of concentrate discharged by the desalination plant. A general schematic of a two-stage RO system is shown in Figure 4.10. In these SWRO systems, typically the entire volume of the concentrate generated by the first stage SWRO system is directed to a second-stage SWRO system for further treatment and enhanced recovery. Permeate from both systems is blended prior to final use.

The main advantage of such SWRO system configuration is that it allows achieving very high level of use (recovery) of the available source seawater and the energy used by the first-stage RO system. For example, while a single-stage SWRO system configuration typically transforms 35 to 50% of the source seawater to potable water, the two-stage SWRO system recovery may reach 60 to 65%. Designing the SWRO plant around higher recovery allows minimizing the size of the plant intake and pretreatment facilities, and the capital expenditures for their construction and operation.

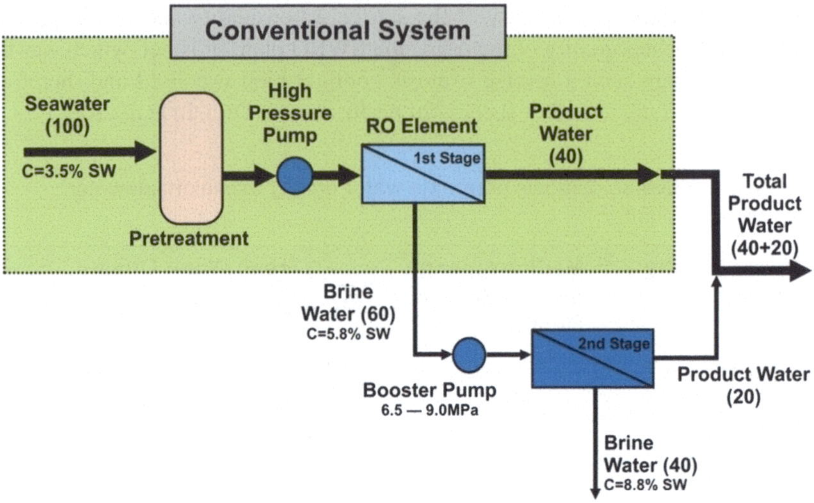

Figure 4.10 Two-stage SWRO system.

However, because of the high salinity of the first-stage concentrate (typically above 55,000 mg/L), the practical implementation of two-stage SWRO systems requires use of high-pressure SWRO membranes, membrane vessels, piping and auxiliary equipment that can withstand and perform well at very high pressures (up to 98 bars/1,400 psi). The cost of this equipment is usually higher than the cost of the same size and type of equipment built to operate at more 'standard' pressures (i.e., below 70 bars/1,000 psi). Therefore, the viability of using two-stage SWRO system has to be carefully assessed based on a comprehensive life-cycle cost analysis for the site specific conditions of a given project. To date, two-stage systems have been mainly used to upgrade capacity and improve energy use of existing conventional single-stage SWRO plants.

Sometimes, two-stage SWRO systems are designed to be operated with lower feed pressure to the first stage, because of the highly fouling nature of the feed water. Since operation at lower feed pressure allows reduced fouling of the first stage/pass elements, this configuration could be beneficial if the pretreatment system is not very robust and the water quality from the SWRO system is of high alluvial organic or particulate content.

4.4.8 Hybrid SWRO systems with multiple passes and stages

The two-pass and two-stage RO system configurations may be combined to achieve an optimum plant design and tailor desalination plant operation to the site specific water source water quality conditions and product water quality goals.

An example of a full-scale two-pass, two-stage SWRO system application is the 170,500 m^3/day Fujairah seawater desalination plant (Sanz *et al.* 2007). A general treatment process flow schematic of this plant is depicted on Figure 4.11. The plant uses Gulf of Oman source seawater.

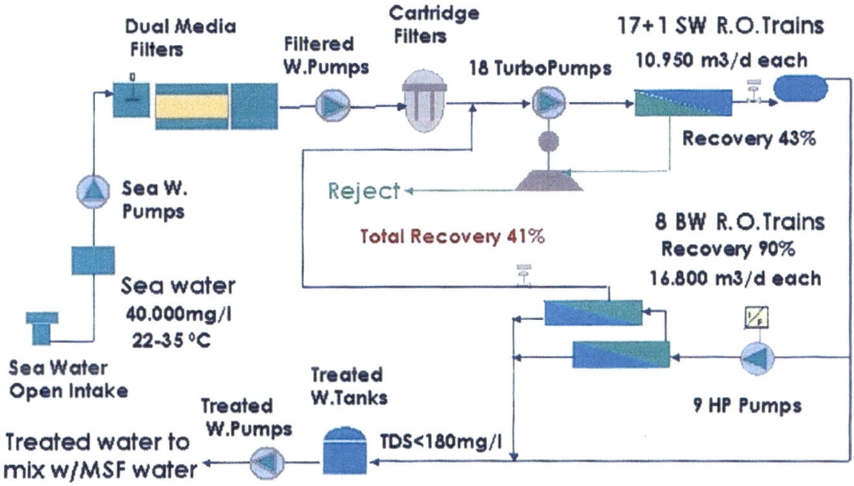

Figure 4.11 Fujairah seawater desalination plant schematic.

The first pass of the Fujairah plant consists of 17 duty and 1 standby RO trains using standard rejection SWRO membranes producing permeate of TDS concentration of 400 to 500 mg/L. The overall recovery of the first pass SWRO system is 43%. The second pass has 8 BWRO trains with a total recovery rate of 90%. The second-pass BWRO train has two stages and treats approximately 80% of the first-pass permeate to a TDS concentration of 10 to 20 mg/L. The rest (i.e., 20%) of the first-pass permeate is blended with the second-pass permeate to produce finished water with a TDS concentration of less than 120 mg/L. Concentrate produced by the second-pass RO system has salinity lower than that of the source seawater and is recycled to the feed of the first pass RO system (see Figure 4.11). Despite its complexity, the two-pass, two-stage RO system in Fujairah performs very reliably.

Another example of a two-pass/two-stage SWRO plant is the 146,000 m^3/day (38 MGD) Point Lisas facility in Trinidad (Figure 4.12). This plant produces high-quality desalinated seawater with TDS concentration of 85 mg/L or less which is predominantly used for industrial applications. The first pass of the Point Lisas SWRO system consists of six two-stage RO units. Each of the first stage RO trains uses SWRO membranes and is coupled with Pelton wheel energy recovery device. The second stage trains are equipped with brackish water RO elements.

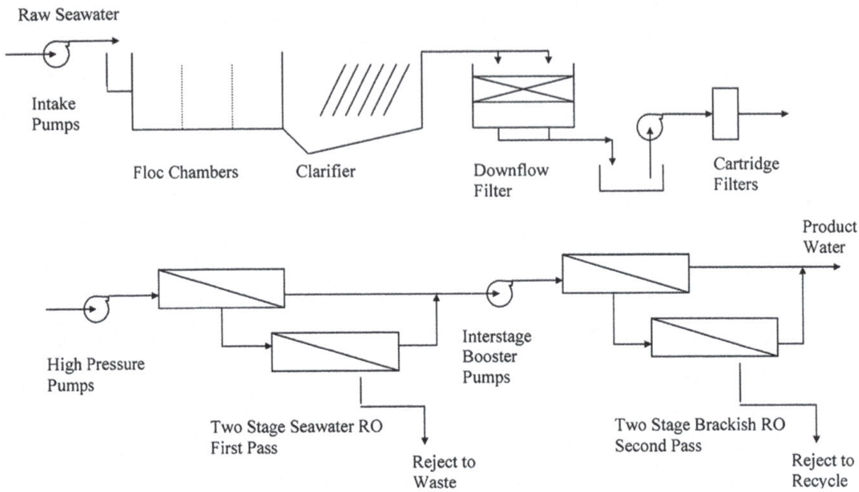

Figure 4.12 Point Lisas seawater desalination plant schematic.

The entire volume of permeate from the first pass of the Point Lisas SWRO system is further treated in a second pass RO system to meet the final product water quality specifications. The second pass system also consists of two stages – each equipped with BWRO membranes. The Point Lisas seawater desalination plant has the same number of first-pass and second-pass RO membrane trains.

Because of the high fouling potential of its source seawater, this two pass-two stage configuration has proven to be beneficial. The high pressure pumps of the first stage are designed for 80% lower feed pressure than those of typical single stage systems, which significantly reduces the fouling on the RO elements of this pass. The permeate from the first pass is treated through a second pass system and an inter-stage booster pump is installed to deliver the remaining 20% of the feed pressure needed to produce water similar to that in a single pass RO. Overall, this configuration produces better water quality, higher recovery and results in less membrane fouling than a typical single-pass system for the same source water quality.

4.5 THREE-CENTER RO SYSTEM CONFIGURATION

As indicated previously, a typical conventional SWRO system is configured in individual equally sized RO trains each of which is serviced by a separate transfer pump, cartridge filter vessel, high pressure feed pump, and energy recovery equipment dedicated to this RO train (see Figure 4.13). The size of the individual RO trains depends of the overall production capacity of the SWRO plant and typically varies between 2,000 m^3/day (0.5 MGD) and 21,000 m^3/day (5.5 MGD). The main advantage of this RO train-based configuration is that it is modular and allows for relatively easy flow distribution and servicing of the individual trains.

Since typically the size of the individual RO trains does not exceed 10 to 20% of the total plant production capacity, train shut down for maintenance (membrane cleaning and replacement, and equipment service) is handled either by using a standby RO train or by temporary increase of the production capacity of the RO trains remaining in service.

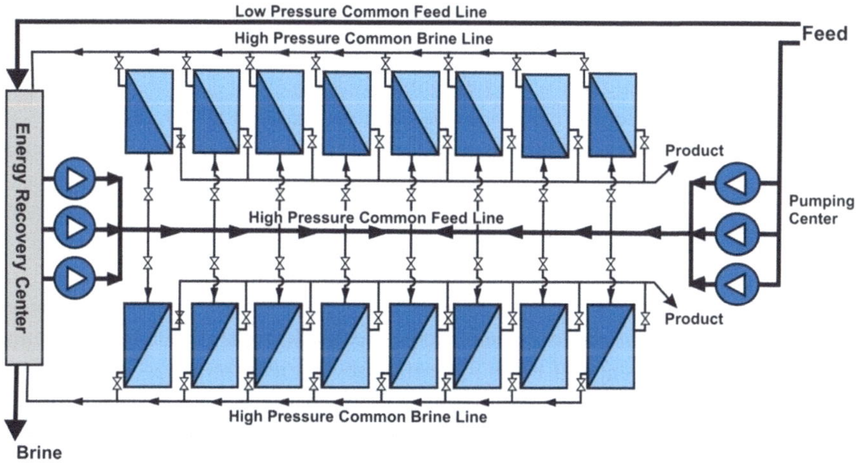

Figure 4.13 Three-center SWRO system configuration.

The RO-train based configuration is suitable when the SWRO plant is designed and intended to operate at a constant production output. At present, most of the existing large municipal SWRO plants worldwide are designed to supplement existing conventional water supply sources rather than to be the primary or the only source of water supply for a given area. Therefore, the operation of these SWRO plants does not need to have the flexibility to follow the actual diurnal and monthly fluctuations in product water demand and most of the existing plants are designed to operate at constant production capacity.

In the future, SWRO is likely to become a prime rather than a supplemental source of water supply for many coastal communities pressured by population growth, especially for large urbanized or industrial areas with limited traditional local sources of fresh water supply (i.e., groundwater or river or lake water). The SWRO systems servicing such areas have to be designed to have the operational flexibility of matching desalination plant production with product water demand fluctuations.

Shift of the SWRO plant operational paradigm from constant to variable production flow requires a change of the typical SWRO configuration from one that is most suitable for constant production output to one that is most cost-effective for delivery of varying permeate production. A response to such a shift of the desalination plant operating paradigm is the three-center RO system configuration in Ashkelon, Israel (see Figure 4.13). Under this configuration, the

RO membrane vessels, high-pressure pumps and energy recovery equipment are no longer separated in individual RO trains, but are rather combined in three functional centers – a high-pressure RO feed pumping center, a membrane center and an energy recovery center (Liberman, 2002). The three functional centers are interconnected via service piping.

The RO feed pumping center includes only a few large-capacity high-pressure pumps that deliver seawater to the RO membrane center. The main benefit of using few large-capacity high-pressure pumps rather than a large number of small capacity units is the gain in overall pumping efficiency.

Typically, the smaller the ratio between the pressure and the flow delivered by a given pump, the better the pump efficiency and the 'flatter' the pump curve (i.e. the pump efficiency is less dependent of the variation of the delivered flow). Therefore, pump efficiency can be improved by either reducing the pressure delivered by the pump or by increasing pump flow. Since the pump operating pressure decrease is limited by the RO system target salt separation performance, the main approach to improve pump efficiency is to increase unit pump flow. While a conventional size high pressure RO feed pump of small capacity would typically have maximum total energy use efficiency of 80 to 85%, the use of a ten times larger size pump may allow to increase the pump efficiency to 88 to 92%, especially for large SWRO plants. This beneficial feature of the three-center design is very valuable in the case of systems delivering varying flow.

While in a conventional RO train design the membrane vessels are typically grouped in 100 to 200 units per train and in 2 to 20 RO trains, the membrane center configuration contains two to four times larger number of RO vessel groups (banks) and a smaller number of membrane vessels per bank. Under this configuration, the individual vessel banks are directly connected to the high-pressure pump feed lines and can be taken off service one at a time for membrane replacement and cleaning.

Although the feed water distribution piping for such membrane center configuration is more elaborate and costly than that use of individual RO trains that contain two to three times more vessels per train, what is lost in capital expenditure is gained in overall system performance reliability and availability. A reliability analysis completed for a 95,000 m^3/day (25 MGD) SWRO plant (Liberman, 2005) indicates that the optimum number of vessels per bank for this scenario is 54 and number of RO banks per plant is 20. A typical RO-train based configuration would include two to four times more (108 to 216) vessels per RO train and two to four times less (5 to 10) RO trains. According to this analysis, the use of the three center configuration instead of the conventional RO-train based approach improves RO system availability from 92 to 96% (avg. 95%) to 98%, which is a significant benefit in terms of the additional amount of water delivered to the customers and improvement in water supply reliability.

The centralized energy recovery system included in the three-center configuration (Figure 4.13) uses high-efficiency pressure-exchanger based energy recovery technology. The proposed configuration improves the overall energy recovery efficiency of the RO system and reduces system power, equipment and construction costs. While typically the energy recovery of conventional Pelton wheel systems drops significantly when the overall SWRO plant recovery decreases, the energy recovery efficiency of the pressure-exchanger systems improves for lower plant recoveries. This allows cost effective operation of the SWRO plant while delivering variable product water flow. For example, if the SWRO plant output has to be reduced by 40% to accommodate low diurnal demand, a SWRO system with RO train-based configuration has to shut down 40% of its trains and if this low demand persists, it has to flush these trains to prevent membrane degradation and to prepare them for the next start up.

An RO system with a three-center configuration would only need to lower its overall recovery to achieve the same reduction of the diurnal demand. Although temporary operation at lower recovery would result in elevated costs for pumping and pretreatment of larger volumes of source water, these extra operational expenses are typically compensated for by the improved energy recovery efficiency that results from operating the SWRO the system at a lower water recovery ratio. In addition to providing flexibility by operating cost-effectively the desalination plant at varying production flows, the three-center design also yields low-fouling RO plant operations of very high availability factor.

4.6 REFERENCES

Bartels G., Lai K. and Wilf M. (2007). New Generation Low Fouling Nanofiltration Membranes, Proceedings of April 2007 Conference of the European Desalination Society, Halkidiki, Greece.

Dreizin Y., Tenne A. and Hoffman D. (2008). Integrating large scale seawater desalination plants within Israel's water supply system. *Desalination*, **220**, 132–149.

Gorenflo A., Brusilovsky M., Faigon M. and Liberman B. (2007). High pH operation in seawater reverse osmosis permeate: first results from the world's largest SWRO plant in Ashkelon. *Desalination*, **203**, 82–90.

Greenlee L. F., Lawler D. F., Freeman B. D., Marrot B. and Moulin P.(2009). Reverse osmosis desalination: water sources, technology, and today's challenges. *Water Research*, **43**, 2317–2348.

Liberman B. (2002). The Importance of Energy Recovery Devices in Reverse Osmosis Desalination, The future of Desalination in Texas – Volume 2, Technical Papers, Case Studies Desalination Technology Resources (report # 363), Texas Water Development Board, December.

Liberman B. and Wilf M. (2005). Evolution of Configuration of RO Seawater Desalination Elements, Proceedings of IDA World Congress – Singapore 2005, Ref: SP05–059.

Sanz M. A., Bonnelye V. and Cremer G. (2007). Fujairah reverse osmosis plant: 2 years of operation, *Desalination*, **203**, 91–99.

Wilf M., Awerbuch L., Bartels C., Mickley M., Pearce G. and Voutchkov (2007). The Guidebook To Membrane Desalination Technology: Reverse Osmosis, Nanofiltration And Hybrid Systems Process Design, Applications And Economics. Balaban Publishers, Rehovot.

Chapter 5

Fouling in reverse osmosis

Brian Bolto, Thuy Tran and Manh Hoang*

CSIRO Materials Science and Engineering, Private Bag 33, Clayton South MDC, Victoria 3169, Australia

**Corresponding author: brian.bolto@csiro.au*

ABSTRACT

Analysis of the foulants isolated from fouled reverse osmosis (RO) membranes reveals simple scale deposits and, more frequently, composites of colloids, organic compounds, particulates and biomass that are difficult to remove. The common scales appear to be limited to calcium carbonate, calcium fluoride and the sulphates of calcium, strontium and barium, plus phosphate when present. Scaling is generally a problem with ground waters of varying mineral content, and is usually not crucial for seawater feeds. Biofouling, on the other hand, can be a serious threat to all RO operations. Biomass accounts for the major part of the organic material in foulants. Pre-treatment is usually essential for the removal of turbidity and dissolved inorganic and organic matter to reduce the fouling potential of the feed water. Biological growth and attachment conditions must be minimised. The most effective control of biofouling is prevention of the colonisation or proliferation of microorganisms by disinfection. Anti-fouling membranes have a role, but as yet the products are not the sole answer as their protective properties are temporary, indicating the need for further research in the area.

5.1 INTRODUCTION

Membrane fouling is a major problem in membrane filtration processes, causing deterioration in both the quantity and quality of the treated water, and a reduced membrane life, which results in higher treatment costs. It is an interfacial phenomenon, arising from the adsorption of materials on the membrane exterior, as the membranes are non-porous. The amount of fouling varies as a function of feed water quality, with membrane replacement costs as a function of total water cost ranging from minimal, at ~5% (Adham, 2007), to substantial levels comparable with energy costs (~23%) (Malaeb & Ayoub, 2011).

While very early publications virtually ignored fouling as an issue for RO, it is notable that one of the first pilot plant studies of a cellulose acetate membrane system reported that the membrane became fouled with iron-containing deposits (Sieveka, 1966). The issue has been much highlighted since then, with the appearance of many reports and reviews (Goosens *et al.* 2004; Malaeb & Ayoub, 2011; Peña *et al.* 2013). The types of material that can cause fouling include inorganic salts, colloids, organic compounds and microbial material.

5.2 INORGANIC SPECIES

Inorganic salts can precipitate out of the concentrate stream as the various solubility limits are exceeded so that they deposit on the membrane surface as scale. This type of fouling is tackled by limiting water recoveries, adjusting pH or adding suitable anti-scalants (Ning, 1999; Sahachaiyunta *et al.* 2002; Kumar *et al.* 2006; Shirazi *et al.* 2010). Most scaling is by calcium carbonate, silica, sulphates of calcium, strontium and barium, and calcium fluoride. The deposition of crystals is inhibited by anti-scalants that suppress crystallisation. Scale control systems are well advanced following substantial effort with regard to boilers, evaporators and cooling towers. Anionic polymers, polyphosphates and organophosphorus compounds are used at the 1–5 mg/L level. They bind to the surface of growing crystal nuclei and retard the rate of crystallisation from supersaturated solutions, modifying the crystal packing. Because of the very short residence time in RO (of the order of a few seconds), the concentration of seed crystals is very low and the temperature is constant, so that high levels of supersaturation without crystallisation are possible. However, there can be interference from other inorganic or organic material in the water. Also, anionic polymers employed as anti-scalants and dispersants can be deposited on the membrane surface and concentrate metal ions of higher valence, such as calcium, iron and aluminium species. Scaling is common with brackish water feeds, but is not generally a problem for seawater and very high salinity waters.

Silica can be a serious problem in the treatment of brackish ground waters, about which many articles have been written (Chen *et al.* 2009). It is recommended that silica levels for safe operating practice of RO plants be kept below 120–155 mg/L (Darton, 1999). Plants have been operated using anti-scalants even when the silica level in the RO concentrate is high as 240–250 mg/L. The type and extent of silica fouling also depends on the presence of other species (Sahachaiyunta *et al.* 2002). It has been shown that iron greatly affects the scale structure on the membrane surface compared to that which occurs when manganese, nickel or barium are present.

A study has been made of synthetic solutions of inorganic scalants arising from $CaCl_2$, $CaCO_3$, $Ca_3(PO_4)_2$, $CaSO_4$ and $MgSO_4$ and their effect on an NF membrane, which is presumed to be an aromatic polyamide membrane because of its chlorine intolerance (Jarusutthirak *et al.* 2007). The membrane had a molecular weight cut off of 150–300 Da. Fouling was dependant on cation valency and on anion type, whether chloride, carbonate, sulphate or phosphate, with the latter being the worst offender. Of the cations, calcium caused a higher flux decline than magnesium because of a greater concentration of precipitated species. For calcium chloride the interaction between the cation and surface water NOM played an important role in flux decline, attributed to cake formation on the membrane surface.

5.3 COLLOIDAL MATERIAL

Colloids can be organic, inorganic or composites. They are introduced by natural drainage and erosion, and consist of sand, silt, clay and the like. The colloidal matter that can be present in natural waters is listed in Table 5.1 (Ning, 1999). Ferric, aluminium and manganese hydroxides and silica grow by polymerisation and crosslinking with organic and inorganic polymers into particles of increasing size that become gels and amorphous foulants. Biotic debris such as polysaccharides and cellular fragments contribute greatly to such fouling. Colloidal matter may agglomerate and adhere to membranes because of increased concentration, salinity, compaction, flocculation, surface interactions and other physical and chemical factors. The particles so formed may be large enough to be removed with 1–5 μm membrane pre-filters. Anticoagulants and anti-deposition agents show promise in inhibiting fouling by this material. Membrane pre-treatments to remove colloids and high MW organics have been studied extensively, with attention being given to other membrane technology such as ultrafiltration (UF) as well as the addition of coagulants.

Table 5.1 Colloidal matter in natural waters (Ning, 1999).

Microorganisms
Biological debris (plant and animal)
Polysaccharides (gums, slimes, plankton, fibrils)
Lipoproteins (secretions)
Clay (hydrous aluminium and iron silicates)
Silt
Oils
Kerogen (aged polysaccharides, marine snow)
Humic acids, lignins, tannins
Iron and manganese oxides
Calcium carbonate
Sulphur and sulphides

5.4 ADSORPTION OF ORGANICS

The adsorption on the membrane surface of the organic compounds present in natural organic matter (NOM) causes a flux decline that is dependant on pH and ionic strength (Braghetta *et al.* 1997). Experiments have been carried out with a nanofiltration (NF) hollow fibre membrane composed of a porous polysulphone support of 100–300 μm thickness onto which was superimposed a skin layer of sulphonated polysulphone of 0.1 μm thickness. The molecular weight (MW) cut off was 1,000 Da. Some sulphonic acid groups were present which provided a negatively charged membrane surface intended to repel the attachment of anionic organic material. Altering the pH caused a change in the pure water permeate flux, which decreased with decreasing pH. This was taken to indicate that membrane compaction occurred as a result of reduced electrostatic charge repulsion within the membrane matrix. There was a decreased rejection of both inorganic ions and NOM at lower pH and higher ionic strength. This also caused a shift in molecular size to smaller dimensions. The NOM accumulated at the membrane surface and impeded water flow. NOM is notorious for fouling low-pressure membranes, especially those that are hydrophobic (Amy *et al.* 2001). NOM can influence the rejection of dissolved matter by steric and electrostatic exclusion, and can cause fouling by precipitation or adsorption. A number of workers have attempted to explain membrane fouling in terms of the nature of the offending NOM.

Experiments on the fouling of an aromatic polyamide thin-film composite NF membrane by three humic acids from Suwannee River water, from peat and an acid-cleaned commercial soil product (of MW 1,000–5,000, 10,000–30,000 and >50,000 Da respectively) demonstrated that fouling increased with the addition of calcium ions, with increasing electrolyte concentration and decreasing pH (Hong & Elimelech, 1997). The aquatic humic acid caused less fouling than the other two, which were almost the same in their flux decline pattern. The zeta potential of the membrane became less negative with increasing divalent ion concentration, for both Ca^{++} and Mg^{++}, rising from –23 to –14.5 mV at 10^{-5} and 10^{-2} M at pH 8. The divalent cations interacted specifically with humic acid carboxylate groups, lowering the humic acid charge and the electrostatic repulsion between the membrane and the humic acid molecules. The addition of a strong chelating agent to the feed water reduced fouling considerably, and also improved the cleaning efficiency of fouled membranes dramatically.

Pilot work on the removal of dissolved matter from groundwater with RO spiral wound membranes after UF pre-treatment showed a flux decline after 100 days (Gwona *et al.* 2003). The removal of inorganic matter was 76–95%, and of organic matter 80–84%. Further investigation showed that the major foulants appeared to be Si bound for organic matter, which formed the first fouling layer close to the membrane, and Ca bound for inorganic matter, which was layered on top of the organic material.

5.5 BIOFOULING

Despite the high salinity of seawater, there is an abundance of biotic life. The growth and anchoring of microorganisms are a prominent source of fouling (Ning, 1999). The moderate temperatures and nutrient levels in feed waters can support an explosive growth of microorganisms at times. Numerous species of bacteria, algae, fungi, and other marine microorganisms grow in a favourable environment. The type, concentration and growth potential of the biological species are determined by the temperature, sunlight, pH, dissolved oxygen content and the presence of organic and inorganic nutrients (Saad, 1992). Bacteria secrete exocellular polymeric substances (EPS) in a unique mechanism facilitating the formation of a difficult-to-remove biofilm that can be a serious threat to RO operation. Fouling by EPS is a complex phenomenon, covering their secretion, transformation, release, and adsorption, and has been reviewed recently (Wang *et al.* 2013). Biomass accounts for the major part of the organic material in recovered foulants. Modification of the membrane surface chemistry to reduce the adsorption of EPS can produce a more fouling-resistant membrane, but much of the details of these developments are propriety (Redondo, 1999; Cohen & Sehn, 2000).

Laboratory studies of RO biofouling by *Pseudomonas aeruginosa* PA01 have been undertaken to elucidate the mechanism that causes reduced performance (Herzberg & Elimelech, 2007). The lower flux and increased salt passage were ascribed to the growth of a biofilm made up of bacterial cells and their EPS, and the resultant increase in trans-membrane osmotic pressure, plus the additional hydraulic resistance. It was concluded that the former played a dominant role in RO biofouling. The adsorption of polysaccharides derived from EPS was found to be more tenacious than that of proteins, and was increased by the addition of calcium ions to the feed (Herzberg *et al.* 2009). This was attributed to specific EPS-calcium interactions, which must therefore have a major role in the biofouling of RO membranes. Tests on the fouling of RO membranes by the protein bovine serum albumin have similarly shown that the phenomenon is significantly enhanced by calcium ions the feed, which has implications for wastewater reclamation (Ang & Elimelech, 2007). The calcium effect outlined here is reminiscent of the situation with fouling by inorganics and NOM, and indicates that a softening pre-treatment would be beneficial for very hard feed waters.

The significance of the deposition of transparent EPS particles of size 0.05–0.4 μm in fouling has been highlighted and methods of their prior removal tested (Villacorte *et al.* 2009). The deposited polymer particles may initiate biological fouling by providing food stock for microbes, which build up a film when they excrete more organic material. These sticky substances can enhance the deposition of more colloids to the surface and further aggravate fouling. Tests on five full-scale and one pilot system on various surface waters indicated that UF was the best pre-treatment for removing particulate matter >0.4 μm, but neither it nor conventional pre-treatment were an efficient for removal of smaller-sized colloidal matter.

One sure way of preventing biofouling is to eliminate or minimise the favourable growth conditions, but many of them cannot be easily controlled, as detailed in a review from which the following is taken (Saad, 1992). The microorganisms can be introduced into the water system by either air or water. Aerobic bacteria find a healthy environment in warm, shallow, sunlit surface waters with adequate dissolved oxygen, a pH of 6.5–8.5 and an abundance of organic and inorganic nutrients. Anaerobic bacteria are usually present in closed water systems with little or no dissolved oxygen, but can be activated if there are sufficient nutrients are present. Both types of bacteria can exist at different locations in the same system. The simplest form of biofouling is initiated by bacterial attachment to inner piping and component surfaces, especially in low-flow dead-ended regions, bends, elbows, etc. There the first layers of biofilm are formed, and as bacteria start to multiply, thriving partly on the organic matter that accumulates within the matrix of the biofilm from dead cells, more and more organic debris is formed and the attachment of the thickened film to surfaces gets stronger and more complex. The deposited biofilm continues to grow, attracting free-floating microorganisms, organic matter and nutrients that are present in the water. The deposits become more cohesive in structure and tougher to dislocate, even with biocides. Eventually, some of the biofilm is sloughed off and the bacterial population explodes into all system components, including the membrane. An accelerated build up of differential pressure across the membrane starts and grows in an exponential manner. Within a few days the flux and salt rejection are adversely affected, generally before any significant pressure drop occurs in spiral-wound thin film composite membranes. The opposite is typically seen in hollow fibre systems, and if the build up is allowed to continue there may be irreversible mechanical damage such as web material tearing; in spiral wound systems telescoping of cartridges can take place.

Sources of an environment conducive to bacterial colonisation are (Saad, 1992):

- Media filters that are run for extended periods between backwashes, without adequate biostat control in the product water, and which are not disinfected during backwash
- Polyelectrolyte and anti-scalant solution tanks which provide more than three days storage and are not routinely washed out, including dosing lines
- Carbon filter beds if not sanitised at least weekly
- Open tanks used for backwash water storage if not maintained with sufficient biostat residual
- Cartridge filters operated for extended times without sufficient biostat added to the product water.

Pre-treatment of feed water, adequate maintenance of upstream unit operations, a continuous flow of water through the system, good monitoring and sanitising during operation and the use of preservatives during downtime are important. Because of the complex nature of microbiological growth mechanisms and the detrimental, often irreversible, effects of biofouling, effective controls must be incorporated in the RO system design, operation and maintenance.

Biofouling control for polyamide RO membranes is restricted because the membranes are vulnerable to chlorine, which eventually decreases salt rejection and shortens membrane life. The use of an alternative disinfectant in the form of dichloro-isocyanurate has been investigated for biofouling control of a polyamide membrane (Yu, J. *et al.* 2013). There was a 99% salt rejection, whereas for chlorine-exposed membranes this was decreased to 80%, while the biofilm grown on the membrane was inactivated to the same extent as with chlorine.

5.6 ADVANCES IN MEMBRANE AUTOPSY

Autopsies of fouled membranes have been carried out in order to better understand the physico-chemical processes governing the fouling (Butt *et al.* 1997; Speth *et al.* 1998; Sahachaiyunta *et al.* 2002; Vrouwenvelder & van der Kooij, 2002; Gwona *et al.* 2003). The methods of chemical and structural analyses used in these studies include inductively coupled plasma mass spectrometry, gas chromatography/mass spectrometry, Fourier transform infra-red spectroscopy (FTIR) and X-ray diffraction (XRD). They provide the average composition of the surface deposits. Because these deposits are complex and heterogeneous, information on average composition is of limited value in elucidating the fouling mechanisms. Direct observations using optical and electron microscopic methods, including scanning electron microscope (SEM) and associated energy-dispersive X-ray spectroscopy (SEM/EDXS), often focus on the top surface deposits, but not on the underlying deposit layers. This leads to an incomplete understanding of the deposition kinetics of various foulants, and therefore of the fouling mechanisms, particularly where thicker deposits have been developed.

The fouling of a spiral wound RO membrane after nearly one year of service in a brackish water treatment plant has been investigated (Tran *et al.* 2007). Both the top surface and the cross-section of the fouled membrane were analysed to monitor the development of the fouling layer. The study used a range of complementary analytical techniques, including IR spectroscopy and SEM/EDXS. It was found that the extent of fouling was uneven across the membrane surface, with regions underneath or in the vicinity of the strands of the feed spacer being more severely affected. The fouling appeared to have been stage-wise, as it consisted of an initial thin fouling layer of an amorphous matrix with embedded particulate matter. SEM/EDXS analysis suggested that the amorphous matrix comprised C–O–Al–P complexes and the particulate matter was mostly aluminium silicates. Subsequently, as the fouling layer reached a thickness of about 5 µm, further amorphous material, probably extracellular polymeric substances such as polysaccharides, started to deposit on top of the existing fouling layer. SEM/EDXS analysis of the secondary amorphous material showed only C, O signals. There was no indication of any particulate matter or any inorganic elements within this sub-layer. It acted as a substrate upon which aluminium silicate crystals grew exclusively in the absence of other foulants such as NOM. The inherent limitations of previous spectroscopic procedures prevented further elucidation of the chemical nature and distribution of the organic materials in this fouling layer.

A further assessment of long-term fouling in RO membranes has been investigated by extensive membrane autopsy using similar analytical techniques (Melian-Martel *et al.* 2012). The membranes were taken from a seawater desalination plant after four years of operation. Characterization performed using visual observation, optical microscopy, SEM-EDXS, XRD, FTIR, X-ray fluorescence, inductively coupled plasma optical emission spectrometry, atomic absorption spectroscopy and ion chromatography. The fouling layer was found to consist of particulate matter embedded in an apparently amorphous matrix, which was unevenly distributed over the membrane surface, with the greatest accumulation in the valley areas. Inorganic and organic foulants were identified. The inorganic foulants consisted mainly of hydrogen aluminosilicates and halite. The main organic substances were proteins, polysaccharides and NOM. Chemical analysis revealed that Si, Al, Cl, Ca, Na, Mg and K were the main elements present. Elements present in minor amounts were Ti, Zn, Fe, Cu, Cr, Mn and Ni.

Although the SEM/EDXS method had adequate spatial resolution, it could only provide elemental information, not structural information, of the surface deposits. FTIR can be used to perform structural analysis, but instruments using a conventional IR source encounter a signal-to-noise ratio limitation when apertures confine the IR to an area of 20–30 μm in diameter. A synchrotron infrared source, with highly-collimated, polarised beam about 100–1,000 times more intense than conventional IR source, can achieve spatial resolution down to about 3 microns in the mid-IR range (Levenson *et al.* 2008; Holman *et al.* 2010). The use of synchrotron FTIR microspectroscopy (FTIRM) to analyse the cross-sections of fouled RO membrane has allowed smaller regions to be probed and has led to the identification of different organic fractions present through the depth of the fouling layer (Tran, 2012). Figure 5.1 shows an example of synchrotron IR chemical mapping of a cross-section of fouled RO membrane.

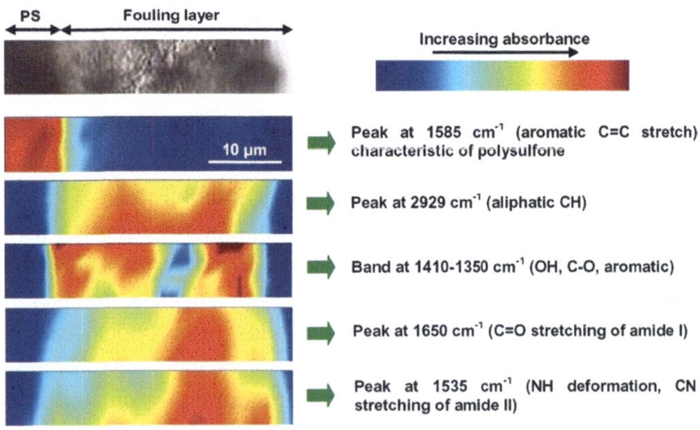

Figure 5.1 Synchrotron IR chemical mapping of a cross-section of fouled RO membrane (PS = polysulfone support).

It can be seen that the characteristic peak of polysulfone at 1585 cm^{-1} (aromatic C=C stretch) is confined to the RO membrane. The segregation of different functional groups into separate regions in the fouling layer is also evident. In particular, the more hydrophilic OH and CO species (1,410–1,350 cm^{-1}) are more pronounced in the vicinity of the membrane surface, compared to the hydrophobic aliphatic CH species (2,929 cm^{-1}) which exhibit high intensity in regions with low OH and CO levels. These data are consistent with a preferential adsorption of more hydrophilic species onto the membrane surface, whereas hydrophobic aliphatic species form distinct parts of the fouling layer but are not directly bound to the surface.

In addition, the characteristic amide I and amide II peaks of proteinaceous compounds exhibit high intensity particularly in the outer region of the fouling layer, rather than distributing evenly across the depth the layer. This is consistent with the deposition of proteinaceous compounds at a later phase of the RO operation and suggests a restricted diffusion of these compounds through the organic matrix to the membrane surface.

The high spatial resolution of synchrotron FTIRM can provide detailed insights into the distribution of different organic and proteinaceous compounds across the depth of the fouling layer. This knowledge would facilitate a more comprehensive understanding of the mechanism of membrane fouling and the development of more effective fouling control strategies.

5.7 PRE-TREATMENTS

The need for pre-treatment for RO has been emphasised and reviewed (Kumar *et al.* 2006; Vedavyasan, 2007; Greenlee *et al.* 2009; Valavala *et al.* 2011). A good example of what can happen without appropriate pre-treatment is the situation at Tampa Bay, Florida (Siverns *et al.* 2004). The largest seawater desalination facility in the United States, it was intended to produce an initial 95 ML/d, with a future planned expansion of 37 ML/d to help reduce the growing demand on the area's aquifers. It first produced water in 2003, but had to be placed in standby mode nearly a year later. The plant used a pre-treatment system comprised of a two-stage dual sand filtration to remove turbidity, algae, organic material, and other particulate matter from the incoming raw seawater. The system was not able to consistently meet the silt density index target set to protect the RO system. As a result, the filters and membranes clogged up quickly, the membranes fouled rapidly and the plant could not achieve peak water production. It faced more frequent membrane cleaning, significantly increasing operating costs because of higher energy consumption, increased chemical usage, and more frequent membrane replacement. As a consequence it only operated intermittently and produced less drinking water than designed. The plant required major repair and remediation work, including modifications to the chemical facilities and the pre-treatment system, which now makes use of ferric sulphate addition, flocculation, and sedimentation. It was finally fully operational in early 2008, some six years behind schedule.

Pre-treatment has been found to be essential for at least half of the major RO desalination plants that have been installed around the world. It is prudent in any RO or NF application to carry out the removal of turbidity and dissolved inorganic and organic matter and so reduce the fouling potential of the feed water. It can be the most critical item in the process train. This is discussed extensively in Chapter 2, especially with regards to the engineering aspects of granular media filtration and coagulation, whereas here the emphasis in this chapter is on the situation at the molecular level. Adequate pre-treatment is highly important in seawater desalination because of the marine flora, algal blooms and phytoplankton that cause high turbidity. The quality of raw seawater is exceedingly variable because of geophysical conditions such as currents, evaporation rates and weather. It can be affected by the local environment because of storm water runoff or pollution in general, notably from hydrocarbons in some areas, which can lead to increased sedimentation or extra loads of nutrients, pathogens, pollutants or salinity (El-Azizi & Omran, 2002).

The types of methods used are determined by the chemical nature of the membrane, the feed water composition and the system configuration. Generally, pre-treatment approaches include scale control and pH adjustment, conventional coagulation/sedimentation/filtration, microfiltration (MF) or UF membranes, combinations of conventional and MF/UF, or simply the use of cartridge filtration (Adham *et al.* 2006; Burbano *et al.* 2007). A survey has been carried out on 62 plants worldwide, mostly in the United States, with 84% of them RO plants and 16% NF. It was confined to plants of 4 ML/day or larger constructed since 1990. They were used for seawater (11%), surface water (6%) and brackish water (60%) desalination, and wastewater reclamation (23%). Simple cartridge filtration was universal, and was sometimes the sole treatment. The percentages of the various methods over all applications are given in Table 5.2.

Table 5.2 Pretreatment methods.

Cartridge Filtration Only	Cartridge Filtration Plus Other	Membrane Filtration, via MF	Coagulation/ Sedimentation/ Filtration	Combination of Conventional and MF
55%	45%	21%	13%	11%

5.7.1 Scale control

Chemicals are often added to protect against the formation of scale, made up of $CaCO_3$, $CaSO_4$, $BaSO_4$, $SrSO_4$, CaF_2 or SiO_2, particularly for ground water, and this is usually followed by filtration through a cartridge filter (Bodzek, 2000). Acidification to a lower pH level and the addition of scale inhibitors such as sodium hexametaphosphate follows, decreasing the formation of carbonate and sulphate deposits. The deposition of silica is minimised by operating at pH >8 or <7 and raising the temperature. Eutrophication problems with phosphates have meant that they are largely replaced by polymers (Fritzmann *et al.* 2007). Another example is the addition of a trimethylglycine composition that eliminates the build up of phosphate and sulphate scale (Morris, 2011).

5.7.2 Conventional coagulation/sedimentation/filtration

Conventional coagulation/sedimentation/filtration is involved in most pre-treatment for seawater applications, with ferric chloride as the preferred coagulant in 100% of the plants surveyed, although alum is certainly used elsewhere. Some 85% of the plants made use of dual media filters. The pre-treatment is optimised to enhance the efficiency of the RO membranes by removing as much as possible of the particulate, microbial and organic matter. Chlorine is typically dosed at 0.5 mg/L upstream of the coagulant to reduce biofilm formation in the downstream filters; copper sulphate is also employed. The 140 ML/day RO plant in Perth, Western Australia makes use of an iron-based coagulant, followed by single stage dual media filtration and 5 μm cartridge filters (McCann, 2007).

Polymer coagulant may be dosed with the inorganic coagulant to help destabilise colloidal matter, but the practice varies with location. Thus in one study the addition of 1 mg/L of a cationic polymer as coagulant aid was tested, but in the first year of operation, optimising the coagulation process eliminated the need for polymer altogether (Irwin & Thompson, 2003). Other work reported significant advantages from the use of 0.1 mg/L of cationic polymer because it reduced the coagulant dose, making for worthwhile cost savings (Al-Sheikh, 1997). The selection of the optimum polymer type has been explored (El-Maharawy & Hafez, 2000). Comparison of cationic, anionic and non-ionic polymers showed that a non-ionic polymer gave the best results, although all organic polymers improved the coagulation process significantly. In general, it has been found that such polymers result in lower doses, longer filter runs and better solids retention, but have issues associated with backwashing (Hagen & Comstock, 2004, quoted by Adham *et al.* 2006; Patterson *et al.* 2006; Franks *et al.* 2006). Inorganic coagulants, on the other hand, offered better filtrate quality, were effective over a broader dose range and gave greater organics removal.

The effects of coagulant residuals on polyamide RO membrane performance have been investigated (Gabelich *et al.* 2006). A wide range of inorganic and organic polymeric additives was examined. It was found that aluminium residuals, most notably from alum coagulation, caused colloidal fouling because of interactions with silica to form aluminium silicates, and with anti-scalants. The effectiveness of the latter reagents was also reduced. Although ferric chloride lowered the tendency towards colloidal fouling, the presence of ferric ions enhanced the oxidation of polyamide membranes by chloramines. Aluminium chlorohydrate also minimised colloidal fouling and reduced aluminium/anti-scalant interactions, but seemed to enhance oxidation of the membrane surface. The fouling of membranes by organic polymers appeared to be minimal.

The effects of pre-treatment with coagulants on UF performance with turbid surface waters vary with the mode of operation, depending on whether coagulation is conventional, forms a dynamic layer on the membrane, or is in-line (Sharp *et al.* 2006). The dynamic or secondary, coagulant-based layer achieved higher steady state flux values than unassisted UF, while in-line coagulation gave the worst flux decline. Later work on optimised coagulation with poly(aluminium chloride) on UF behaviour with river water has shown that it is advantageous to form the strongest possible flocs as they bind more particulates in the water and are less likely to foul because the formation of a metal hydroxide gel layer on the membrane surface (Dixon *et al.* 2013). The best results were obtained at medium agitation with optimum or enhanced doses.

The NOM removal efficiency of polysilicato-iron (PSI) coagulants and the fouling potential of PSI pre-treated waters have been studied using polyvinylidene fluoride (PVDF) and polypropylene (PP) MF membranes (Tran *et al.* 2006). The results showed that it was more effective than conventional coagulants in improving the fouling potential. Aluminium-based coagulants, particularly aluminium chlorohydrate, worked best at lower coagulant dose. Increasing the coagulant dose to improve NOM removal led to increased membrane fouling, possibly due to the increased level of unsettleable flocs and pore blocking. For PSI with its larger floc size, the advantage of increased NOM removal was not overridden by the adverse effect of pore blocking. PSI shows a much better coagulation performance than ferric sulphate (Matilainen *et al.* 2010). In a similar way, the use of an *in situ* Fe(III) coagulant formed by reacting Fe(II) and $KMnO_4$ prior to a PVDF UF membrane gave larger flocs and caused a lower rate of fouling compared with conventional dosing of an Fe(III) coagulant (Yu *et al.* 2013). This was ascribed to the formation of a thinner and more porous cake layer with the *in situ* coagulant.

5.7.3 Dissolved air flotation

Because of the high organic content of some seawater it has been necessary to use dissolved air flotation (DAF) as a pre-treatment after dosing with a coagulant such as ferric chloride or aluminum sulphate to flocculate the suspended matter (Veolia, 2012). Removal is achieved by dissolving air in the seawater under pressure and then releasing the air at atmospheric pressure in a flotation tank or basin. The released air forms very small bubbles to which the suspended matter adheres, causing it to float to the surface of the water where it is removed by a skimming device. After filtration the treated water passes to the membranes. DAF is particularly suitable particularly for algal laden waters and supplies that have low turbidity and high colour. It is frequently used on industrial wastewaters, and achieves superior results at higher flow and contaminant loadings than can be achieved with conventional coagulation/sedimentation/filtration (Edzwald, 1985). For surface water containing cyanobacteria a dual DAF/MF system was more effective than DAF alone (Amaral *et al.* 2013).

The basic water chemistry and physics when DAF is used as a pre-treatment of seawater for RO have been explored recently, with a focus on both the dissolving of air and its subsequent efficient release as fine micro-bubbles, to ensure the required level of clarification is achieved ahead of filtration (Amato *et al.* 2013). The differences with seawater as a feed compared to freshwater have been addressed in adapting DAF to seawater (Haarhoff & Edzwald, 2013). The most important are the lower solubility of air (at 20°C, 0.64 mol/kL for seawater versus 0.84 mol/kL for freshwater) and the air transfer efficiency, that for a typical air saturator with seawater being only 74% of the freshwater value. These shortfalls can be corrected by increasing the recycle rate or operating the saturator at a higher pressure.

5.7.4 Ion exchange

There have been many studies of NOM removal by ion exchange, usually with the aim of lowering the formation of disinfection by-products (Bourke & Slunjski, 1999; Bolto *et al.* 2002; Tan *et al.* 2005). Mostly a charge mechanism applies, but there may be others such as physical adsorption and hydrogen bonding. There are major differences in the types of compounds removed relative to coagulation, which preferentially removes hydrophobic material. Ion exchange processing of natural surface waters has a substantial impact on NOM removal. However, when magnetic ion exchange is used before a PVDF MF membrane there is only a moderate decrease in fouling which is not replicated in the case of a PVDF UF membrane (Huang *et al.* 2012). This has been ascribed to the preference of the resin for medium MW organic compounds, rather than the high MW material that is relevant to fouling. Yet another study reports that magnetic ion exchange gives better NOM removal and decreased fouling of a polysulphone UF membrane, attributed to the good removal of the low MW organic compounds responsible for membrane fouling (Huang *et al.* 2009; Kabsch-Korbutowicz *et al.* 2006).

Ion exchange has been found to remove most of the hydrophilic and low MW organic compounds and a significant amount of the hydrophobic compounds from a biologically treated secondary effluent (Zhang *et al.* 2006). When used as a pre-treatment before a submerged membrane reactor incorporating a hydrophilic polyethylene MF membrane, magnetic ion exchange significantly reduces membrane fouling. Fouling can be further reduced by addition of a small dose of powdered activated carbon to the membrane reactor.

Ion exchange is especially appropriate for the selective removal of inorganic ions that may cause membrane scaling or other problems (Lipnizki *et al.* 2012). The scale-forming compounds are mainly calcium, barium and strontium sulphates, calcium carbonate and silica. Iron fouling because of the negative charge on the membrane can occur if ferric chloride has been the coagulant added in a prior step. Metal cations can be removed by a high recovery softening with a weak acid resin of the carboxylic acid type, typically used before RO in treating brackish waters. Iminodiacetic acid chelating resins are used for Sr^{2+} and Ba^{2+} removal from

RO concentrates ahead of further RO treatment, while to prevent other scaling problems an aminophosphonic acid resin may be employed. Wastewaters from the production of gas from unconventional sources are another feedstock that can be challenging to treat.

5.7.5 Low pressure MF or UF membranes

An integrated or hybrid membrane approach has been reviewed, making use of some of the membrane technologies summarised in Table 5.3 (Adham *et al.* 2006), which gives a general indication of the species that are rejected by membranes of increasing tightness.

Table 5.3 Contaminants generally rejected by various membrane processes (Adham *et al.* 2006).

Process	Monovalent Ions	Divalent Ions	Micro-Organisms	Suspended Solids
MF	–	–	–	√
UF	–	–	√	√
NF	–	√	√	√
RO	√	√	√	√

Using low pressure (70–600 kPa) MF or UF membrane technology offers several benefits. The pre-treated seawater then has a higher quality in terms of suspended solids and biological content (Henthorne & Quigley, 2003), so there is

- less cleaning of the RO membranes required
- less fouling so lower pressure drops
- longer RO membrane life
- increased flux
- reduced footprint
- lower overall chemical usage
- less sludge handling.

It has been found that the use of MF prior to an RO system can significantly reduce the fouling potential, as first demonstrated with a PP MF membrane used on secondary effluent (Del Pino & Durham, 1999). The permeate flux, cleaning frequency and lifetime of membranes in water reclamation applications can be improved with the use of a PVDF MF membrane prior to RO (Xiao *et al.* 2013). The system had a low operating pressure of below 40 kPa and a high permeate flux of 52–60 L/m^2/h. The recovery 86%–92% and the permeability could be easily recovered.

An example of UF before RO where operation of a UF pilot plant in a high recovery (80%), low flux (60 L/m^{2}h) mode was found to be the best pre-treatment for a seawater application (Xu *et al.* 2007). From the original feed having a total suspended solids content of 4.8–8.9 mg/L and a chemical oxygen demand of 1.15–1.76 mg/L, the feed for the RO plant had a turbidity 0.01 NTU, with no chemicals being used for disinfection, flocculation, enhanced chemical backwashing or cleaning. In another seawater RO example it was found necessary when there were algal blooms for two months of the year to add in-line coagulation with ferric chloride prior to UF (Schurer *et al.* 2013).

Removal of the biodegradable organic content of seawater by means of a fluidised bed biological activated carbon or membrane bioreactor has been tested (Visvanathan *et al.* 2002). It gave good removals and a high flux improvement.

5.7.6 High pressure NF membranes

NF membranes are used for preliminary softening and organic removal, taking out material that has MW >200 Da. The reduction of NaCl is usually about 50%, and that of $CaSO_4$ about 90%. The pressures required are 0.3–3 MPa, versus 4–7 MPa for RO. The application of NF before RO has been investigated as a means of reducing the concentration of divalent ions that are responsible for membrane scaling (Hassan *et al.* 2000). It provides excellent feed water for RO, allowing operation at much higher fluxes and recovery rates. Other viewpoints however consider that the approach merely transfers the scaling problem further upstream (El-Maharawy & Hafez, 2000).

NF and RO for the desalination of brackish water is reported, with the emphasis on the effects of di- and mono-valent cations on operation (M'nif *et al.* 2007). At a pressure of only 0.6 MPa, when run in sequence with RO before NF, the brine reject from the RO unit was desalted with respect to divalent ions, so that the recovery of 40% for RO was increased to 80% with the coupled RO/NF. When done in the reverse order, and conducted in parallel with recirculation so that the divalent ions were reduced by NF, the osmotic pressure to the RO decreased and the RO flux and recovery were also increased. However, such pre-treatment with NF has not proven to be economically feasible (Ericsson, 2005, quoted by Fritzmann *et al.* 2007). Higher recoveries can in special circumstances make up for the additional investment cost, but these are unlikely to occur in the practical operation of a desalting plant.

5.7.7 Combined technologies

The prior addition of a coagulant before UF or MF can help in the removal of natural and synthetic organic compounds. In extreme cases such as in Jeddah, Saudi Arabia, the quality of the seawater may be so poor because of the facility being within a port area for example, where frequent ship movement and algal blooms can upset the quality, so a combined approach such as coagulation, dual media filtration and UF may be necessary (Buisson, 2006).

A large RO plant in China is to desalt highly variable poor quality seawater for use as boiler feed water and for potable purposes. It will use enhanced coagulation and clarification, followed by immersed membrane UF to produce 76.8 ML/day of exceptional quality feed water (Siverns, 2006).

In a review of the efficiency of different MF and UF membranes as pre-treatment for the RO of Red Sea water, it was observed that the prior injection of 1 mg/L of $FeCl_3$ before UF was the preferred option (Al-Mashharawi *et al.* 2013).

5.7.8 Comparison of conventional and membrane pre-treatment

The water quality obtained from MF pre-treatment is better than that from conventional approach. One example is on high-fouling seawater from Gibraltar that is subject to seasonal algal blooms, and has an SDI of 13–25. UF pre-treatment with coagulation at low dose gave an SDI of <1, whereas the conventional method failed to reduce the SDI below 2.5 (Brehant *et al.* 2002). A higher flux rate and higher recovery were obtained. RO membranes desalting seawater from the Gulf of Mexico require cleaning after 980 h operation on conventionally pre-treated water, whereas those exposed to membrane pre-treated water needed no cleaning after the 1300 h evaluation period (Henthorne & Quigley, 2003).

Membrane pre-treatment with hollow fibre membranes is claimed to be cost competitive with conventional technology in one study (Wilf & Scheirach, 2001). The higher flux rate and higher recovery following MF make for better reliability and overall economics because of lower investment and operating costs, which would result in about a 10% reduction in total water costs. Another investigation found pre-treatment with hollow fibre UF to be more expensive than the conventional, but feasible for sites that require very extensive conventional pre-treatment, or where there are wide fluctuations in raw water quality (Glueckstern *et al.* 2002).

Overall, the advantages of membrane pre-treatment over conventional are (Henthorne & Quigley, 2003; Kumar *et al.* 2006):

- better feed water quality, especially the suspended solids and biological content
- less RO membrane cleaning, so less chemicals usage
- lower pressure drop in RO arising from fouling
- longer RO membrane life
- increased flux rates and smaller plant footprint, and therefore lower capital investment
- lower overall chemical and sludge handling costs
- better recovery of product water if NF precedes RO (M'nif *et al.* 2007; El-Zanati & El-Khatib, 2007).

Summaries of the situation have been published recently, as shown in Table 5.4 (Vedavyasan, 2007). Data for a specific example at San Francisco Bay are included (Frenkel *et al.* 2006). Many more publications have appeared that substantiate these claims (Bonnelye *et al.* 2004; Pearce *et al.* 2004; Franks *et al.* 2006; Patterson *et al.* 2006; El-Zanati & El-Khatib, 2007; Pearce, 2007a; Bonnelye *et al.* 2008).

A comparison of the cost of conventional and membrane-based pre-treatment has been published (Côté *et al.* 2005). Estimated capital costs for a 75 ML/d plant desalting seawater were 944 and 1,005 \$/kL/d respectively for media filtration and immersed UF as pre-treatment, and the total life cycle costs were 0.615 and 0.637 \$/kL respectively. Thus for UF there was a premium of 6.5% and 3.6% for capital costs and total life cycle costs over that for conventional pre-treatment. However, there is the potential to reduce costs for the all-membrane arrangement taking into account the better and more reliable pre-treatment and consequent operation of the RO at increased flux, with higher recovery and longer membrane life. The overall smaller footprint is another advantage.

Table 5.4 Comparison of benefits, conventional versus membrane pre-treatment (Vedavyasan, 2007).

Item	Conventional	MF/UF
Capital costs	Competitive with MF/UF	Although 0–25% higher, life cycle costs are comparable
Foot print	Larger	30–50% smaller foot print
Energy requirements	Lower, usually gravity flow	Higher
Chemical costs*	High; coagulant and polymer always required	Low, depending on raw water quality; coagulant only for organics reduction
Recovery*	96%	93–95%
Average turbidity*	0.13 NTU†	0.05 NTU†
Cartridge replacement*	2–3 months	3–4 months
RO flux	Lower	Higher
RO capital cost	Higher	Less as flux can be 20% higher
RO operating costs	Fouling potential greater, more frequent cleaning	Less fouling potential and longer membrane life

* For San Francisco Bay (Frenkel *et al.* 2006).
† Nephelometric turbidity units.

A capital cost comparison for feed water from the eastern Mediterranean Sea having a salinity of 38,000 mg/L is shown in Table 5.5. A similar conclusion can be drawn. The chemical cleaning of the RO membranes receiving coagulated seawater that has been MF/UF pre-treated are minimal, and this reduced frequency of membrane cleaning is the main advantage. The major benefit of MF/UF is that it provides a more reliable and stable system, tolerant of variations in the feed water quality and of upsets in the operating system. It is estimated that some 40% of UF/MF systems sold are for RO pre-treatment (Pearce, 2007b).

Table 5.5 Capital cost comparison, conventional versus membrane pre-treatment at 20 ML/day (Pearce, 2007a).

Item	Conventional in kUS$	MF/UF in kUS$
RO	2,000	2,000
Pretreatment	800	1,700
Process equipment, pumps, site works	14,050	1,850
Indirect	590	5,900
Total	22,750	23,450

5.8 ANTI-FOULING MEMBRANES

The commercial membranes used in desalination plants are frequently manufactured from hydrophobic polymers because of their good thermal, mechanical and chemical properties. A major disadvantage of these membranes is their poor performance because of the relatively strong adsorption of pollutants to the membrane surface, causing significant irreversible fouling. A common strategy to mitigate the fouling of water treatment membranes is to modify their surface properties, whilst preserving their desirable bulk properties. Methods of altering the membrane surface have been extensively reviewed recently (Rana & Matsuura, 2010; Nady *et al.* 2011; Kang & Cao, 2012). Various surface modification methods have been employed including increasing surface hydrophilicity, a reduction in the surface roughness, changing surface electrical properties, exploitation of steric repulsion effects, or incorporating biocides to kill bacteria that have already become attached to the surface. Also, the addition of inorganic particles to polymeric membranes is a recent development and the hybrid organic/inorganic RO membranes so formed show great promise as they have good permeability and antifouling plus self-cleaning properties. These various modifications may lead to an improvement in membrane performance. Anti-fouling membranes are much further advanced for the porous membranes used in MF and UF, and less so for non-porous ones, as used for RO and NF, with one review citing more than 160 examples for the former, and only about seven for the latter (Rana & Matsuura, 2010).

5.8.1 Increased hydrophilic character

Surface modification can be by a physical method such as adsorption or coating, or by chemical methods of improving the hydrophilic properties by radical grafting, chemical coupling, plasma-induced polymerisation or chemical vapour deposition (Kang & Cao, 2012). Generally a covalent bond linkage between the membrane and modifier is superior to the physical route and has greater practical utility. A more indirect approach is to make use of quite different reactants in the polymerisation process.

Many types of hydrophilic entities have been used, some examples being listed in Table 5.6, which shows instances of MF, UF, RO and NF membranes being modified. PP MF membranes, initially hydrophobic, have been made hydrophilic by first introducing peroxide groups onto the membrane surface, and then grafting 2-hydroxyethyl methacrylate by means of redox initiation (Wang *et al.* 2000). Bovine serum albumin (BSA) was less readily adsorbed on the membrane and flux recovery was greater after the protein adsorption, showing that the fouling layer could be removed readily because of the hydrophilic nature of the modified membranes. Similar MF membranes have been rendered hydrophilic by UV-initiated grafting with the monomer poly(ethylene glycol)$_{200}$ monomethacrylate (Ma *et al.* 2000, 2001).

Earlier work had shown that PES membranes could be modified by UV-induced grafting and polymerisation of hydrophilic monomers onto their surface. Most work was done on 2-hydroxyethyl methacrylate, but a large range of monomers was examined (Yamagishi *et al.* 1995a). Moderate (20–27%) increases in flux and retention of BSA were observed relative to the ungrafted membrane. With monomers such as *N*-vinyl-2-pyrrolidone the products had a 25% increase in hydrophilicity, a 49% decrease in BSA fouling, and a 4% increase in the protein's retention compared to the ungrafted membrane (Pieracci *et al.* 1999).

A regenerated cellulose membrane had lower fouling and protein adsorption, but the modified polyethersulphone membrane had the best combination of low fouling and high flux. Irradiation alone led to increased protein fouling, even though the surface was hydrophilised. This was attributed to pore enlargement. PES UF membranes prepared with small quantities (2 wt.%) of polyvinyl-pyrrolidone (PVP) of MWs 40,000 and 360,000 Da present have been prepared to increase the membrane permeability and to check the fouling behaviour (Marchese *et al.* 2003). It was found that at pH 4.9 the original and modified membranes were fouled externally by BSA, and internally by DL-histidine, with the BSA fouling occurring more rapidly. Pore blocking of the original membrane by BSA was 1.3 to 3.3 times greater than for the two modified products, with the lower MW PVP having the greater effect. This was explained by the hydrophobic attractive forces between the protein and the membrane being greater for the unmodified membrane. PES UF membranes prepared with small quantities (2 wt.%) of polyvinylpyrrolidone (PVP) of MWs 40,000 and 360,000 Da present have been prepared to increase the membrane permeability and check the fouling behaviour (Marchese *et al.* 2003). It was found that at pH 4.9 the original and modified membranes were fouled externally by BSA, and internally by DL-histidine, with the BSA fouling occurring more rapidly. Pore blocking of the original membrane by BSA was 1.3 to 3.3 times greater than for the two modified products, with the lower MW PVP having the

Table 5.6 Examples of membranes modified by hydrophilic species.

Process	Membrane Polymer	Modifier	Outcome	Reference
UF	PES	2-Hydroxyethyl-methacrylate	Better flux, protein retention	Yamagishi *et al.* (1995a)
UF	PES	N-vinyl-2-pyrrolidone	Less protein fouling	Pieracci *et al.* (1999)
NF	PES	N-Vinyl-2-pyrrolidone	Less NOM fouling	Kilduff *et al.* (2000)
MF	PP	Hydrophilic non-ionic polymers	Generally beneficial	Wang *et al.* (2000)
MF	PP	Poly(ethylene glycol) methacrylate	Generally beneficial	Ma *et al.* (2000, 2001)
UF	PES	Polyvinylpyrrolidone	Less protein fouling	Marchese *et al.* (2003)
UF	Polysulphone	Hydrophilic polymers blends	Higher flux	Yoo *et al.* (2003)
RO, NF	Aromatic polyamide	Poly(vinyl alcohol)	Less fouling by dyes	Myung *et al.* (2005)
NF	PVDF	Poly(oxyethylene methacrylate)	Less fouling in bioreactor	Asatekin *et al.* (2006)
NF	PVDF	Amphiphilic copolymer	Less fouling by humic acid, protein or alginate	Asatekin *et al.* (2007)
UF	PAN	Poly(ethylene glycol) acrylate	Less fouling generally	Mayes & Asatekin (2007)
UF	Cellulose acetate	Poly(ethylene glycol)	Lower cake accumulation	Gullinkala & Escobar, (2008)
RO	Aromatic polyamide	PEG acrylate, polyHEMA	Less flux decline	Sagle *et al.* (2009)
UF	PAN	Polyacrylonitrile-*graft*-poly(ethylene oxide)	Less bacterial adhesion	Adout *et al.* (2010)
RO	Aromatic polyamide	Poly(N-isopropylacrylamide-co-acrylamide),	Less protein fouling	Yu *et al.* (2011)
NF	Aromatic polyamide	PolyHEMA and Fe_3O_4 nanoparticles	Fouling resistant in a magnetic field	Yang *et al.* (2013)
NF	Poly(arylene ether sulfone)	Poly(2-methacryloyl-oxyethyl phosphorylcholine)	Less irreversible protein adsorption	Liu *et al.* (2013)
UF	Polysulphone	N-succinyl chitosan blend	Less protein fouling, higher flux	Kumar *et al.* (2013)
FO	Cellulose triacetate	3-(3,4-dihydroxyphenyl)- L-alanine	Improved fouling resistance	Nguyen *et al.* (2013)

greater effect. This was explained by the hydrophobic attractive forces between the protein and the membrane being greater for the unmodified membrane.

Bisphenol-A polysulphone, widely used in UF membranes because of its excellent chemical and physical properties, is limited in its application because of its hydrophobic nature. Methods of altering the surface chemistry, such as adsorption of surfactant, plasma treatment and the surface grafting of hydrophilic species are limited because they are difficult to apply to hollow fibre formats. Blends of the polymer with hydrophilic water insoluble polymers on the other hand overcome this disadvantage (Yoo *et al.* 2003). By employing copolymers such as poly(1-vinylpyrrolidone-co-styrene) miscible blends with polysulphone can be obtained that show improved performance over the original polysulphone; for example the rejection of poly(ethylene glycol) or PEG of MW 40 kDa is doubled. A UV-irradiation method has also been used to produce highly hydrophilic surfaces on polyarylsulphone UF membranes (Yamagishi *et al.* 1995a, 1995b). The membranes were intrinsically photoactive by virtue of free radical cleavage of the chains, so that no photo-initiators were required. Most experiments were done on the 2-hydroxyethyl methacrylate-grafted products. The polyarylsulphone membranes were less sensitive to UV irradiation than the PES version. A higher flux with similar retentions to the ungrafted membrane were observed when the membranes were challenged with a 0.1% BSA solution.

Other examples include coating an aromatic polyamide RO or NF membrane with poly(vinyl alcohol) (Myung *et al.* 2005), or with poly(ethylene glycol) acrylate (PEGA) (Sagle *et al.* 2009), and modifying a PVDF membrane by a similar approach (Asatekin *et al.* 2006). The latter workers made a PVDF NF membrane from a graft with poly(oxyethylene methacrylate) which showed no irreversible fouling in 10-day dead-end filtration studies of model organic foulants such as bovine serum albumin, sodium alginate and humic acid at concentrations of 1,000 mg/L and above. Analogous results have been obtained with UF membranes containing polyacrylonitrile-*graft*-poly(ethylene oxide) comb copolymers, with the comment that the antifouling performance was exceptional (Asatekin *et al.* 2007). The preparation of UF membranes with poly(ethylene glycol)-*graft*-polyacrylonitrile copolymers has been reported (Su *et al.* 2009). Another UF example includes poly(ethylene glycol) grafts to commercially available membranes that show lower cake accumulation patterns (Gullinkala & Escobar, 2008). Chemically initiated grafting onto polyacrylonitrile (PAN) membranes of hydrophilic side-chains based on PEGA has been reported, to yield nanochannels of tunable width which may result in permselective properties and/or anti-fouling character. (Mayes & Asatekin, 2007).

Poly(2-hydroxyethylmethacrylate) or (polyHEMA) chains have been grafted onto the surface of thin film composite polyamide NF membranes (Yang *et al.* 2013). Super-paramagnetic Fe_3O_4 nanoparticles were then successfully attached to the grafts. In treating feed water containing 500 mg/L $CaCl_2$ and 2,000 mg/L $MgSO_4$ the modified membranes displayed a lower permeate flux and increased salt rejection compared to unmodified membranes, due to the added resistance of the grafted nanolayer. However, in the presence of an oscillating magnetic field they gave both an increased permeate flux and increased salt rejection. The increases were greater for higher densities of attached nanoparticles, although only a fraction of the chain ends were terminated with nanoparticles. The magnetically responsive polyHEMA chains appeared to act as micro-mixers; this novel system may also represent a new class of easier cleaned and fouling resistant membranes. Blending of a water soluble chitosan derivative, N-succinyl chitosan, with a polysulphone membrane has yielded a UF membrane of improved flux and resistance to fouling by BSA (Kumar *et al.* 2013).

An NF membrane prepared by interfacial polymerisation with the zwitterionic monomer N-aminoethyl piperazine propane sulphonate has been described which can resist bacterial adsorption and protein fouling during a 288 h test (An *et al.* 2013). It was felt that the improved hydrophilicity and binding of more free water were responsible. A similar UF membrane has been made by grafting a zwitterionic polymer, poly(2-methacryloyloxyethyl phosphorylcholine) onto a poly(arylene ether sulfone) membrane containing hydroxyl groups (Liu *et al.* 2013). Irreversible protein adsorption was significantly reduced, suggesting superior antifouling performance because of the more hydrophilic character of the membrane. A further example is the deposition of L-DOPA, the poly amino acid 3-(3,4-dihydroxyphenyl)- L-alanine, onto the surface of a forward osmosis (FO) membrane (Nguyen *et al.* 2013). Fouling resistance was improved by 30% for synthetic waters. Yet another case related to FO membranes is the grafting of a polyalkyne zwitterion onto a polyamide membrane, making it more hydrophilic (Yu, H.-Y. 2013). There was no lowering of the water flux, and an excellent antifouling capability was claimed. This was attributed to the shielding of binding sites on the membrane surface and strong hydrophilic repulsion by charge-induced hydration forces, plus steric repulsion by the brush-like grafted zwitterion chains.

An interesting variation on the theme is the modification of RO membranes with a thermo-responsive polymer, poly(N-isopropylacrylamide-co-acrylamide), which has a low critical solution temperature (CST) (Musale & Zeiher, 2010; Musale, 2010; Yu *et al.* 2011). The higher surface hydrophilicity improved the RO performance and improved the fouling resistance to bovine serum albumin, used as a model foulant. The flux recovery and salt rejection after cleaning at 50°C, above the CST, was 89.5% and 98.5% respectively, versus 68.0% and 97.5% for the unmodified membrane (Yu *et al.* 2011). Much lower results were obtained if the washing was carried out below the CST. Poly(N-isopropylacrylamide) or PNIPAM is a well explored thermo-sensitive polymer that exhibits a CST of ~32° in aqueous solution, below which it adopts an extended random coil transformation. On warming, hydrogen bonds weaken to release the water of hydration. Hydrophobic interactions overcome hydrophilic interaction, which leads to a coil to globule transition and ultimately to phase separation. Membranes containing it have a permeability that is temperature sensitive, the biggest change occurring in the region of 32° (Ying *et al.* 2003). This has been ascribed to a change in the conformation of the PNIPAM domains, which assume an extended conformation on the surface, including the pores, at the lower temperature. On warming, the polymer chains exhibit less hydrophilic interactions and shrink via hydrophobic association, opening the pores to give a higher flow rate and facilitating membrane cleaning.

5.8.2 Charge modification

Examples of the modification of various membrane types by the introduction of charged groups are shown in Table 5.7. PP MF membranes have also been modified by UV-initiated grafting with dimethylaminoethyl methacrylate or acrylic acid (Ma *et al.* 2000, 2001). MF experiments showed that in the filtration of carboxylate-modified latex particles, fouling was not strongly dependent on membrane surface chemistry when there was no back pulsing. With backpulsing, however, relative to the unmodified membrane, a 10% increase in permeate volume over one hour was achieved with a negatively charged modified membrane, versus a 20% decrease when the membrane had an opposite positive charge to that on the latex particles.

Photo-induced grafting of 4-vinylpyridine on PP and polyethylene MF membranes has produced composite membranes with the graft polymer located both inside the pores and on the surface (Mika *et al.* 1995). The modified membranes were found to be extremely vulnerable to compression under pressure. In effect, the membranes were converted to NF or RO systems. The reversible swelling of the graft was controllable by pH, allowing the same membrane to show both MF and RO behaviour.

A UV-irradiation method has been used to graft methacrylic acid onto polyarylsuphone UF membranes (Yamagishi *et al.* 1995a, 1995b). The grafted chains could be further functionalised by post-graft reactions. Grafting proceeded to considerable depths into the membranes. Polyarylsulphone membranes were less sensitive to UV irradiation than PES membranes. A higher flux with similar retentions to the ungrafted membrane were observed when the membranes were challenged with a 0.1% BSA solution, indicating less fouling by the protein.

Low temperature plasma induced surface modifications have been carried out on polyacrylonitrile and also polysulphone UF membranes (Ulbricht & Belfort, 1996). There was a drastic and almost permanent increase in surface hydrophilicity, and in the case of polyacrylonitrile the pore structure was also changed. Polymer peroxides created by plasma excitation were used to thermally initiate graft polymerisation of acrylic and methacrylic acids. Modified polysulphone membranes gave improved performance in protein UF because pore etching effects were compensated for by the grafted layer. Modified polyacrylonitrile membranes showed a clear dependency of water permeability on the amount of graft polymer, and there was significantly reduced fouling due to protein adsorption, plus better performance in the UF of protein solutions. The work has been extended to include grafting of sulphonic acid, dimethylamino and trimethylammonium species onto polyacrylonitrile UF membranes by photo initiation (Ulbricht *et al.* 1998). The pore structure was not damaged. The approach was seen as a way of developing low-fouling systems.

Table 5.7 Examples of membranes modified by charged species.

Process	Membrane Polymer	Modification	Outcome	Reference
MF	PP & polyethylene	Polymerisation of cationic monomers in pores	Achieved RO & NF performance	Mika *et al.* (1995); Childs *et al.* (2001, 2003)
UF	Polysulphone	UV irradiation to graft anionic & non-ionic polymers	Higher flux for more grafted membranes	Yamagishi *et al.* (1995b)
UF	Polysulphone	Anionic, cationic & non-ionic polymer grafts	Better performance in protein UF	Ulbricht & Belfort, (1996); Ulbricht *et al.* (1998)
RO	Polyamide	Methacrylic acid, 3-sulphopropyl methacrylate	Less organic fouling, more easily cleaned	Belfer *et al.* (1999); Gilron *et al.* (2001)
MF	PP	Grafting of cationic, anionic & non-ionic polymers	Charge critical in latex particle	Ma *et al.* (2001)
MF	PP	Grafting of acrylic acid	Better wettability, less fouling, but lower flux	Xu *et al.* (2002); Abu Seman *et al.* (2010)
MF	PP	Grafting of cationic, anionic & non-ionic polymers	Graft types show little real differences	Carroll *et al.* (2002)
UF	Polysulphone	Oxygen plasma treatment	Higher flux of gelatin, less fouling	Kim *et al.* (2002)
MF	PP	Grafting of anionic polymer	Less flux decline	Meier-Haack *et al.* (2003)
Not stated	PES	Plasma grafting of poly(acrylic acid)	Less fouling by protein solution	Wavhal & Fisher, (2002)
UF	PES	Grafting of neutral, weakly and strongly acidic groups	Weakly acidic good; neutral worse fouling	Taniguchi *et al.* (2003)
MF	PES	Grafting of cationic & anionic polymers	Potential for less biofouling	Hilal *et al.* (2003)
NF	Sulphonated PES	Highly negative NF membrane	Performs very well in organics removal	Bartels *et al.* (2003)
UF	PES	Permeation of sodium poly(styrene sulphonate)	Better antifouling, higher flux	Reddy *et al.* (2005)
NF	PES	Acrylic acid	Less irreversible NOM fouling	Abu Seman *et al.* (2010)
NF	Polysulphone	N-aminoethyl piperazine propane sulphonate	Resists protein fouling	An *et al.* (2013)

Oxygen plasma treatment can modify polysulphone UF membranes to convert them to hydrophilic membranes (Kim *et al.* 2002). In the process, the isoelectric point increased from pH 3 to 4.5 and the oxygen to carbon ratio from 33 to 50%. The flow rates for water and gelatin increased and there was less fouling relative to the original membrane.

Grafting conditions for the attachment of acrylic acid to a hollow fibre PP MF membrane by peroxide initiation have been studied (Xu *et al.* 2002). A certain amount of a crosslinking agent in the form of divinylbenzene was found to be advantageous, with grafting being discerned on the surface and within the pores of the membrane. Excellent wetting properties were bestowed on the membrane, but a high water flux was not achievable because of pore blocking. The pores of similar membranes have been filled by *in situ* crosslinking of poly-4-vinylpyridine or poly(vinylbenzyl chloride) gels, both later reacted to incorporate quaternary ammonium groups, and *in situ* polymerising of acrylic acid and crosslinkers (Childs *et al.* 2001). The modification in effect changed the membranes from MF to NF or RO systems by closing up the pores. The approach has been extended to incorporate a crosslinked cationic polymer made from 3-acrylamidopropyltrimethylammonium chloride and *N,N*'-methylene-*bis*-acrylamide (Childs *et al.* 2003). The membrane produced had good water softening characteristics and could be operated at high recovery. It was also reported to be low fouling in the presence of humic acid.

PP MF membranes have been modified also by grafting of a cationic or anionic polymers from dimethylaminoethyl methacrylate or acrylic acid (Carroll *et al.* 2002). The flux in pure water experiments was dependent on pH for the grafted charged membranes, but not for the ungrafted, which was attributed to changes in the configuration of the grafted polymers to consist of extended chains rather than coils when in the charged state, thus causing a decline in pore size. When NOM was present at near neutral pH the throughput flux for the cationic membrane having an 11.3% graft was twice that of the ungrafted membrane. This was at least partly attributed to binding of the humate anions to the positively charged surface, causing a conformational change and compaction of the grafted chains and a widening of the pores. With higher grafts though this effect was reversed, the flux decreasing most for a graft of 20.7%, as the larger amount of organic anions now bound to the membrane surface led to pore blockage. The cationic grafts resulted in flux declines caused by NOM fouling that were up to 50% lower than for the ungrafted membrane. Work on an\ similkar MF membrane modified by a 7.2% w/w poly(acrylic acid) graft confirmed that there was less flux decline than with an ungrafted membrane (Meier-Haack *et al.* 2003). The effect depended on raw water characteristics such as hardness, particulate properties and pre-treatment strategies, with soft water having a high NOM content and low turbidity being the most amenable.

Grafting of methacrylic acid to a polyamide RO membrane by redox initiation results in less contaminant adsorption and increased ease of rinsing (Belfer *et al.* 1999). The modification improves the hydrophilicity and maintains or slightly improves flux and rejection, but beyond a certain point the modified RO membrane becomes more permeable to salts. The work was extended to grafting of non-ionic and sulphonic acid monomers, with the outcome that the increase in hydrophilicity lowered the amount of organic material adsorbed and made it easier to clean the membranes (Gilron *et al.* 2001).

A permanent hydrophilic modification of PES membranes has been achieved by low temperature plasma-induced grafting of acrylic acid in the vapour phase (Wavhal & Fisher, 2002). The pore sizes of the grafted membranes were slightly decreased at high graft levels, but the membranes were less susceptible to protein fouling than the original membranes, and the pure water flux was tremendously increased. The grafted products were easier to clean and required little caustic to recover the flux. It was postulated that the plasma treatment alone attacked the carbon in the phenyl group and caused graphitisation of the surface occurred to some extent. There was no physical damage to the polymer surface after treatment, with both sides of the membrane becoming highly hydrophilic. The lack of permanency in the plasma-alone product was because of a structural rearrangement of the polymer chains, but the post-plasma grafting to form poly(acrylic acid) chains locked in hydrophilicity for the entire membrane cross-section.

Surface modification of polyamide brackish water RO membranes with negatively charged and/or neutral hydrophilic groups has been carried out by *in situ* redox polymerisation of acrylate monomers containing —SO_3H, —CO_2H or –$CONH_2$ groups (Reddy *et al.* 2005). Desalting tests on the unmodified membranes showed that after 30 days the salt rejection decreased from 92 to 83% and the flux was halved. For the modified membranes the change in the salt rejection was marginal, but there was a flux reduction of nearly a third. The modified membranes showed good fouling resistant properties in brackish water desalination.

Further NF membranes have been prepared from polyether sulphone (PES) and its sulphonated analogue by UV-initiated graft polymerization of N-vinyl-2-pyrrolidone (Kilduff *et al.* 2000). Fouling by NOM was reduced significantly, but there was also decreased NOM rejection, which was attributed pore blockage by graft polymer chains. Lower irradiation doses gave a compromise of some fouling protection without loss of NOM rejection. Similar products from NF membranes based on PES grafted with acrylic acid by UV-irradiation gave products with an increased rejection of humic acid and less irreversible fouling (Abu Seman *et al.* 2010).

5.8.3 Antibacterial surfaces

Commercial PES MF membranes have been modified by a photo-initiated grafting of polymers containing carboxylic or sulphonic acid groups (Hilal *et al.* 2003). Membranes containing quaternary ammonium groups posed a strong anti-microbial action towards *E. coli*, indicating that they were potentially more resistant to biofouling.

Polyacrylonitrile UF membranes incorporating an amphiphilic copolymer, polyacrylonitrile-*graft*-PEO have been tested for adsorption of *E. coli* K12 (Adout *et al.* 2010). Static tests revealed no binding of the bacteria, even with Ca^{2+} ions present. Under crossflow filtration conditions bacterial adsorption increased with higher ionic strength and the addition of Ca^{2+}, but 97–100% were readily removed on increasing the crossflow and eliminating permeate flow, so that there was no trans-membrane pressure, in contrast with the original unmodified membrane, with less than 25–50% removal. The exceptional resistance of the treated membrane to bacterial adhesion was attributed to steric repulsion by the dense brush layer of poly PEO chains.

A novel polysulphone UF membrane has been blended with a ternary polymer containing capsaicin moieties as an anti-fouling and anti-bacterial agent (Xu *et al.* 2013). The copolymer was made from N-(4-hydroxy-3-methoxybenzyl)acrylamide, methyl methacrylate and acrylic acid. Capsaicin is the active ingredient of chili peppers. The product membranes exhibited excellent anti-bacterial activity and improved anti-fouling properties towards organic foulants.

Graphene oxide modified with hyperbranched polyethyleneimine has been blended into a PES UF membrane (Yu, L. *et al.* 2013). The graphene oxide consisted of highly packed carbon atoms that gave the material unique properties, although there was a slightly lower water flux. They gave better anti-fouling performance and showed effective anti-bacterial properties towards *E. coli.*

Polysulphone MF membranes modified with poly(allylamine hydrochloride)/poly(acrylic acid) polyelectrolyte multilayers have shown significantly improved anti-adhesive behaviour and as a consequence good anti-bacterial characteristics in tests with *E. coli* (Tang *et al.* 2013). This was attributed to the highly swollen and hydrated polyelectrolyte multilayers inhibiting the direct contact or close approach of the bacteria.

A polyamide RO membrane has modified with covalently bonded antibacterial enzymes (Saeki *et al.* 2013). The bonding of lysozyme onto the membranes to the extent of 0.1 $\mu g/cm^2$ resulted in a decrease in a halving of the water flux, but 90% of the salt rejection was maintained. When the membrane was exposed to feed water containing Gram-positive bacteria *Micrococcus lysodeikticus* and *Bacillus subtilis*, the formation of a biofilm was prevented, with the activity remaining after five months storage at 5°C. It was felt that the water flux could be improved by controlling the surface density of the immobilised enzyme, and such membranes could be useful for industrial applications.

Copper coating of an NF membrane by vapour deposition has been shown to give a resistance to microbial attack that is greater than that of the untreated membrane, with a reduced flux loss (Isloor *et al.* 2013). Cultures used in the study were *Bacillus cereus, Staphylococcus aureus, Pseudomonas aeruginosa* and *Escherichia coli.* The inhibition of bacterial growth was ascribed to the release of copper species into the surrounding medium, which however raises drinking water quality concerns. The flux was 13.0 L/m^2h for the treated membrane and 13.3 L/m^2h for the untreated, while the flux decline after 10 days was 9% for the modified membrane, versus 27% for the original.

5.8.4 Challenges

It has been pointed out that there are some significant challenges facing anti-fouling RO membranes: decreased flux, complicated production, higher cost, loss of modifiers – especially on cleaning – and not least, fouling cannot be completely prevented even with anti-fouling membranes. There are at present no membranes that are completely free from fouling (Rana & Matsuura, 2010). Proper pre-treatment, a modified module design and appropriate membrane cleaning also need to be optimised, as well as the trade-off between flux reduction and anti-fouling properties (Kang & Cao, 2012).

5.9 CONCLUSIONS

The most frequent foulants of RO membranes are composites of colloids, particulates and biomass that are difficult to remove. Scaling is a frequent problem with ground waters, but is usually not crucial for seawater feeds. But biofouling is a menace to all RO operations, and accounts for the major part of the organic material in fouling material. Hence pre-treatment is essential for the removal of turbidity and dissolved inorganic and organic matter. Physicochemical or membrane technology can be efficient for this purpose, the choice depending on the benefit and the cost. Biological growth and attachment conditions should be eliminated, the most effective control being prevention of the growth of microorganisms by disinfection.

Anti-fouling membranes have a role, but as yet fouling cannot be completely prevented, so there is no complete answer. Much further research is required in this area. From the point of view of membrane design, there are three parameters that need to be considered: pore size, surface charge and hydrophilic/hydrophobic balance. There has been a great concentration of thinking on pore characteristics and size exclusion effects, and also some attention to the influence of charge and hydrophilicity. The importance of not having an entirely hydrophobic character to the membrane surface is highlighted in the treatment of waters high in strongly hydrophobic compounds: thus a membrane containing aromatic as well as negatively charged groups is more effective in NOM removal than a hydrophilic cellulosic membrane, despite the latter having pores of smaller size. However, increased hydrophilicity is not the complete answer in preventing adsorption on membranes. There should be a focus on the matching of the polymer used in membrane construction with the type of water to be treated; that is, an emphasis on the chemistry behind the interactions, especially hydrophobic forces.

With regard to membrane cleaning, a desirable arrangement would be a membrane that has anionic groups on the entire membrane surface, including the pores, so that there are some potentially negatively charged surface groups present throughout. During an alkaline wash the increased charging of these groups should, because of repulsive forces, cause the pores to widen and release accumulated material. Grafted chains have the disadvantage of expanding and blocking pore access and transmission through the pores. A more futuristic design would be a reversibly cross-linked membrane, the matrix of which that could be made to selectively shrink and open its pores under cleaning conditions. The pH variability of polyion complexes of weakly basic and weakly acidic polymers, with the latter in excess, would be a possible way of achieving this, as would thermally reversible polymer crosslinks based on electrostatic interactions or hydrogen bonding. Membranes containing PNIPAM domains have pores that open up on heating because of this phenomenon. Attention needs to be given to the trade-off between antifouling properties and flux reduction brought about by changes to the membrane.

5.10 REFERENCES

Abu Seman M. N., Khayet M., Bin Ali Z. I. and Hilal N. (2010). Reduction of nanofiltration membrane fouling by UV-initiated graft polymerisation technique. *J. Mem. Sci.,* **355**, 133–141.

Adham S., Burbano A., Chiu K.-P. and Kumar M. (2006). Development of a reverse osmosis/nanofiltration (RO/NF) knowledge base. Final Report for California Energy Commission, Montgomery Watson Harza, San Diego.

Adham S. (2007). Desalination. Proc. Membrane Specialty Conf. II Melbourne, Australian Water Assoc., Paper 12.

Adout A., Kang S., Asatekin A., Mayes A. M. and Elimelech M. (2010). Ultrafiltration membranes incorporating amphiphilic comb copolymer additives prevent irreversible adhesion of bacteria. *Env. Sci. Tech.,* **44**, 2406–2411.

Al-Mashharawi S. K., Ghaffour N., Al-Ghamdi M. and Amy G. L. (2013). Evaluating the efficiency of different microfiltration and ultrafiltration membranes used as pretreatment for Red Sea water reverse osmosis desalination. *Desal. Wat. Treat.,* **51**, 617–626.

Al-Sheikh A. H. H. (1997). SWRO pre-treatment with an emphasis on the Jeddah plant operation experience. *Desal.,* **110**, 37–48.

Amaral P. A. P. d., Coral L. A., Nagel-Hassemer M. E., Belli T. J. and Lapolli F. R. (2013). Association of dissolved air flotation (DAF) with microfiltration for cyanobacterial removal in water supply. *Desal. Wat. Treat.,* **51**, 1664–1671.

Amato T., Park K.-S., Yim W. and Kim T. (2013). SWRO pre-treatment design using high-rate dissolved air flotation including preliminary pilot-scale results. *Desal. Wat. Treat.,* **51**, 1804–1816.

Amy G., Cho J., Yoon Y., Wright S., Clark M. M., Molis E., Combe C., Wang Y., Lucas P., Lee Y., Kumar M., Howe K., Kim K.-S., Pelligrino J. and Irvine S. (2001). NOM rejection by, and fouling of, NF and UF membranes. AwwaRF Report, Awwa Research Foundation, Denver.

An Q.-F., Sun W.-D., Xhao Q., Ji Y.-L. and Gao C.-J. (2013). Study on a novel nanofiltraion membrane prepared by interfacial polymerisation with zwitterionic amine monomers. *J. Mem. Sci.,* **431**, 171–179.

Ang W. S. and Elimelech M. (2007). Protein (BSA) fouling of reverse osmosis membranes: implications for wastewater reclamation. *J. Mem. Sci.,* **296**, 83–92.

Asatekin A., Menniti A., Kang S., Elimelech M., Morgenroth E. and Mayes A. M. (2006). Antifouling nanofiltration membranes for membrane bioreactors from self-assembling graft copolymers. *J. Mem. Sci.,* **285**, 81–89.

Asatekin A., Kang S., Elimelech M. and Mayes A. M. (2007). Antifouling ultrafiltration membranes containing polyacrylonitrile-*graft*-poly(ethylene oxide) comb copolymer additives. *J. Mem. Sci.,* **298**, 136–146.

Bartels C., Wilf M., Long J. and Campbell J. (2003). Selective colour removal nanofiltration membrane for the 7 MGD irvine ranch water treatment project. *Proc. AWA Federal Conv.* Perth. CD ROM.

Belfer S., Gilron J. and Kedem O. (1999). Characterisation of commercial RO and UF modified and fouled membranes by means of ATR/FTIR. *Desal.,* **124**, 175–180.

Bodzek M. (2000). Membrane techniques in water treatment and renovation. In: *Water Management, Purification and Conservation in Arid Climates,* M. F. A. Goosen and W. H. Shayya (eds), Technomic, Lancaster, Pennsylvania, pp. 45–101.

Bolto B. A., Dixon D. R., Eldridge R. J. and King S. J. (2002). Removal of natural organic matter by ion exchange. *Wat. Res.,* **36**, 5057–5065.

Bonnelye V., Sanz M. A., Durand J.-P., Plasse L., Gueguen F. and Mazounie P. (2004). Reverse osmosis on open intake seawater: pre-treatment strategy. *Desal.* **167**, 191–200.

Bonnelye V., Guey L. and Del Castillo J. (2008). UF/MF as RO pre-treatment: the real benefit. *Desal.,* **222**, 59–65.

Bou-Hamad S., Abdel-Jawad M., Ebrahim S., Al-Mansour M. and Al-Hijji A. (1997). Performance evaluation of three different pre-treatment systems for SWRO technique. *Desal.,* **110**, 85–92.

Bourke M. and Slunjski M. (1999). MIEX® DOC process launched in Western Australia. *Wat.,* **26**(6), 17–20.

Brehant A., Bonnelye V. and Perez M. (2002). Comparison of MF/UF pre-treatment with conventional filtration prior to RO membranes for surface seawater desalination. *Desal.,* **144**, 353–360.

Braghetta A., DiGiano F. A. and Ball W. P. (1997). Nanofiltration of natural organic matter: pH and ionic strength effects. *J. Env. Eng.,* 628–640.

Buisson H. (2006). Pre-treatment experiences and lessons learnt from over 25 years of constructing large SWRO plants. *Proc. AMTA Conf. Amer. Mem. Tech. Assoc.*, Anaheim.

Burbano A. A., Adham S. S. and Pearce W. R. (2007). The state of full-scale RO/NF desalination – results from a worldwide survey. *J. AWWA* **99**(4), 116–127.

Butt F. H., Rahman F. and Baduruthamal U. (1997). Characterisation of foulants by autopsy of RO desalination membranes. *Desal.,* **114**, 51–64.

Carroll T., Booker N. A. and Meier-Haack J. (2002). Polyelectrolyte-grafted microfiltration membranes to control fouling by natural organic matter in drinking water. *J. Mem. Sci.,* **203**, 3–13.

Chen S.-S., Cheng H.-H. and Yang S.-R. (2009). In-Line Coagulation/Ultrafiltration for Silica Removal from Brackish Water as RO Membrane Pretreatment. *Separation & Purif. Technol.,* **70**, 112–117.

Childs R. F., Mika A. M., Pandey A. K., McCrory C., Mouton S. and Dickson J. M. (2001). Nanofiltration using pore-filled membranes: effect of polyelectrolyte composition on performance. *Separation & Purif. Technol.* **22–23**, 507–517.

Childs R. F., Mika A. M. and Zhou J. (2003). Evaluation of the nanofiltration properties of charged gel-filled porous membranes. *Proc. Internat. Membrane Science & Technology Conf.* Sydney, Paper 94.

Cohen S. and Sehn P. (2000). Four years field experience with fouling resistant reverse osmosis membranes. *Desal.* **132**, 211–215.

Côté P., Siverns S. and Monti S. (2005). Comparison of membrane-based solutions for water reclamation and desalination. *Desal.,* **182**, 251–257.

Darton E. G. (1999). RO plant experiences with high silica waters in the Canary Islands. *Desal.* **124**, 33–41.

Del Pino M. P. and Durham B. (1999). Wastewater reuse through dual-membrane processes: opportunities for sustainable water resources. *Desal.,* **124**, 271–277.

Dixon M., Staaks C., Fabris R., Vimonses V., Chow C. W. K., Panglisch S., van Leeuwen J. A. and Drikas M. (2013). The impact of optimised coagulation on membrane fouling for coagulation/ultrafiltration process. *Desal. Wat. Treat.,* **51**, 2718–2725.

Edzwald J. K. (1985). Principles and applications of dissolved air flotation. *Wat. Sci. Tech.,* **31**, 1–23.

El-Azizi I. M. and Omran A. A. M. (2002). Design criteria of 10,000 m^3/d SWRO desalination plant of Tajoura, Libya. *Desal.,* **153**, 273–279.

El-Maharawy S. and Hafez A. (2000). Technical management of RO system. *Desal.,* **131**, 173–188.

El-Zanati E. and El-Khatib K. M. (2007). Integrated membrane-based desalination system. *Desal.,* **205**, 15–25.

Ericsson P. (2005). Evaluation of nanofiltration as pre-treatment to reverse osmosis in seawater desalination. *Proc. IDA World Congress*, Singapore.

Franks R., Wilf M., Voutchkov N., Murkute P. and Kizer J. (2006). A pilot study using seawater reverse osmosis membranes in combination with various pre-treatments to meet the challenges of Pacific seawater desalination. *Proc. AMTA Conf. Amer. Mem. Tech. Assoc.*, Anaheim.

Frenkel V. S., Reynolds T. K., Kennedy P. E. and MacInnis G. (2006). Membrane (UF/MF) vs. conventional pre-treatment for SWRO in California. *Proc. AMTA Conf. Amer. Membrane Tech. Assoc.*, Anaheim.

Fritzmann C., Löwenberg J., Wintgens T. and Melin T. (2007). State-of-the-art of reverse osmosis desalination. *Desal.,* **216**, 1–76.

Gabelich C. J., Franklin J. C., Cohen Y. and Suffet I. H. (2006). Reverse osmosis pre-treatment: challenges with conventional treatment. In: Membrane Treatment for Drinking Water and Reuse Applications K. J. Howe (ed.), Amer. Waterworks Assoc., Denver, pp. 149–168.

Gilron J., Belfer S., Väissänen P. and Nyström M. (2001). Effects of surface modification on antifouling and performance properties of reverse osmosis membranes. *Desal.,* **140**, 167–179.

Glueckstern P., Priel M. and Wilf M. (2002). Field evaluation of capillary UF technology as a pretreatment for large seawater RO plants. *Desal.,* **147**, 55–62.

Goosens M. F. A., Sablani S. S., Al-Hinai H., Al-Obeidani S., Al-Belushi R. and Jackson D. (2004). Fouling of reverse osmosis and ultrafiltration membranes: a critical review. *Sep. Sci. Tech.,* **39**(10), 2261–2297.

Greenlee L. F., Lawler D. F., Freeman B. D., Marrot B. and Moulin P. (2009). Reverse osmosis desalination: water sources, technology, and today's challenges. *Wat. Res.,* **43**, 2317–2348.

Gullinkala T. and Escobar I. (2008). Study of the hydrophilic-enhanced ultrafiltration membrane. *Env. Prog.,* **27**, 210–217.

Gwona E.-M., Yu M.-J., Oh H.-K. and Ylee Y.-H. (2003). Fouling characteristics of NF and RO operated for removal of dissolved matter from groundwater. *Wat. Res.,* **37**, 2989–2997.

Haarhoff J. and Edzwald J. K. (2013). Adapting dissolved air flotation for the clarification of seawater. *Desal.,* **311**, 90–94.

Hagen D. and Comstock D. (2004). Granular media filtration pre-treatment. *Proc. IDA Conf. Congress: RO Pre-treatment Workshop*, San Diego. Int. Desal. Assoc., Topsfield, Mass.

Hassan A. M., Farooque A. M., Jamaluddin A. T. M., Al-Amoudi A. S., Al-Sofi M. A. K. and Al-Rubaian A. F. (2000). A demonstration plant based on the new NF-SWRO process. *Desal.,* **131**, 157–171.

Henthorne L. and Quigley R. (2003). Evaluation of MF, UF and conventional pre-treatment for seawater RO applications. *Proc. IDA World Congress*, Sep. 28-Oct. 3, Paradise Island, Bahamas. International Desalination Assoc., Topsfield, Mass.

Herzberg M. and Elimelech M. (2007) Biofouling of reverse osmosis membranes: role of biofilm-enhanced osmotic pressure. *J. Mem. Sci.,* **295**, 11–20.

Herzberg M., Kang S. and Elimelech M. (2009). Role of extracellular polymeric substances (EPS) in biofouling of reverse osmosis membranes. *Env. Sci. Tech.* **43**, 4393–4398.

Hilal N., Al-Khatib L., Atkin B. P., Kochkodan V. and Potapchenko V. (2003). Photochemical modification of membrane surfaces for (bio)fouling reduction: a nano-scale study using AFM. *Desal.,* **158**, 65–72.

Holman H.-Y. N., Bechtel H. A., Hao Z. and Martin M. C. (2010). Synchrotron IR spectromicroscopy: chemistry of living cells. *Anal. Chem.,* **82**, 8757–8765.

Hong S. and Elimelech M. (1997). Chemical and physical aspects of natural organic matter (NOM) fouling of nanofiltration membranes. *J. Mem. Sci.,* **132**, 159–181.

Huang H., Schwab K. J. and Jacangelo J. G. (2009). Pretreatment for low pressure membranes in water treatment: a review. *Env. Sci. Tech.,* **43**, 3011–3019.

Huang H., Cho H.-H., Schwab K. J. and Jacangelo J. G. (2012). Effects of magnetic ion exchange pretreatment on low pressure membrane filtration of a natural surface water. *Wat. Res.,* **46**, 5483–5490.

Irwin K. J. and Thompson J. D. (2003). Orinoco fluctuations fail to make filters fail. *Int. J. Desal. & Wat. Reuse*, **13**(3), 12–16.

Isloor A. M., Ganesh B. M., Isloor S. M., Ismail A. F., Nagaraj H. S. and Pattabi M. (2013). Studies on copper coated polysulfone/modified poly isobutylene alt-maleic anhydride blend membrane and its antibiofouling property. *Desal.,* **308**, 82–88.

Jarusutthirak C., Mattaraj S. and Jiraratananon R. (2007). Influence of inorganic scalants and natural organic matter on nanofiltration membrane fouling. *J. Mem. Sci.,* **287**, 138–145.

Kang G.-D. and Cao Y.-M. (2012). Development of antifouling reverse osmosis membranes for water treatment: a review. *Wat. Res.,* **46**, 584–600.

Kilduff J. E., Mattaraj S., Pieracci J. P. and Belfort G. (2000). Photochemical modification of poly(ether sulfone) and sulfonated poly(sulfone) nanofiltration membranes for control of fouling by natural organic matter. *Desal.,* **132**, 133–142.

Kabsch-Korbutowicz M., Bilyc A. and Molczan M. (2006). The effect of feed water pretreatment on ultrafiltration membrane performance. *Polish J. Env. Stud.,* **15**, 719–725.

Kim K. S., Lee K. H., Cho K. and Park C. E. (2002). Surface modification of polysulphone ultrafiltration membrane by oxygen plasma treatment. *J. Mem. Sci.,* **199**, 135–145.

Kumar M., Adham S. and Pearce W. R. (2006). Investigation of seawater reverse osmosis fouling and its relationship to pre-treatment type. *Env. Sci. Tech.,* **40**, 2037–2044.

Kumar R., Isloor A. M., Ismail A. F. and Matsuura T. (2013). Performance improvement of polysulfone ultrafiltration membrane using N-succinyl chitosan as additive. *Desal.,* **318**, 1–8.

Levenson E., Lerch P. and Martin M. C. (2008). Spatial resolution limits for synchrotron-based spectromicroscopy in the mid- and near-infrared. *J. Synch. Rad.,* **15**, 323–328.

Lipnizki J., Adams B., Okazaki M. and Sharpe A. (2012). Combining reverse osmosis and ion exchange. *Filtr. + Sep.,* **49**, 30–33.

Liu Y., Zhang S. and Wang G. (2013). The preparation of antifouling ultrafiltration membrane by surface grafting zwitterionic polymer onto poly(arylene ether sulfone) containing hydroxyl groups membrane. *Desal.,* **316**, 127–136.

Ma H., Davis R. H. and Bowman C. N. (2000). A novel sequential photinduced living graft polymerisation. *Macromol.,* **33**, 331–335.

Ma H. M., Hakim L. F., Bowman C. N. and Davis R. H. (2001). Factors affecting membrane fouling reduction by surface modification and backpulsing. *J. Mem. Sci.,* **189**, 255–270.

Malaeb L. and Ayoub G. M. (2011). Reverse osmosis technology for water treatment: state of the art review. *Desal.,* **267**, 1–8.

Matilainen A., Vepsäläinen M. and Sillanpää M. (2010). Natural organic matter removal by coagulation during drinking water treatment: a review. *Adv. Colloid & Interface Sci.,* **159**, 189–197.

Mayes A. M. and Asatekin A. (2007). Fouling resistant membranes formed with polyacrylonitrile graft copolymers. *PCT Patent Appln.* 2007120631. *Chem. Abs.,* **147**, 487248.

McCann W. (2007). Perth marks Australia's progress to desalination. *Water 21*, June 2007, 17–18.

Meier-Haack J., Booker N. A. and Carroll T. (2003). A permeability-controlled microfiltration membrane for reduced fouling in drinking water treatment. *Wat. Res.,* **37**, 585–588.

Melian-Martel N., Sadhwani J. J., Malamis S. and Ochsenkuhn-Petropoulou M. (2012). Structural and chemical characterization of long-term reverse osmosis membrane fouling in a full scale desalination plant. *Desal.,* **305**, 44–53.

Mika A. M., Childs R. F., Dickson J. M., McCarry B. E. and Gagnon D. R. (1995). A new class of polyelectrolyte-filled microfiltration membranes with environmentally controlled porosity. *J. Mem. Sci.,* **108**, 37–56.

M'nif A., Bouguecha S., Hamrouni B. and Dhahbi M. (2007). Coupling of membrane processes for brackish water desalination. *Desal.,* **203**, 331–336.

Morris W. H. (2011). Method for removing and reducing scaling. *US Patent*, 2011/0155,664 and WO2011/082,285A2

Musale D. A. (2010). Method of cleaning fouled and/or scale membranes. *US Patent*, 7,674,382, B2.

Musale D. A. and Zeiher E. H. K. (2010). Method of cleaning fouled or scale membranes. *US Patent*, 7,662,289 B2.

Myung S.-W., Choi I.-H., Lee S.-H., Kim I.-C. and Lee K.-H. (2005). Use of fouling resistant nanofiltration and reverse osmosis membranes for dyeing wastewater effluent treatment. *Wat. Sci. Tech.,* **51**(6–7), 159–164.

Nady N., Franssen M. C. R., Zuilhof H., Eldin M. S. M., Boom R. and Schroën K. (2011). Modification methods for poly(arylsulphone) membranes: a mini-review focusing on surface modification. *Desal.,* **275**, 1–9.

Nguyen A., Azari S. and Zou L. (2013). Coating zwitterionic amino acid L-DOPA to increase the fouling resistance of forward osmosis membrane. *Desal.,* **312**, 82–87.

Ning R. (1999). Process chemistry relevant to the gulf. *Desal.,* **123**, 157–163.

Patterson A. W., Johnson W. T., Sundstrom G. P. and Gallagher P. M. (2006). Seawater pretreatment using low-pressure membranes. *Proc. AMTA Conf. Amer. Membrane Technol. Assoc.*, Anaheim.

Pearce G., Talo S., Chida K., Basha A. and Gulamhusein A. (2004). Pre-treatment options for large scale SWRO plants: case studies of UF trials at Kinsada, Saudi Arabia, and conventional pre-treatment in Spain. *Desal.,* **167**, 175–189.

Pearce G. (2007a). The case for UF/MF pre-treatment to RO in seawater applications. *Desal.,* **203**, 286–295.

Pearce G. (2007b). Introduction to membranes: manufacturer's comparison. Part 1. *Filtr. + Sep.* October, pp. 36–38.

Peña N., Gallego S., del Vigo F. and Chesters S. P. (2013). Evaluating impact of fouling on reverse osmosis membranes performance. *Desal. Wat. Treat.,* **51**, 958–968.

Pieracci J., Crivello J. V. and Belfort G. (1999). Photochemical modification of 10 kDa polyethersulfone ultrafiltration membranes for reduction of biofouling. *J. Mem. Sci.,* **156**, 223–240.

Rana D. and Matsuura T. (2010). Surface modifications for antifouling membranes. *Chem. Rev.,* **110**, 2448–2471.

Reddy A. V. R., Trivedi J. J., Devmurari C. V., Mohan D. J., Singh P., Rao A. P., Joshi S. V. and Ghosh P. K. (2005). Fouling resistant membranes in desalination and water recovery. *Desal.,* **183**, 301–306.

Redondo J. A. (1999). Improve RO system performance and reduce operating cost with FILMTEC fouling-resistant (FR) elements. *Desal.,* **126**, 249–259.

Saad M. A. (1992). Biofouling prevention in RO polymeric membrane systems. *Desal.,* **88**, 85–105.

Saeki D., Nagao S., Sawada I., Ohmukai Y., Maruyama T. and Matsuyama H. (2013). Development of antibacterial polyamide reverse osmosis membrane modified with a covalently immobilised enzyme. *J. Mem. Sci.,* **428**, 403–409.

Sagle A. C., Van Wagner E. M., Ju H., McCloskey B. D., Freeman B. D. and Sharma M. M. (2009). PEG-coated reverse osmosis membranes: desalination properties and fouling resistance. *J. Mem. Sci.,* **340**, 92–108.

Sahachaiyunta P., Koo T. and Sheikholeslami R. (2002). Effect of several inorganic species on silica fouling in RO membranes. *Desal.,* **144**, 373–378.

Schurer R., Tabatabai A., Villacorte L., Schippers J. C. and Kennedy M. D. (2013). Three years operational experience with ultrafiltration as SWRO pre-treatment during algal bloom. *Desal. Wat. Treat.,* **51**, 1034–1042.

Sharp M. M. and Escobar I. C. (2006). Determining the effects of coagulant pre-treatment on ultrafiltration. In: Membrane Treatment for Drinking Water and Reuse Applications K. J. Howe (ed.), Amer. Waterworks Assoc., Denver, pp.169–186.

Shirazi S., Lin C.-J. and Chen D. (2010). Inorganic fouling of pressure-driven membrane processes – a critical review. *Desal.,* **250**, 236–248.

Sieveka E. H. (1966). Reverse osmosis plot plants. In: Desalination by Reverse Osmosis U. Merten (ed.), MIT Press Cambridge, Massachusetts, pp. 239–270.

Siverns S., Maness R. and Movahed B. (2004). Alternative pre-treatment tests with UF yield excellent results at Tampa Bay. *J. AWWA* **96**(12), 30–32.

Siverns S. (2006). Using ultrafiltration as a pre-treatment before RO. *Ultrapure Wat.,* **23**(3), 36–39.

Speth T. F., Summers R. S. and Gusses A. M. (1998). Nanofiltration foulants from a treated surface water. *Env. Sci. Tech.*, **32**, 3612–3617.

Su Y. L., Cheng W., Li C. and Jiang Z. Y. (2009). Preparation of antifouling ultrafiltration with poly(ethylene glycol)-graft-polyacrylonitrile copolymers. *J. Mem. Sci.*, **329**, 246–252.

Tan Y., Kilduff J. E., Kitis M. and Karanfil T. (2005). Dissolved organic matter removal and disinfection byproduct formation control using ion exchange. *Desal.*, **176**, 189–200.

Taniguchi M., Kilduff J. E. and Belfort G. (2003). Low fouling synthetic membranes by UV-assisted graft polymerisation: monomer selection to mitigate fouling by natural organic matter. *J. Mem. Sci.*, **222**, 59–70.

Tang L., Gu W., Yi P., Bitter J. L., Hong J.-Y., Fairbrother D. H. and Chen K.-L. (2013). Bacterial anti-adhesive properties of polysulphone membranes modified with polyelectrolyte multilayers. *J. Mem. Sci.*, **446**, 201–211.

Tran T., Gray S., Naughton R. and Bolto B. (2006). Polysilicato-iron for improved NOM removal and membrane performance. *J. Mem. Sci.*, **280**, 560–571.

Tran T., Bolto B., Gray S., Hoang M. and Ostarcevic (2007). An autopsy study of a fouled reverse osmosis membrane element used in a brackish water treatment plant. *Wat. Res.*, **41**, 3915–3923.

Tran T. (2012). Personal communication.

Ulbricht M. and Belfort G. (1996). Surface modification of ultrafiltration membranes by low temperature plasma II. graft polymerisation onto polyacrylonitrile and polysulfone. *J. Mem. Sci.*, **111**, 193–215.

Ulbricht M., Richau K. and Kamusewitz (1998). Chemically and morphologically defined ultrafiltration membrane surfaces prepared by heterogeneous photo-initiated graft polymerisation. *Coll. & Surf. A: Physicochem. & Eng. Aspects*, **138**, 353–366.

Valavala R., Sohn J., Han J., Her N. and Yoon Y. (2011). Pretreatment in reverse osmosis seawater desalination: a short review. *Env. Eng. Res.*, **16**, 205–212.

Vedavyasan C. V. (2007). Pre-treatment trends – an overview. *Desal.*, **203**, 296–299.

Veolia (2012). Veolia Water to upgrade namibian desalination plant. *Filtr. + Sep.*, **49**(6), 8.

Villacorte L. O., Kennedy M. D., Amy G. L. and Schippers J. C. (2009). The fate of transparent exopolymer particles in integrated membrane systems: removal through pre-treatment processes and deposition on reverse osmosis membranes. *Wat. Res.*, **43**, 5039–5052.

Visvanathan C., Boonthanon N., Sathasivan A. and Jegatheesan V. (2002). Pre-treatment of seawater for biodegradable organic content removal using membrane bioreactor. *Desal.*, **153**, 133–140.

Vrouwenvelder J. S. and van der Kooij D. (2002). Diagnosis of fouling problems of NF and RO membrane installations by a quick scan. *Desal.*, **153**, 121–124.

Wang Y., Kim J.-H., Choo K.-H., Lee Y.-S. and Lee C.-K. (2000). Hydrophilic modification of polypropylene microfiltration membranes by ozone-induced graft polymerisation. *J. Mem. Sci.*, **169**, 269–276.

Wang Z., Han X., Ma J., Wang P., Mei X. and Wu Z. (2013). Recent advances in membrane fouling caused by extracellular polymeric substances: a mini-review. *Desal. Wat. Treat.*, **51**, 5121–5131.

Wavhal D. S. and Fisher E. R. (2002). Hydrophilic modification of polyethersulfone membranes by low temperature plasma-induced graft polymerisation. *J. Mem. Sci.*, **209**, 255–269.

Wilf M. and Scheirach M. K. (2001). Improved performance and cost reduction of RO seawater systems using UF pre-treatment. *Desal.,* **135**, 61–68.

Xiao Y., Liu X. D., Wang D. X., Lin Y. K., Han Y. P. and Wang X. L. (2013). Feasibility of using an innovative PVDF MF membrane prior to RO for reuse of a secondary municipal effluent. *Desal.,* **311**, 16–23.

Xu J., Ruan G., Chu X., Yao Y., Su B. and Gao C. (2007). A pilot study of UF pre-treatment without any chemicals for SWRO desalination in China. *Desal.,* **207**, 216–226.

Xu J., Feng X., Hou J., Wang X., Shan B., Yu U. and Gao C. (2013). Preparation and characterisation of a novel polysulphone UF membrane using a copolymer with capsaicin-mimic moieties for improved anti-fouling properties. *J. Mem. Sci.,* **446**, 171–180.

Xu Z., Wang J., Shen L., Men D. and Xu Y. (2002). Microporous polypropylene hollow fibre membrane. part I. surface modification by the graft polymerisation of acrylic acid. *J. Mem. Sci.,* **196**, 221–229.

Yamagishi H., Crivello J. V. and Belfort G. (1995a). Development of a novel photochemical technique for modifying

Yamagishi H., Crivello J. V. and Belfort G. (1995b). Evaluation of chemically modified polyarylsulfone ultrafiltration membranes. *J. Mem. Sci.,* **105**, 249–259.

Yang Q., Himstedt H. H., Ulbricht M., Qian X. and Wickramasinghe S. R. (2013). Designing magnetic field responsive nanofiltration membranes. *J. Mem. Sci.,* **430**, 70–78.

Ying L., Kang E. T. and Neoh K. G. (2003). Characterisation of membranes prepared from blends of poly(acrylic acid)-graft-poly(vinylidene fluoride) with poly(*N*-isopropylacrylamide) and their temperature- and pH-sensitive microfiltration. *J. Mem. Sci.,* **224**, 93–106.

Yu H.-Y., Kang Y., Liu Y. and Mi B. (2013). Grafting polyzwitterions onto polyamide by click chemistry and nucleophilic substitution on nitrogen: a novel approach to enhance membrane fouling resistance. *J. Mem. Sci.,* **449**, 50–57.

Yu J., Baek Y., Yoon H. and Yoon J. (2013). New disinfectant to control biofouling of polyamide reverse osmosis membrane. *J. Mem. Sci.,* **427**, 30–36.

Yu L., Zhang Y., Zhang B., Liu J., Zhang H. and Song C. (2013). Preparation and characterisation of HPEI-GO/PES ultrafiltration membrane with antifouling and antibacterial properties. *J. Mem. Sci.,* **447**, 452–462.

Yu S., Liu X., Liu J., Wu D., Liu M. and Gao C. (2011). Surface modification of thin-film composite polyamide reverse osmosis membranes with thermo-responsive polymer (TRP) for improved fouling resistance and cleaning efficiency. *Sep. Purif. Tech.,* **76**, 283–291.

Yu W.-Z., Graham N., Liu H.-J., Li H. and Qu J.-H. (2013). Membrane fouling by fe-humic cake layers in nano-scale: effect of in-situ formed Fe(III)coagulant. *J. Mem. Sci.,* **431**, 47–54.

Zhang R., Vigneswaran S., Ngo H. H. and Nguyen H. (2006). Magnetic ion exchange (MIEX®) resin as a pre-treatment to a submerged membrane system in the treatment of biologically treated wastewater. *Desal.,* **192**, 296–302.

Chapter 6

Reverse osmosis beneficiation

Filicia Wicaksana[1,2], Anthony G. Fane[1,3], Tang Chuyang[1,4] and Wang Rong[1,3]*

[1]Singapore Membrane Technology Centre, Nanyang Technological University, Singapore

[2]Department of Chemical and Materials Engineering, the University of Auckland, New Zealand

[3]School of Civil and Environmental Engineering, Nanyang Technological University, Singapore

[4]Department of Civil Engineering, University of Hong Kong, Pokfulam, Hong Kong

**Corresponding author: f.wicaksana@auckland.ac.nz*

6.1 INTRODUCTION

Being the most established desalination technology, reverse osmosis (RO) desalination plants have been installed in many parts of the world, particularly in coastal regions that encounter fresh water scarcity issues. Along with the progress of this technology, numerous endeavours have been assessed to advance the process performance. Primary efforts have been focused on minimising the power requirement as well as counteracting membrane fouling and scaling. In recent years, reverse osmosis has been coupled with various technologies to enhance its performance, to reduce the energy demand as well as to limit brine disposal, including zero liquid discharge. This chapter presents an overview on the improvement of reverse osmosis processes by retrofitting with various emerging technologies (Figure 6.1) such as an integration of membrane distillation (MD) to increase the recovery, a combination with membrane distillation crystallisation (MDC) for treatment of brine concentrate, an implementation of forward osmosis (FO) process with RO brine as draw solution, as well as the use of brine concentrate in pressure retarded osmosis (PRO) for power generation. Several green technologies with renewable energy sources such as the use of wind power and solar energy for RO system operation are also discussed.

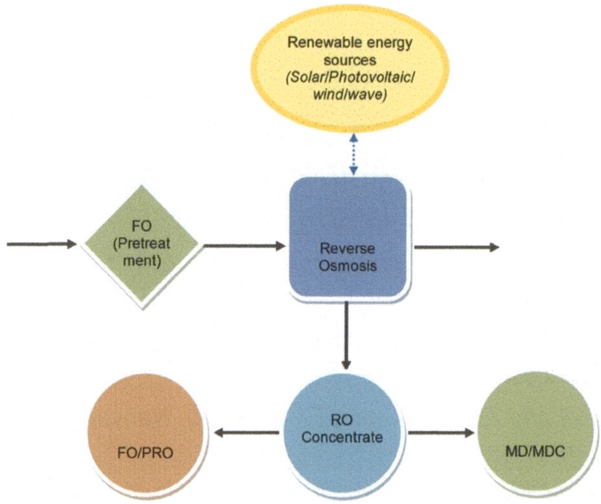

Figure 6.1 Combinations of reverse osmosis and other processes.

6.2 ENGINEERED OSMOSIS PROCESSES

Reverse osmosis, forward osmosis and pressure retarded osmosis can be differentiated based on their driving forces. Unlike reverse osmosis that relies on a high hydraulic pressure to overcome the osmotic pressure of saline water, an osmotically-driven process can be considered as a natural phenomenon since water transport occurs due to an osmotic pressure gradient between two aqueous solutions separated by a membrane.

Forward Osmosis involves the use of a draw solution with greater osmotic pressure than the feed water to induce the water transport across a dense semi-permeable membrane. As a result, water will be separated from the feed side to the draw solution side. This natural phenomenon occurs without applied pressure; hence the energy consumption is low. However, supplementary energy may be required to retrieve the water from the draw solution depending on the intended use of the final product and the type of the draw solution.

Possessing intermediate characteristics between FO and RO, Pressure Retarded Osmosis (PRO) also relies on salinity gradient where fresh water diffuses through a semi-permeable membrane into a pressurised saline solution. Power can be generated by depressurising the diluted saline water via a hydro-turbine or generator set.

The transport of water in FO, PRO and RO can be expressed as:

$$J_w = A(\Delta P - \sigma\Delta\pi) \tag{6.1}$$

where J_w is the water flux, A is the water permeability coefficient of the membrane, ΔP is the applied pressure, σ is the reflection coefficient, and $\Delta\pi$ is the osmotic pressure difference across the membrane. For non-permeating solutes, the reflection coefficient is ~1. $\Delta P = 0$ for FO processes, $\Delta P < \Delta\pi$ for PRO, and $\Delta P > \Delta\pi$ for RO. Figure 6.2 illustrates the relationship between water flux and applied pressure in FO, PRO and RO.

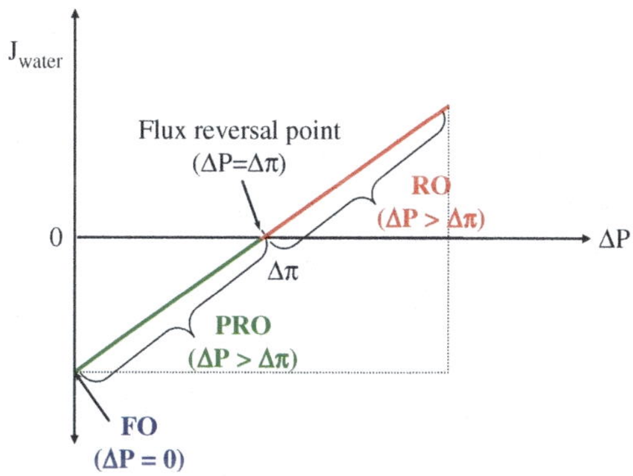

Figure 6.2 The relationship between water flux and applied pressure in FO, RO and PRO. (Adapted from Lee *et al.* 1981).

6.2.1 Osmotically-driven processes: applications, benefits and limitations

FO processes have several potential applications such as for desalination (Cath, 2010; Choi *et al.* 2009; Elimelech 2007; Khaydarov & Khaydarov, 2007; Kravath & Davis, 1975; McCutcheon *et al.* 2006; McGinnis & Elimelech, 2007; Miller & Evans, 2006; Tang & Ng, 2008), wastewater treatment (Catalyx-Inc, Cath *et al.* 2005; Holloway *et al.* 2007), food processing (Babu *et al.* 2006; Dova *et al.* 2007; Nayak & Rastogi, 2010; Petrotos *et al.* 1998) and pharmaceutical industries (Wright *et al.* 2003). A major application of pressure retarded osmosis (PRO) is for power generation (Achilli *et al.* 2009a; Gerstandt *et al.* 2008; Lee *et al.* 1981; Loeb, 1998; Loeb, 2002). Further details are discussed in Section 6.4.2.

In addition to lower energy consumption, FO may have several other benefits. Due to the lack of external pressure, it is believed that FO fouling propensity could be lower than pressure-driven processes (Achilli *et al.* 2009b; Holloway *et al.* 2007). However, further verification is needed to support this statement, although recent results suggest fouling in FO is more reversible (resulting in easier cleaning) (Lee *et al.* 2010; Mi & Elimelech, 2010). With no high pressure or high temperature involved, FO is particularly beneficial for the processing of heat-sensitive and more delicate substances such as food and pharmaceutical products, since degeneration of components in the feed due to high pressure and temperature can be avoided. Despite these advantages, FO membranes can encounter internal concentration polarisation (ICP) phenomenon due to the build up of salt or dilution of draw solutes within its porous sub-structure (Cath *et al.* 2006; McCutcheon & Elimelech, 2008; Mehta & Loeb, 1978). This phenomenon reduces the osmotic pressure gradient, resulting in a lower permeation flux. ICP is exacerbated by membrane support structures that have low porosity and are thick and for these reasons RO membranes perform poorly in FO. The effect of ICP could be minimised by optimising the FO membrane structure (thin and porous support) and performance.

Ideally, FO membranes should exhibit high water flux, have good structural integrity and high salt rejection with minimal cross-contamination of salts. In order to obtain these characteristics, FO membranes are typically asymmetric in structure comprising a dense selective layer and a porous sub-layer. At present, there are only a few commercially available FO membranes: cellulose tri-acetate (CTA) flat sheet membrane (Herron, 2008; Herron *et al.* 1994) and a more recently developed thin film composite (TFC) membrane (Hydration-Technologies-Innovations-Llc 2012) manufactured by US-based Hydration Technologies Innovations Llc (HTI). It was claimed that their TFC FO membranes were highly durable, dimensionally stable with more than double the permeability and flux rate of CTA membranes. Another US-based company, Catalyx, Inc. claims to produce thin and highly porous FO membranes that limits the ICP (Catalyx-Inc 2009). In addition to commercial FO membranes, various types of FO membranes with different structures and materials have been developed by researchers (Amini *et al.* 2013; Cho *et al.* 2013; Emadzadeh *et al.* 2014; Goh *et al.* 2013; Han *et al.* 2012; Hoover *et al.* 2013; Liu *et al.* 2013; Nguyen *et al.* 2013; Qiu *et al.* 2012; Setiawan *et al.* 2011; Su *et al.* 2010; Wang *et al.* 2007; Wang *et al.* 2009; Wang *et al.* 2010; Wei *et al.* 2011; Yang *et al.* 2009; Yu *et al.* 2011). Generally, these membranes have been reported to have high permeate flux, high salt rejection and relatively low ICP.

It appears that both osmotically-driven and pressure-driven processes have their own advantages and disadvantages; thus, it may be possible to enhance the performance of one particular process by combining it with other processes.

6.2.2 Integrated FO and RO processes

The forward osmosis process can be integrated with reverse osmosis systems in several ways. A combined system of forward osmosis and reverse osmosis processes, called manipulated osmosis (MO) technology was developed in 2003 by Adel Sharif and Al-Mayahi to transform seawater to drinking water (Surrey-University). This reportedly low energy and low cost technology, involves the use of forward osmosis process to extract fresh water from seawater followed by reverse osmosis process to regenerate the draw solution for reuse. An illustration of the process is shown in Figure 6.3. In this process, since the draw solute osmotic pressure must exceed that of sea water, the RO pressure needs to be raised. On this basis the energy savings are doubtful unless the FO pre-treatment reduces the overall fouling energy penalties.

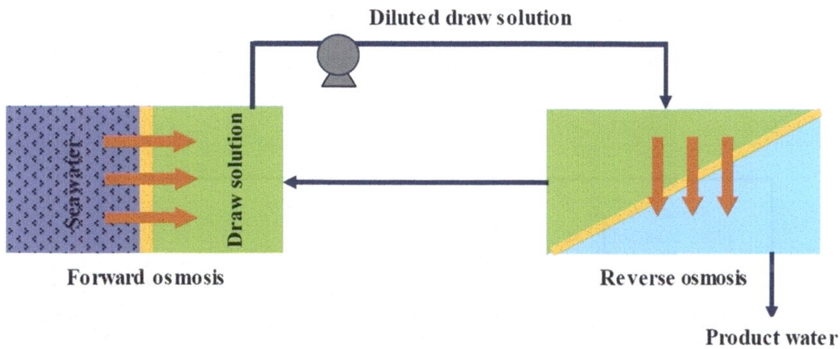

Figure 6.3 Manipulated osmosis (MO) desalination process (Sharif, 2011).

A similar concept has been adapted to produce desalinated water for agricultural irrigation purposes (Shaffer *et al.* 2012). Water for agricultural applications must comply with stringent boron and chloride standards. Hence, the production of desalinated water for agriculture may require a higher energy demand, compared to desalination for potable use; since ancillary post treatment may be required. The total energy consumption to produce desalinated water for agricultural irrigation ranges from 3–7 kWh/m^3 of product water, while pre- and post-treatments require 0.7–0.9 kWh/m^3 of product water. It has been reported that a hybrid FO-RO system could reach water quality standards for agricultural irrigation and required 10–30% lower energy than a conventional double-stage reverse osmosis process (Shaffer *et al.* 2012). In this combined system, the forward osmosis membrane functions as a pretreatment for the reverse osmosis process by substituting conventional pretreatments (coagulation and conventional filtration) or membrane-based pretreatment (microfiltration or ultrafiltration), as well as being a post-treatment for boron and chloride removal such that a second-stage reverse osmosis unit is no longer needed.

A pilot scale hybrid FO-RO configuration has also been utilised to treat impaired water originated from the permeate of a sequencing batch membrane bioreactor (SBMBR) (Hancock *et al.* 2013). The integrated system works synergistically to perform water reclamation and seawater desalination. In this system, seawater is used as the draw solution to extract water from the impaired water. Three RO membrane elements are used to recover the water from the diluted draw solution.

Another approach is to place a RO system in between two FO processes as a multiple-barrier system for water reuse (Bamaga *et al.* 2011, Sharif, 2011). In the first FO stage, saline water (seawater or brackish water concentrate) is used as draw solution to extract water from recycled water. The diluted seawater is re-concentrated with RO system to produce fresh water. A second FO stage is applied to dilute the seawater prior to discharge as well as to further concentrate and decrease the volume of the impaired water for subsequent treatment (Figure 6.4). Since the saline water has been diluted in the first FO stage, the energy demand for the following RO desalination process is lower as compared to a RO process without pre-dilution of feed. The challenge with this application is that the product water is partially reclaimed wastewater, and may not be accepted for direct potable use. It would be well suited to industrial and agricultural use or for indirect potable (subject to local regulations).

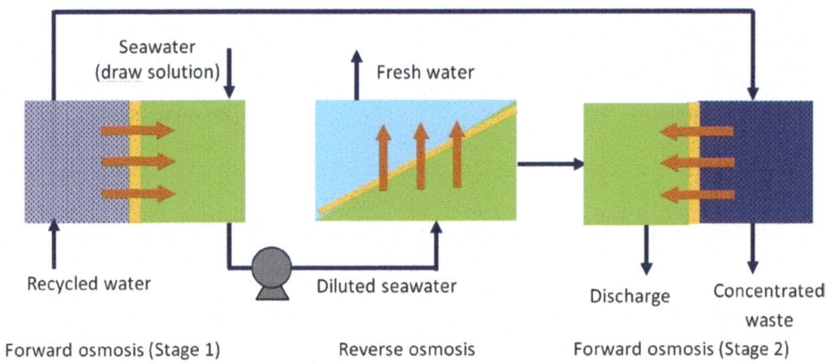

Figure 6.4 Integrated FO-RO system for water reuse (Bamaga *et al.* 2011, Sharif 2011).

The reverse osmosis process is commonly employed to produce potable water from seawater on ships. However, the need for high pressure (40–70 bars) to operate the system increases the energy consumption and affects the cost effectiveness of the RO process. RO can be operated at a lower pressure, but it has to compensate with a lower recovery rate. Requiring no external pressure, forward osmosis can be considered as a low cost process. However, generally, the FO permeate produced from sources with high salinity cannot fulfil the drinking water standard. A forward osmosis and reverse osmosis assembly was invented to generate a high quality potable water with no external energy input (Lampi *et al.* 2005). Forward osmosis can be hydraulically coupled to a reverse osmosis system in a housing to form a forward osmosis pressurised reverse osmosis (FOPRO) device. The housing assembly of the batch system comprise a forward osmosis element at one end and a RO element at the other end. A moveable piston is placed between the FO end and the RO end of the housing to transmit hydraulic pressure generated by the FO element to the RO unit.

6.3 RENEWABLE SOURCES AS ENERGY ENHANCEMENT FACILITIES FOR REVERSE OSMOSIS PROCESS

The idea of utilising renewable energy sources is attractive, particularly for energy intensive process such as a reverse osmosis desalination plant. Various renewable energy sources have been explored as energy enhancement facilities for reverse osmosis processes. The type of renewable energy that can be used is region specific and depends on the scale of the desalination plant. The utilisation of renewable energy resources to operate a reverse osmosis desalination plant may be crucial, particularly in remote regions with abundant natural energy resources such as solar and wind, but which experience fresh water shortage and limited electrical power supply. Typical capacities of reverse osmosis process with selected renewable energy sources are summarised in Table 6.1. Further discussion on these renewable energy sources can be found in the following sections.

Table 6.1 Potential renewable energy sources for reverse osmosis process (Papapetrou *et al.* 2010).

Energy Source	Typical Capacity	Power Requirement	Water Production Cost	Stage of Technology
PV	<100 m³/d	Electrical: BW: 0.5–1.5 kWh/m³ SW: 4–5 kWh/m³	BW: 5–7 €/m³ SW: 9–12 €/m³	Applications/ advanced R&D
Wind	50–2,000 m³/d	Electrical: BW: 0.5–1.5 kWh/m³ SW: 4–5 kWh/m³	under 100 m³/d BW: 3–5 €/m³ SW: 5–7 €/m³ around 1,000 m³/d 1.5–4 €/m³	Applications/ advanced R&D
Wave	1,000–3,000 m³/d	Pressurised water: 1.8–2.4 kWh/m³ Electrical: 2.2–2.8 kWh/m³	0.5–1.0 €/m³ (prospective cost)	Basic research

6.3.1 Wind-powered RO plants

Wind power is one common renewable energy source for desalination (Fadigas & Dias, 2009; Forstmeier *et al.* 2007; García-Rodríguez *et al.* 2001; Miranda & Infield, 2003; Peñate *et al.* 2011). It can be converted as kinetic energy to drive the rotor of a windmill. The output power can be used as energy supply for the desalination plant. Several desalination plants in Australia are operated by wind power. For example, the Kurnell Desalination Plant in south Sydney is powered by 100% renewable energy from The Capital Wind Farm in Bungendore, New South Wales. The desalination plant uses reverse osmosis technology that produces 250 megaliters of drinking water per day and provides up to 15% of Sydney's water supply needs (Desalination-Sydney-Water-Corporation, Kurnell-Desalination-Plant-Australia)

Renewable energy is commonly used as an auxiliary power supply or intermittent energy source, for example as an integrated wind-diesel system to ensure undisrupted supply of electricity during calm periods (Lindemann, 2004). In relation to this matter, there are two types of wind power system: a wind power system that is connected to public power grids is known as an on-grid system, while the system that is not connected to a local electric power supply is classified as off-grid system. Commercial on-grid wind energy devices are currently available with power generation capacities ranging from 0.3 to 3.6 MW (Lindemann, 2004) An off-grid system would be more appropriate in remote regions with an approximate wind speed of 4 m/s and where proper infrastructure is unavailable (Southwest-Windpower). In order to avoid power interruption in off-grid wind power system, a wind turbine can be combined with solar photovoltaic panels such that both systems can be alternated on sunny days with low wind speed and on windy days with less sunlight (Kershman *et al.* 2003, Southwest-Windpower).

The fluctuation in electricity produced by renewable sources would result in discontinuity of operation. Therefore, to overcome this problem, many desalination facilities use renewable energy sources while connected to a smart grid that is able to absorb or supply energy to the plant as required. This strategy has been adopted by the Perth Seawater Desalination Plant in Kwinana. The 48 wind turbines on the Emu Downs Wind Farm generate 80 MW of power, 30% of the power generated (24 MW) is used to operate the desalination plant (Perth-Seawater-Desalination-Plant-Australia).

6.3.2 Solar-powered and photovoltaic

Other renewable energy resources that can be used to supply power to RO process are solar energy and photovoltaic (PV) technology (Abdallah *et al.* 2005; Bilton *et al.* 2011; Gandhidasan & Al-Mojel, 2009; Lindemann 2004; Petersen *et al.* 1981; Qiblawey *et al.* 2011; Thomson & Infield, 2002). Photovoltaic technology involves the use of semiconductors to generate electricity under solar irradiation. This power generation system uses a number of solar cells containing photovoltaic materials. Photovoltaic devices are currently dominated by silicon-based materials such as monocrystalline silicon, polycrystalline silicon and amorphous silicon, other materials that have been used are cadmium telluride, copper indium gallium selenide/sulfide (Unold & Schock, 2011), and gallium arsenide nitride (O'Connor & Eure, 2011). Common PV-RO systems designs use lead-acid batteries. Unfortunately, the use of these batteries in hot climates would shorten their effective life span. In order to address this issue, a more recent PV-RO system with a designed capacity of 3 m^3/day was developed without batteries (Thomson & Infield, 2002). The non-battery system could enhance the energy efficiency and cost effectiveness of the system. Alternatively, the vanadium redox battery developed at the University of New South Wales, Australia offers great benefits due to low cost, large energy storage, and less sensitive to the environmental conditions (Rahman & Skyllas-Kazacos, 2009).

6.3.3 Wave energy

Although the idea of using wave energy for desalination system has been reported since the 1970s (Pleass, 1978; Pleass, 1979), ocean wave energy can be considered as a relatively immature form of renewable energy technology to operate the desalination plant as compared to solar and wind power. Arid regions with abundant wave energy resources such as Kenya, Oman and South Africa would benefit from this technology (Davies, 2005). Ocean waves are naturally available for coastal desalination plants. Harnessing of wave energy, however, can be costly. Therefore, proper design and construction of the wave energy collector device is important. Various designs and technologies have been reported. These include

(1) the use of oscillating buoys (DELBUOY) to drive piston pumps that supply seawater to single-pass RO membranes (Cruz & Salter, 2006; Davies, 2005; Hicks *et al.* 1989; Pleass, 1974);
(2) the desalination duck concept proposed by the Wave Power Group at the University of Edinburgh (Salter, 1974; Salter, 1980; Salter, 1989; Salter *et al.* 1976). The mechanical movement of the duck was converted to electrical power;
(3) McCabe wave pump that comprises a three-section hinged floating barge with two oscillating arms attached symmetrically to a central section (McCormick, 2001; McCormick & Kim, 1997). The mechanical forces were harnessed by pistons that pumped seawater to RO units;
(4) Oscillating water Column (OWC) device installed near Vizhinjam port in India (Indian-Express-Online-Media, 2004; Sharmilla *et al.* 2004). Waves crashing into a concrete caisson would compress and decompress the column of air. These actions generated a pressure variation that rotates a turbine to drive the reverse osmosis desalination plant. The energy generated was reported to be 6–7 kW to produce 7–8 m^3/day of desalinated water and
(5) the use of the water hammer effect to generate sufficient intermittent pressures to drive a reverse osmosis unit (Sawyer & Maratos, 2000).

The world's first wave-powered desalination plant is expected to commence construction in 2014 on Garden Island in Western Australia. The reverse osmosis desalination plant will be integrated with Carnegie's Perth Wave Energy Project (PWEP) infrastructure to produce freshwater (Australia-Carnegie-wave-energy).

6.4 REVERSE OSMOSIS CONCENTRATE

The increasing number of RO desalination plants may impose a negative impact on the environment due to issues associated with concentrate disposal. Major constituents in RO concentrate encompass soluble inorganic salts, low molecular weight soluble organics and biological substances which originate from the feed water (Pérez-González *et al.* 2012; Soliz *et al.* 2011). However, the concentration of these substances can be 1.5–6 times higher than that in the feed water for 35%–85% recovery (Squire, 2000). Coastal desalination plants typically discharge the brine into the ocean creating potential hazards to the marine environment. Inland desalination plants need to source alternative disposal methods since direct discharge to the sea is not possible. Evaporation ponds, deep well injection and surface water discharge are common brine disposal options for inland desalination plants (Ahmed *et al.* 2001; Arnal *et al.* 2005; Gabelich *et al.* 2010; Greenlee *et al.* 2009). However, these disposal methods may create potential risks of leakage and contamination of water bodies. Therefore, there is a critical need for more environmentally friendly brine disposal alternatives. Various innovative treatment options have been assessed which are primarily aimed

(1) to recover valuable matters or remove unwanted/hazardous substances from RO concentrate such as a possible use of electrodialysis for treatment of brine concentrate (Banasiak & Schafer, 2006), the implementation of ion exchange, bipolar membrane electrodialysis, and electrochlorination (Adham *et al.* 2006). Nevertheless, these alternative technologies have not been well-established; hence further assessment and optimisation are required to evaluate their cost effectiveness and viability, particularly for medium and large scale desalination plants;
(2) to reduce the volume of RO brine prior to disposal, such as dewatering of RO brine using forward osmosis process (Adham *et al.* 2007), and
(3) to reuse the RO concentrate, for example as a draw solution in FO. Several promising technologies for further use and processing of brine concentrate are discussed in more detail in the following sections.

6.4.1 Forward osmosis

The selection of draw solution used for a forward osmosis process will depend on its intended product. Proper selection of draw solution is crucial if pure water is the intended product since the draw solution should be easily regenerable. However, if a concentrated feed is the product (for example for protein enrichment or fruit juice concentration), reverse osmosis concentrate can be used as the draw solution.

In this case, the draw solution recovery is not needed and the RO retentate can be disposed of safely upon dilution (Figure 6.5).

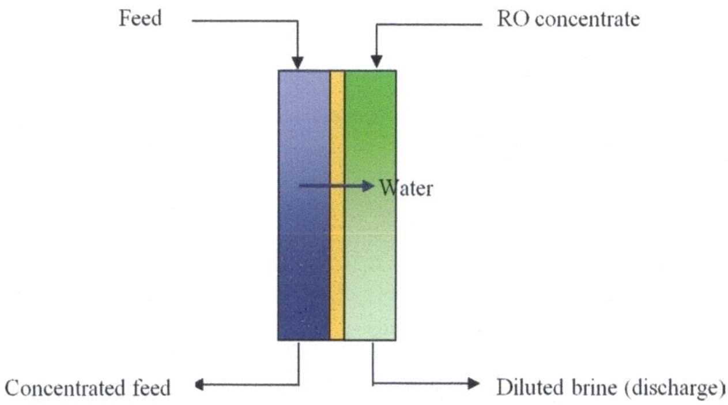

Figure 6.5 The use of RO concentrate as draw solution in FO process.

Catalyx Inc. has developed a patent pending Two Way Osmosis (TWO) system that combines forward osmosis with reverse osmosis for dewatering of sodium sulphate brine without the need for supplementary chemicals or other pre-treatment. It is claimed that the volume of concentrated liquid waste could be reduced by 50% with this system which consequently reduces the brine disposal costs (Catalyx-Inc 2009).

6.4.2 Pressure retarded osmosis

The idea of generating power by mixing fresh water and saline water has been reported as early as in the 1950s (Pattle, 1954; Post *et al.* 2007). Several decades later, Sydney Loeb initiated the use of pressure retarded osmosis for power generation from concentrated brines (Loeb, 1976; Loeb *et al.* 1976) as well as from other high salinity sources such as the Dead Sea (Loeb, 1998) and the Great Salt Lake (Loeb, 2001). In 2009, the world's first osmotic power plant was opened by the Norwegian renewable energy company (Statkraft) in Tofte on Hurum, near Oslo (Statkraft Press Centre, 2009). After operating for more than four years, it was reported in late 2013 that Statkraft will now discontinue its operation (Statkraft, 2013). At present, various studies on PRO have been performed to optimise the membrane and process performances (Chou *et al.* 2012; Hoover *et al.* 2013; She *et al.* 2012; Xu *et al.* 2010; Yip & Elimelech, 2012). The use of a PRO system as an energy recovery for seawater desalination plant has been projected as part of Mega-ton Water System project in Fukuoka (Kurihara, 2012).

The principle of pressure retarded osmosis power plant is depicted in Figure 6.6. When a semi permeable membrane is placed between a pressurised concentrated seawater (draw solution side) and fresh water (feed side), trans-membrane diffusion of water will occur from the feed side to the draw solution side resulting in an increased volume of pressurized solution on the draw solution side. Energy can be harvested by de-pressurising the diluted seawater through a hydroturbine. The magnitude of power that can be generated is determined by the salinity gradient between fresh water and salt water. Higher salinity gradients will generate more power.

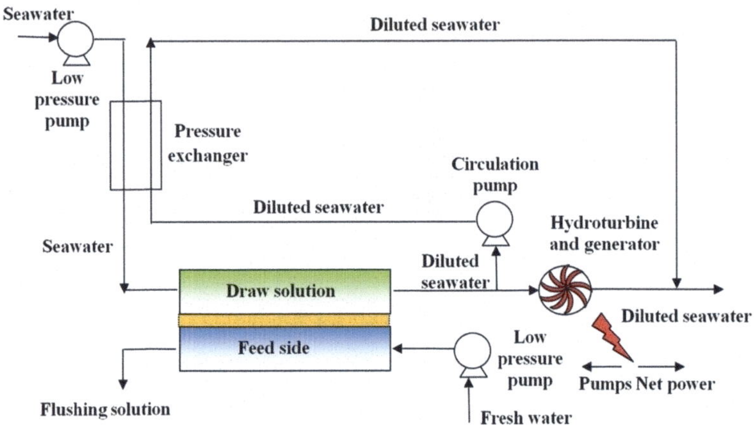

Figure 6.6 Pressure retarded osmosis power plant. Source: Aadapted from Achilli *et al.* (2009a).

A similar concept was proposed by Sharqawy *et al.* on the use of PRO as an energy recovery device for RO desalination plant. It is claimed that the incorporation of PRO unit could reduce the input power by 38% compared to the original RO plant (Sharqawy *et al.* 2011), assuming the PRO could operate without fouling.

In PRO systems the selection of a proper membrane is very important to achieve an economically feasible power generation process. Typical membranes used for PRO are composite membranes; hence the key issues will be the optimisation of membrane skin layer and porous substructure to achieve a desirable water flux with low salt permeability and low structural parameter (Lee *et al.* 1981; Loeb, 2001; Post *et al.* 2007; She *et al.* 2012). In addition, the membranes should be able to withstand the standard operating pressure in a PRO plant ~12–14 bars for seawater (Sivertsen *et al.* 2012) or greater when concentrated brine is used as a draw solution (the optimum pressure is 1/2 the osmotic pressure difference) (Chou *et al.* 2013). Depending on the feed sources and membrane material, the cost of energy produced by PRO was estimated to be approximately \$0.058/kWh (Loeb, 1998) to \$0.15/kWh (Loeb, 2001; Loeb, 2002). Chou *et al.* developed thin-film composite polyethersulfone-based hollow fibre membranes that were specially designed for PRO application (Chou *et al.* 2012). It was reported that a power density of 10.6 W/m^2 could be achieved at 9 bars operating pressure with 1 M NaCl as draw solution and 40 mM NaCl as feed. Recently, TFC hollow fibre PRO membranes capable of 20 W/m^2 have been developed (Chou *et al.* 2013). Hoover *et al.* incorporated electrospun polymeric nanofibres into a thin-film composite membrane to achieve high water flux and salt rejection for potential application in osmotically-driven membrane processes, including PRO (Hoover *et al.* 2013).

6.4.3 Membrane distillation

6.4.3.1 Basic principle of membrane distillation

The membrane distillation (MD) process has been known since the 1960's (Findley, 1967; Findley *et al.* 1969; Weyl, 1967). Membrane distillation is analogous to the conventional distillation process as both processes require energy input to heat the feed solution to provide the latent heat of vaporization, except that MD uses a hydrophobic microporous membrane placed between a heated feed and a cooled permeate streams (Figure 6.7). The temperature gradient between the heated and cooled solutions facing the membrane causes a vapour pressure difference that acts as a driving force for mass transfer across the membrane. The hydrophobic membrane allows passage of water vapour, while liquid water remains on the feed side provided that the liquid pressure does not surpass the minimum entry pressure of the porous membrane. The water vapour is then condensed on the cooled permeate side. Unlike conventional distillation, membrane distillation can generate a high purity product and can be operated under mild operating conditions. The operating temperature is commonly between 40 to 80°C (Cipollina *et al.* 2012; Phattaranawik *et al.* 2003; Schofield *et al.* 1990a; Yang *et al.* 2011)

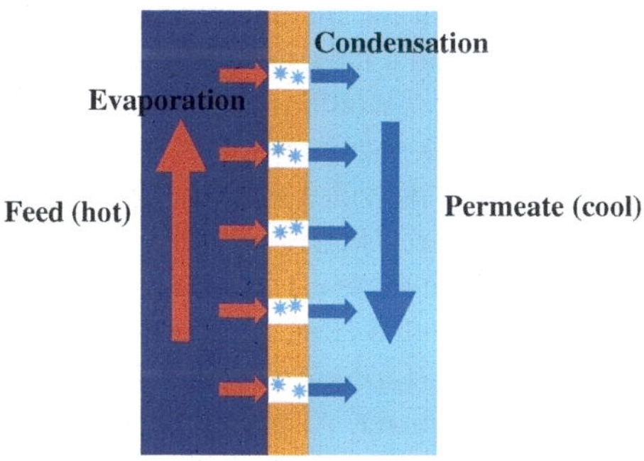

Figure 6.7 Membrane distillation process.

6.4.3.2 MD configurations

Membrane distillation can be operated in various configurations, as a direct contact membrane distillation (DCMD) in which the cold aqueous stream of the permeate side is in direct contact with the membrane (Qu *et al.* 2009b; Song *et al.* 2007), with an air gap in between the condensing surface and the membrane (air gap membrane distillation or AGMD) (Meindersma *et al.* 2006), with an inert gas passing the permeate side to sweep the vapor to an external condenser (sweeping gas membrane distillation or SGMD) (Alkhudhiri *et al.* 2012; Drioli *et al.* 2009; Pangarkar *et al.* 2011; Tomaszewska, 2000), or a vacuum membrane distillation (VMD) in which a vacuum is applied on the permeate side (without passing any liquid through the permeate side of the membrane) (Izquierdo-Gil & Jonsson, 2003; Lawson & Lloyd, 1997; Mericq *et al.* 2010; Safavi & Mohammadi, 2009). AGMD and DCMD are more applicable for separation processes where water is the permeating fluid since these MD configurations do not require an external condenser. Conversely, SGMD and VMD are commonly used for removal of volatile organic substances or dissolved gases from an aqueous stream (Pangarkar *et al.* 2011). However, recently memsys®, a technology company based in Singapore and Germany, has introduced a patented multi-effect VMD technology into water processing applications (Zhao *et al.* 2013).

6.4.3.3 Membrane characteristics

Ideally MD membranes should be highly selective and porous to achieve a desirable permeate flux and product quality. The membrane should not be wetted by liquids to avoid contamination of the product (the contact angle of the hydrophobic surface should be >90°). Edwie and Chung suggested that membrane wetting was strongly influenced by membrane pore size and the morphology beneath the membrane surface, while the wetting was less affected by membrane wall thickness (Edwie & Chung, 2012). A typical pore size of membranes used in MD process ranges from 0.2–0.4 μm (Alkhudhiri *et al.* 2012; Hsu *et al.* 2002; Yun *et al.* 2006), while the porosity varies from 30 to 85% (El-Bourawi *et al.* 2006). However, it was recommended that the membrane porosity should be greater than 70% for optimum performance and a maximum pore diameter of 0.5–0.6 μm was suggested as large membrane pores would increase the risk of membrane wetting (Schneider *et al.* 1988). In DCMD, the energy loss by conduction is decreased by thicker membranes and this also provides mechanical strength. However thicker membranes have greater transport resistance (Schofield *et al.* 1990a; Schofield *et al.* 1990b; Xu *et al.* 1994). To balance these effects, an optimum membrane thickness of 30–60 μm has been recommended (Laganà *et al.* 2000). However, membrane thickness has little effect on the performance of AGMD since the stationary air gap is the main resistance to mass transfer. At present, common membrane materials utilised in MD process are polymeric membranes such as polytetrafluoroethylene (PTFE) (Mericq *et al.* 2010; Saffarini *et al.* 2012), polyvinylidene fluoride (PVDF) (Edwie & Chung, 2012; Khayet *et al.* 2004; Qu *et al.* 2009a; Yang *et al.* 2011) and polypropylene (PP) (Ji *et al.* 2010; Song *et al.* 2007). MD performance can also be limited by temperature polarisation (TP) (Schofield *et al.* 1987), and this can be controlled by careful hydrodynamic design.

6.4.3.4 Energy sources in membrane distillation

The primary energy need in membrane distillation is thermal energy, hence MD can be competitive with other membrane processes in circumstances where various alternative energy sources such as solar power (Cipollina *et al.* 2012; Hogan *et al.* 1991; Qtaishat & Banat, 2013; Saffarini *et al.* 2012; Walton *et al.* 2004), geothermal energy (Goosen *et al.* 2010; Khayet & Matsuura, 2011; Sarbatty & Chiam, 2013) and waste heat (Drioli *et al.* 2009; Walton *et al.* 2004) are accessible.

Geothermal energy is an interesting option. The world's first commercial geothermal power plant was built in Italy in 1911, followed by New Zealand in 1958 when its Wairakei station was commissioned (Goosen *et al.* 2010). Geothermal energy is a substantial part of the power generation capacity in New Zealand. In 2011, geothermal energy contributed up to 19% of New Zealand's Total Primary Energy Supply (Ministry-of-Economic-Development 2012). The top ten countries in electricity generation from geothermal energy are USA, Philippines, Indonesia, Mexico, Italy, New Zealand, Iceland, Japan, El Salvadore and Kenya (Bertani, 2010). It has been reported that the incorporation of geothermal energy into a vacuum membrane distillation plant could reduce the water production cost by up to 60% (Sarbatty & Chiam, 2013).

Industrial waste heat is an energy that is produced from industrial activities without being used, such as hot exhaust gasses released to the environment, high temperature substances produced from exothermic processes such as sulphuric acid production and heat emitted from hot equipment surfaces. Various studies have reported that approximately 20 to 50% of industrial energy consumption would eventually be released as waste heat (BCS, 2008), while up to 740 MWh heat can be produced per 1,000 tonnes of sulphuric acid (Van der Bruggen, 2003). If waste heat can be recovered, it can be used in various applications. In regions where sulphuric acid plants are widely available such as in the Middle East and Northern Africa, the excess heat can be beneficial to membrane distillation plants as an alternative energy source, particularly to heat the feed. Hence, substantial energy saving can be achieved.

6.4.3.5 An integrated membrane distillation and reverse osmosis process

A combined reverse osmosis and membrane distillation system has been used to treat RO brine with a concentration of 75 g/L at 35°C (Drioli *et al.* 1999). This integrated unit could attain a higher recovery factor (86.7%) than those of membrane distillation (77%) and reverse osmosis alone (40%). It was claimed that the amount of water generated by the RO/MD system was more than double the amount of water produced by a RO process alone at the same cost. The MD process alone could produce the same amount of water as the RO/MD unit, but at a higher cost.

A vacuum membrane distillation coupled with a reverse osmosis unit was assessed in a simulation study (Mericq *et al.* 2010) The integrated system could achieve a high recovery factor (up to 89%) with the reduction of brine volume by a factor of 5.5 (the concentration of brine was 7.6 times higher than that of seawater).

6.4.3.6 Membrane distillation crystallization

Membrane distillation can be integrated with a crystallizer to retrieve the solids from the concentrate; this combination is known as membrane distillation crystallization (MDC) (Edwie & Chung, 2012; Gryta, 2002a; Guan *et al.* 2012; Ji *et al.* 2010; Mariah *et al.* 2006; Tun *et al.* 2005). MDC provides an alternative solution for the brine disposal problem to achieve zero discharge of hazardous substances to waterways. The separation of water from saline water is governed by thermodynamics; hence RO can only reach a certain brine concentration level due to a limit associated with its osmotic pressure value. Since MD is a thermally driven membrane process, it allows further concentration of RO brine to approach its saturation level (Ji *et al.* 2010). Salts can be recovered by passing the saturated brine through a crystalliser. Typical sizes of NaCl crystals formed ranged from 20 to 200 μm (Edwie & Chung, 2012; Ji *et al.* 2010). It was reported that the degree of salt recovery and the crystal size distribution were determined by the degree of supersaturation or cooling rate. A faster cooling rate would result in a higher salt recovery and a smaller crystal size (Edwie & Chung, 2012).

One of the challenges of an integrated membrane distillation and crystallisation process is to control the crystal formation so that it will not occur before the saturated feed reaches a crystallizer. In the MD process, the saline solution passes through the membrane module and is eventually concentrated. The formation of salt crystals on the membrane surface could occur during this process. The presence of salt crystals on the membrane surface would stimulate the wettability of the membrane (Gryta, 2002b). The salt build up would also cause module blockage. However, this problem can be avoided by applying a high feed flowrate (Guan *et al.* 2012) with possible higher energy demand as a trade-off.

The energy consumption of MDC system with hollow fibre membranes at moderate temperatures (40–50°C) was estimated to be around 15–20 kWh/m^3 (Macedonio *et al.* 2007). This energy requirement can be considered as relatively high. However, it is still 30–50% lower than conventional evaporative equipment for NaCl crystallization processes such as force circulation crystallisers and draft-tube-baffle crystallisers which are commonly operated at temperatures higher than 70°C (Bennett *et al.* 1973; Ji *et al.* 2010). A combined MDC and RO system could become very attractive, particularly if the production of salt crystals is sufficiently high. Results from economic evaluation suggested that it is possible for the gain from the sale of collected salt to cover the entire cost of the desalination process (Macedonio *et al.* 2007). In addition, since a significant portion of cost in the MD process originates from the cost of energy for heating the feed, further cost saving can be obtained if thermal energy such as low grade heat source, solar or geothermal energy is available in the plant. Finally thermally-driven MD has a benefit because it can be driven by non-fossil fuel energy and is intrinsically a low GHG process.

6.5 CONCLUSIONS

The incorporation of various technologies such as FO, PRO and MD in RO processes offers potential benefits. FO can be used as a pretreatment prior to RO process, resulting in substantial reduction of RO fouling and power consumption. The use of renewable energy sources can transform a RO process to become less energy intensive, particularly for regions with abundant natural resources such as solar, wind and wave energy. Integrated RO-FO, RO-PRO, and RO-MDC systems provide more environmentally friendly alternative solutions in contrast to common desalination practices for brine disposal. Zero liquid discharge can be achieved by re-using the concentrated brine as draw solution in FO, for power generation in PRO, and by transforming the brine concentrate to salt crystals as by-product.

6.6 REFERENCES

Abdallah S., Abu-Hilal M. and Mohsen M. S. (2005). Performance of a photovoltaic powered reverse osmosis system under local climatic conditions. *Desalination*, **183**(1–3), 95–104.

Achilli A., Cath T. Y. and Childress A. E. (2009a). Power generation with pressure retarded osmosis: an experimental and theoretical investigation. *Journal of Membrane Science*, **343**(1–2), 42–52.

Achilli A., Cath T. Y., Marchand E. A. and Childress A. E. (2009b). The forward osmosis membrane bioreactor: a low fouling alternative to MBR processes. *Desalination*, **239**(1–3), 10–21.

Adham S., Oppenheimer J., Kumar M., Webb L. and Kottenstette R. 2006. Innovative approaches to RO concentrate management: beneficial reuse and concentrate minimization in WEFTEC. Water Environment Foundation, Dallas, Texas.

Adham S., Oppenheimer J., Liu L. and Kumar M. (2007). Dewatering reverse osmosis concentrate from water reuse applications using forward osmosis. Water Reuse Foundation. Alexandria.

Ahmed M., Shayya W. H., Hoey D. and Al-Handaly J. (2001). Brine disposal from reverse osmosis desalination plants in Oman and the United Arab Emirates. *Desalination*, **133**(2), 135–147.

Alkhudhiri A., Darwish N. and Hilal N. (2012). Membrane distillation: a comprehensive review. *Desalination*, **287**(Special issue), 2–18.

Amini M., Jahanshahi M. and Rahimpour A. (2013). Synthesis of novel thin film nanocomposite (TFN) forward osmosis membranes using functionalized multi-walled carbon nanotubes. *Journal of Membrane Science*, **435**, 233–241.

Arnal J. M., Sancho M., Iborra I., Gozálvez J. M., Santafé A. and Lora J. (2005). Concentration of brines from RO desalination plants by natural evaporation. *Desalination*, **182**(1–3), 435–439.

Australia-Carnegie-Wave-Energy Receives First Payment for CETO Seawater Desalination Plant. https://www.offshorewind.biz/2013/05/08/australia-carnegie-wave-energy-receives-first-payment-for-ceto-seawater-desalination-plant/ (accessed 24 May 2013)

Babu B. R., Rastogi N. K. and Raghavarao K. S. M. S. (2006). Effect of process parameters on transmembrane flux during direct osmosis. *Journal of Membrane Science*, **280**(1–2), 185–194.

Bamaga O. A., Yokochi A., Zabara B. and Babaqi A. S. (2011). Hybrid FO/RO desalination system: preliminary assessment of osmotic energy recovery and designs of new FO membrane module configurations. *Desalination*, **268**(1–3), 163–169.

Banasiak L. J. and Schafer A. I. 2006. Electrodialysis for Reverse Osmosis Concentrate Treatment. Available from: http://www.see.ed.ac.uk/IIE/research/membrane/Posters/Electrodialysis%20for%20reverse%20osmosis%20concentrate%20treatment.pdf (accessed, 2013)

BCS (2008). Waste heat recovery: technology and opportunities in US industry.

Bennett R. C., Fiedelman H. and Randolph A. D. (1973). Crystallizer-influenced nucleation. *Chemical Engineering Progress*, **69**(7), 86–93.

Bertani R. 2010. Geothermal power generation in the world, 2005–2010 update report. In World Geothermal Congress. Bali, Indonesia.

Bilton A. M., Wiesman R., Arif A. F. M., Zubair S. M. and Dubowsky S. (2011). On the feasibility of community-scale photovoltaic-powered reverse osmosis desalination systems for remote locations. *Renewable Energy*, **36**(12), 3246–3256.

Catalyx-Inc http://www.catalyxinc.com/forward-osmosis.html (accessed March 2010)

Catalyx-Inc (2009). Contract signals commercial deployment of forward osmosis system. *Membrane Technology*, **2009**(4), 2.

Cath T. Y. (2010). Osmotically and thermally driven membrane processes for enhancement of water recovery in desalination processes. *Desalination and Water Treatment*, **15**(1–3), 279–286.

Cath T. Y., Childress A. E. and Elimelech M. (2006). Forward osmosis: principles, applications, and recent developments. *Journal of Membrane Science*, **281**(1–2), 70–87.

Cath T. Y., Gormly S., Beaudry E. G., Flynn M. T., Adams V. D. and Childress A. E. (2005). Membrane contactor processes for wastewater reclamation in space: Part I. direct osmotic concentration as pretreatment for reverse osmosis. *Journal of Membrane Science*, **257**(1–2), 85–98.

Cho Y. H., Han J., Han S., Guiver M. D. and Park H. B. (2013). Polyamide thin-film composite membranes based on carboxylated polysulfone microporous support membranes for forward osmosis. *Journal of Membrane Science*, **445**, 220–227.

Choi Y. J., Choi J. S., Oh H. J., Lee S., Yang D. R. and Kim J. H. (2009). Towards a combined system of forward osmosis and reverse osmosis for seawater desalination. *Desalination*, **247**(1–3), 239–246.

Chou S., Wang R. and Fane A. G. (2013). Robust and high performance hollow fiber membranes for energy harvesting from salinity gradients by pressure retarded osmosis. *Journal of Membrane Science*, **448**, 44–54.

Chou S., Wang R., Shi L., She Q., Tang C. and Fane A. G. (2012). Thin-film composite hollow fiber membranes for pressure retarded osmosis (PRO) process with high power density. *Journal of Membrane Science*, **389**, 25–33.

Cipollina A., Di Sparti M. G., Tamburini A. and Micale G. (2012). Development of a membrane distillation module for solar energy seawater desalination. *Chemical Engineering Research and Design*, **90**(12), 2101–2121.

Cruz J. and Salter S. 2006. Update on the design of an offshore wave powered desalination device. In OWEMES. Civitavecchia (Rome).

Davies P. A. (2005). Wave-powered desalination: resource assessment and review of technology. *Desalination*, **186**(1–3), 97–109.

Desalination-Sydney-Water-Corporation http://archive-au.com/page/275941/2012-09-05/ http://www.sydneywater.com.au/Water4Life/Desalination/ (accessed January 2014)

Dova M. I., Petrotos K. B. and Lazarides H. N. (2007). On the direct osmotic concentration of liquid foods. part I: impact of process parameters on process performance. *Journal of Food Engineering*, **78**(2), 422–430.

Drioli E., Criscuoli A. and Molero L. P. (2009). Water and wastewater treatment technologies-vol III-membrane distillation, in encyclopedia of life support systems(EOLSS).

Drioli E., Laganà F., Criscuoli A. and Barbieri G. (1999). Integrated membrane operations in desalination process. *Desalination*, **122**(2–3), 141–145.

Edwie F. and Chung T.-S. (2012). Development of hollow fiber membranes for water and salt recovery from highly concentrated brine via direct contact membrane distillation and crystallization. *Journal of Membrane Science*, **421–422**, 111–123.

El-Bourawi M. S., Ding Z., Ma R. and Khayet M. (2006). A framework for better understanding membrane distillation separation process. *Journal of Membrane Science*, **285**(1–2), 4–29.

Elimelech M. (2007). Yale constructs forward osmosis desalination pilot plant. *Membrane Technology*, **2007**(1), 7–8.

Emadzadeh D., Lau W. J., Matsuura T., Rahbari-Sisakht M. and Ismail A. F. (2014). A novel thin film composite forward osmosis membrane prepared from PSf–TiO2 nanocomposite substrate for water desalination. *Chemical Engineering Journal*, **237**, 70–80.

Fadigas E. A. F. A. and Dias J. R. (2009). Desalination of water by reverse osmosis using gravitational potential energy and wind energy. *Desalination*, **237**(1–3), 140–146.

Findley M. E. (1967). Vaporization through porous membrane. *Industrial & Engineering Chemistry Process Design and Development*, **6**(2), 226–237.

Findley M. E., Tanna V. V., Rao Y. B. and Yeh C. L. (1969). Mass and heat transfer relations in evaporation through porous membranes. *AIChE Journal*, **15**(4), 483–489.

Forstmeier M., Mannerheim F., D'Amato F., Shah M., Liu Y., Baldea M. and Stella A. (2007). Feasibility study on wind-powered desalination. *Desalination*, **203**(1–3), 463–470.

Gabelich C. J., Xu P. and Cohen Y. (2010). Chapter 10 concentrate treatment for inland desalting, in Sustainability Science and Engineering 2. Elsevier, pp. 295–326.

Gandhidasan P. and Al-Mojel S. A. (2009). Effect of feed pressure on the performance of the photovoltaic powered reverse osmosis seawater desalination system. *Renewable Energy*, **34**(12), 2824–2830.

García-Rodríguez L., Romero-Ternero V. and Gómez-Camacho C. (2001). Economic analysis of wind-powered desalination. *Desalination*, **137**(1–3), 259–265.

Gerstandt K., Peinemann K. V., Skilhagen S. E., Thorsen T. and Holt T. (2008). Membrane processes in energy supply for an osmotic power plant. *Desalination*, **224**(1–3), 64–70.

Goh K., Setiawan L., Wei L., Jiang W., Wang R. and Chen Y. (2013). Fabrication of novel functionalized multi-walled carbon nanotube immobilized hollow fiber membranes for enhanced performance in forward osmosis process. *Journal of Membrane Science*, **446**, 244–254.

Goosen M., Mahmoudi H. and Ghaffour N. (2010). Water desalination using geothermal energy. *Energies*, **3**(8), 1423–1442.

Greenlee L. F., Lawler D. F., Freeman B. D., Marrot B. and Moulin P. (2009). Reverse osmosis desalination: water sources, technology, and today's challenges. *Water Research*, **43**(9), 2317–2348.

Gryta M. (2002a). Concentration of NaCl solution by membrane distillation integrated with crystallization. *Separation Science and Technology*, **37**(15), 3535–3558.

Gryta M. (2002b). Direct contact membrane distillation with crystallization applied to NaCl solutions. *Chemical Papers*, **56**(1), 14–19.

Guan G., Wang R., Wicaksana F., Yang X. and Fane A. G. (2012). Analysis of membrane distillation crystallization system for high salinity brine treatment with zero discharge using aspen flowsheet simulation. *Industrial & Engineering Chemistry Research*, **51**(41), 13405–13413.

Han G., Zhang S., Li X., Widjojo N. and Chung T.-S. (2012). Thin film composite forward osmosis membranes based on polydopamine modified polysulfone substrates with enhancements in both water flux and salt rejection. *Chemical Engineering Science*, **80**, 219–231.

Hancock N. T., Xu P., Roby M. J., Gomez J. D. and Cath T. Y. (2013). Towards direct potable reuse with forward osmosis: technical assessment of long-term process performance at the pilot scale. *Journal of Membrane Science*, **445**, 34–46.

Herron J. (2008). US Patent 7,445,712 B2. Asymmetric forward osmosis membranes.

Herron J. R., Beaudry E. G., Jochums C. E. and Medina L. E. (1994). US patent 5,281,430. Osmotic concentration apparatus and method for direct osmotic concentration of fruit juices.

Hicks D. C., Mitcheson G. R., Pleass C. M. and Salevan J. F. (1989). Delbouy: ocean wave-powered seawater reverse osmosis desalination systems. *Desalination*, **73**, 81–94.

Hogan P. A., Sudjito, Fane A. G. and Morrison G. L. (1991). Desalination by solar heated membrane distillation. *Desalination*, **81**(1–3), 81–90.

Holloway R. W., Childress A. E., Dennett K. E. and Cath T. Y. (2007). Forward osmosis for concentration of anaerobic digester. *Water Research*, **41**(17), 4005–4014.

Hoover L. A., Schiffman J. D. and Elimelech M. (2013). Nanofibers in thin-film composite membrane support layers: enabling expanded application of forward and pressure retarded osmosis. *Desalination*, **308**, 73–81.

Hsu S. T., Cheng K. T. and Chiou J. S. (2002). Seawater desalination by direct contact membrane distillation. *Desalination*, **143**(3), 279–287.

Hydration-Technologies-Innovations-Llc (2012). HTI starts producing thin-film composite FO membrane. *Membrane Technology*, **2012**(7), 5–6.

Indian-Express-Online-Media (2004). Desalination plant is wave-powered. *Membrane Technology*, **2004**(3), 4.

Izquierdo-Gil M. A. and Jonsson G. (2003). Factors affecting flux and ethanol separation performance in vacuum membrane distillation. *Journal of Membrane Science*, **214**(1), 113–130.

Ji X., Curcio E., Al Obaidani S., Di Profio G., Fontananova E. and Drioli E. (2010). Membrane distillation-crystallization of seawater reverse osmosis brines. *Separation and Purification Technology*, **71**(1), 76–82.

Kershman S. A., Rheinländer J. and Gabler H. (2003). Seawater reverse osmosis powered from renewable energy sources – hybrid wind/photovoltaic/grid power supply for small-scale desalination in Libya. *Desalination*, **153**(1–3), 17–23.

Khaydarov R. A. and Khaydarov R. R. (2007). Solar powered direct osmosis desalination. *Desalination*, **217**(1–3), 225–232.

Khayet M., Khulbe K. and Matsuura T. (2004). Characterization of membranes for membrane distillation by atomic force microscopy and estimation of their water vapor transfer coefficients in vacuum membrane distillation process. *Journal of Membrane Science*, **238**(1–2), 199–211.

Khayet M. and Matsuura T. (2011). Chapter 14 – membrane distillation hybrid systems, in Membrane Distillation, Principles and Applications. Elsevier, Amsterdam, pp. 399–427.

Kravath R. E. and Davis J. A. (1975). Desalination of sea water by direct osmosis. *Desalination*, **16**(2), 151–155.

Kurihara M. (2012). Membrane research for water treatment facing the age of global mega-competition & collaboration, in Advanced Membrane Technology V: Membranes for Sustainable Water, Energy and the Environment. An ECI Conference Series, Engineering Conferences International, Singapore.

Kurnell-Desalination-Plant-Australia http://www.water-technology.net/projects/kurnell-desalination/ (accessed January 2014)

Laganà F., Barbiari G. and Drioli E. (2000). Direct contact membrane distillation: modelling and concentration experiments. *Journal of Membrane Science*, **166**(1), 1–11.

Lampi K., Beaudry E. and Herron J. (2005). US Patent 6,849,184 B1. Forward osmosis pressurized device and process for generating potable water.

Lawson K. W. and Lloyd D. G. (1997). Review. membrane distillation. *Journal of Membrane Science*, **124**(1), 1–25.

Lee K. L., Baker R. W. and Lonsdale H. K. (1981). Membranes for power generation by pressure-retarded osmosis. *Journal of Membrane Science*, **8**(2), 141–171.

Lee S., Boo C., Elimelech M. and Hong S. (2010). Comparison of fouling behavior in forward osmosis (FO) and reverse osmosis (RO). *Journal of Membrane Science*, **365**(1–2), 34–39.

Lindemann J. H. (2004). Wind and solar powered seawater desalination applied solutions for the mediterranean, the middle east and the gulf countries. *Desalination*, **168**, 73–80.

Liu X., Qi S., Li Y., Yang L., Cao B. and Tang C. Y. (2013). Synthesis and characterization of novel antibacterial silver nanocomposite nanofiltration and forward osmosis membranes based on layer-by-layer assembly. *Water Research*, **47**(9), 3081–3092.

Loeb S. (1976). Production of energy from concentrated brines by pressure-retarded osmosis : I. Preliminary technical and economic correlations. *Journal of Membrane Science*, **1**, 49–63.

Loeb S. (1998). Energy production at the dead sea by pressure-retarded osmosis: challenge or chimera? *Desalination*, **120**(3), 247–262.

Loeb S. (2001). One hundred and thirty benign and renewable megawatts from great salt lake? the possibilities of hydroelectric power by pressure-retarded osmosis. *Desalination*, **141**(1), 85–91.

Loeb S. (2002). One hundred and thirty benign and renewable megawatts from great salt lake? the possibilities of hydroelectric power by pressure-retarded osmosis with spiral module membranes: desalination, 141 (2001) 85–91. *Desalination*, **142**(2), 207.

Loeb S., Van Hessen F. and Shahaf D. (1976). Production of energy from concentrated brines by pressure-retarded osmosis: II. experimental results and projected energy costs. *Journal of Membrane Science*, **1**, 249–269.

Macedonio F., Curcio E. and Drioli E. (2007). Integrated membrane systems for seawater desalination: energetic and exergetic analysis, economic evaluation, experimental study. *Desalination*, **203**(1–3), 260–276.

Mariah L., Buckley C. A., Brouckaert C. J., Curcio E., Drioli E., Jaganyi D. and Ramjugernath D. (2006). Membrane distillation of concetrated brines – roles of water activities in the evaluation of driving force. *Journal of Membrane Science*, **280**(1–2), 937–947.

McCormick M. E. (2001). Wave-powered reverse osmosis desalination. *Sea Technology*, **42**, 37–39.

McCormick M. E. and Kim Y. C. 1997. Ocean wave powered desalination. in Proceedings of the International Association for Hydraulic Research Congress.

McCutcheon J. R. and Elimelech M. (2008). Influence of membrane support layer hydrophobicity on water flux in osmotically driven membrane processes. *Journal of Membrane Science*, **318**(1–2), 458–466.

McCutcheon J. R., McGinnis R. L. and Elimelech M. (2006). Desalination by ammonia-carbon dioxide forward osmosis: influence of draw and feed solution concentrations on process performance. *Journal of Membrane Science*, **278**(1–2), 114–123.

McGinnis R. L. and Elimelech M. (2007). Energy requirements of ammonia-carbon dioxide forward osmosis desalination. *Desalination*, **207**(1–3), 370–382.

Mehta G. D. and Loeb S. (1978). Internal polarization in the porous substructure of a semi-permeable membrane under pressure-retarded osmosis. *Journal of Membrane Science*, **4**, 261–265.

Meindersma G. W., Guijt C. M. and de Haan A. B. (2006). Desalination and water recycling by air gap membrane distillation. *Desalination*, **187**(1–3), 291–301.

Mericq J.-P., Laborie S., Cabassud C. (2010). Vacuum membrane distillation of seawater reverse osmosis brines. *Water Research*, **44**(18), 5260–5273.

Mi B. and Elimelech M. (2010). Organic fouling of forward osmosis membranes: fouling reversibility and cleaning without chemical reagents. *Journal of Membrane Science*, **348**(1–2), 337–345.

Miller J. E. and Evans L. R. (2006). Forward osmosis: a new approach to water purification and desalination, in sandia report. Sandia National Laboratories, California.

Ministry-of-Economic-Development, New Zealand Energy Data File. (2012).

Miranda M. S. and Infield D. (2003). A wind-powered seawater reverse-osmosis system without batteries. *Desalination*, **153**(1–3), 9–16.

Nayak C. A. and Rastogi N. K. (2010). Forward osmosis for the concentration of anthocyanin from garcinia indica choisy. *Separation and Purification Technology*, **71**(2), 144–151.

Nguyen T. P. N., Yun E.-T., Kim I.-C. and Kwon Y.-N. (2013). Preparation of cellulose triacetate/cellulose acetate (CTA/CA)-based membranes for forward osmosis. *Journal of Membrane Science*, **433**, 49–59.

O'Connor M. and Eure J. (2011). New photovoltaic material could deliver twice the solar power. medill reports,

Pangarkar B. L., Sane M. G. and Guddad M. (2011). Reverse osmosis and membrane distillation for desalination of groudwater: a review. *International Scholarly Research Network Materials Science*, **2011**.

Papapetrou M., Wieghaus M. and Biercamp C. (2010). Roadmap for the development of desalination powered by renewable energy, prodes project report. Intelligent Energy for Europe programme, Fraunhofer Verlag, Germany.

Pattle R. E. (1954). Production of electric power by mixing fresh and salt water in the hydroelectric pile. *Nature*, **174**(4431), 660.

Peñate B., Castellano F., Bello A. and García-Rodríguez L. (2011). Assessment of a stand-alone gradual capacity reverse osmosis desalination plant to adapt to wind power availability: a case study. *Energy*, **36**(7), 4372–4384.

Pérez-González A., Urtiaga A. M., Ibáñez R. and Ortiz I. (2012). State of the art and review on the treatment technologies of water reverse osmosis concentrates. *Water Research*, **46**(2), 267–283.

Perth-Seawater-Desalination-Plant-Australia. water-technology.net http://www.water-technology.net/projects/perth/ (accessed January 2014)

Petersen G., Fries S., Mohn J. and Müller A. (1981). Wind and solar powered reverse osmosis desalination units – design, start up, operating experience. *Desalination*, **39**, 125–135.

Petrotos K. B., Quantick P. and Petropakis H. (1998). A study of the direct osmotic concentration of tomato juice in tubular membrane – module configuration. I. the effect of certain basic process parameters on the process performance. *Journal of Membrane Science*, **150**(1), 99–110.

Phattaranawik J., Jiraratananon R. and Fane A. G. (2003). Effect of pore size distribution and air flux on mass transport in direct contact membrane distillation. *Journal of Membrane Science*, **215**(1–2), 75–85.

Pleass C. M. 1974. The use of wave powered seawater desalination systems. in Proceeding of the International Symposium on Waves and Energy. Canterbury.

Pleass C. M. 1978. The use of wave powered systems for desalination. in Proc. International Symposium on Wave and Tidal Energy. Cranfield, UK.

Pleass C. M. (1979). Seawater desalting using wave powered reverse osmosis systems, in Proc. ACS Conference on Chemical and Economic Resources. pp. 385–401.

Post J. W., Veerman J., Hamelers H. V. M., Euverink G. J. W., Metz S. J., Nymeijer K. and Buisman C. J. N. (2007). Salinity-gradient power: evaluation of pressure-retarded osmosis and reverse electrodialysis. *Journal of Membrane Science*, **288**(1–2), 218–230.

Qiblawey H., Banat F. and Al-Nasser Q. (2011). Performance of reverse osmosis pilot plant powered by photovoltaic in Jordan. *Renewable Energy*, **36**(12), 3452–3460.

Qiu C., Setiawan L., Wang R., Tang C. Y. and Fane A. G. (2012). High performance flat sheet forward osmosis membrane with an NF-like selective layer on a woven fabric embedded substrate. *Desalination*, **287**, 266–270.

Qtaishat M. R. and Banat F. (2013). Desalination by solar powered membrane distillation systems. *Desalination*, **308**, 186–197.

Qu D., Wang J., Fan B., Luan Z. and Hou D. (2009a). Study on concentrating primary reverse osmosis retentate by direct contact membrane distillation. *Desalination*, **247**(1–3), 540–550.

Qu D., Wang J., Wang L., Hou D., Luan Z. and Wang B. (2009b). Integration of accelerated precipitation softening with membrane distillation for high-recovery desalination of primary reverse osmosis concentrate. *Separation and Purification Technology*, **67**(1), 21–25.

Rahman F. and Skyllas-Kazacos M. (2009). Vanadium redox battery: positive half-cell electrolyte studies. *Journal of Power Sources*, **189**(2), 1212–1219.

Safavi M. and Mohammadi T. (2009). High-salinity water desalination using VMD. *Chemical Engineering Journal*, **149**(1–3), 191–195.

Saffarini R. B., Summers E. K., Arafat H. A. and Lienhard V J. H. (2012). Economic evaluation of stand-alone solar powered membrane distillation systems. *Desalination*, **299**, 55–62.

Salter S. H. (1974). Wave power. *Nature*, **249**(5459), 720–724.

Salter S. H. (1980). Recent progress on ducks. *IEE Proceedings, Part A*, **127**(5), 308–319.

Salter S. H. (1989). World progress in wave energy-1988. *International Journal of Ambient Energy*, **10**(1), 3–24.

Salter S. H., Jeffrey D. C. and Taylor J. R. M. (1976). The architecture of nodding duck wave power generators. *Naval Architect*, (1), 21–24.

Sarbatty R. and Chiam C. K. (2013). Evaluation of geothermal energy in desalination by vacuum membrane distillation. *Applied Energy*, **112**, 737–746.

Sawyer R. A. and Maratos D. F. 2000. Wave power for desalination using unsteady incompressible duct flow (water hammer). in Proceedings of Mediterranean Conference on Policies and Strategies for Desalination and Renewable Energies. Santorini, Greece.

Schneider K., Holz W., Wollbeck R. and Ripperger S. (1988). Membranes and modules for transmembrane distillation. *Journal of Membrane Science*, **39**(1), 25–42.

Schofield R. W., Fane A. G. and Fell C. J. D. (1987). Heat and mass transfer in membrane distillation. *Journal of Membrane Science*, **33**(3), 299–313.

Schofield R. W., Fane A. G. and Fell C. J. D. (1990a). Gas and vapor transport through microporous membranes. II. membrane distillation. *Journal of Membrane Science*, **53**(1–2), 173–185.

Schofield R. W., Fane A. G., Fell C. J. D. and Macoun R. (1990b). Factors affecting flux in membrane distillation. *Desalination*, **77**, 279–294.

Setiawan L., Wang R., Li K. and Fane A. G. (2011). Fabrication of novel poly(amide–imide) forward osmosis hollow fiber membranes with a positively charged nanofiltration-like selective layer. *Journal of Membrane Science*, **369**(1–2), 196–205.

Shaffer D. L., Yip N. Y., Gilron J. and Elimelech M. (2012). Seawater desalination for agriculture by integrated forward and reverse osmosis: improved product water quality for potentially less energy. *Journal of Membrane Science*, **415–416**, 1–8.

Sharif A. 2011. Low energy production of fresh water from the sea: manipulated osmosis. in Water Security Symposium. Muscat, Oman.

Sharmilla N., Jalihal P., Swamy A. K. and Ravindran M. (2004). Wave powered desalination system. *Energy*, **29**(11), 1659–1672.

Sharqawy M. H., Zubair S. M., Lienhard V. J. H. (2011). Second law analysis of reverse osmosis desalination plants: an alternative design using pressure retarded osmosis. *Energy*, **36**(11), 6617–6626.

She Q., Jin X. and Tang C. Y. (2012). Osmotic power production from salinity gradient resource by pressure retarded osmosis: effects of operating conditions and reverse solute diffusion. *Journal of Membrane Science*, **401–402**, 262–273.

Sivertsen E., Holt T., Thelin W. and Brekke G. (2012). Modelling mass transport in hollow fibre membranes used for pressure retarded osmosis. *Journal of Membrane Science*, **417–418**, 69–79.

Soliz D., Glenn E. P., Seaman R., Yoklic M., Nelson S. G. and Brown P. (2011). Water consumption, irrigation efficiency and nutritional value of atriplex lentiformis grown on reverse osmosis brine in a desert irrigation district. *Agriculture, Ecosystems and Environment*, **140**(3–4), 473–483.

Song L., Li B., Sirkar K. K. and Gilron J. L. (2007). Direct contact membrane distillation-based desalination: novel membranes, devices, larger-scale studies, and a model. *Industrial & Engineering Chemistry Research*, **46**(8), 2307–2323.

Southwest-Windpower what is the difference between on-grid and off-grid wind turbines? http://www.windenergy.com/community/blog/what-difference-between-grid-and-grid-wind-turbines (accessed January 2013)

Squire D. (2000). Reverse osmosis concentrate disposal in the UK. *Desalination*, **132**(1–3), 47–54.

Statkraft (2013). Statkraft halts osmotic power investments, http://www.statkraft.com/presscentre/news/statkraft-halts-osmotic-power-investments.aspx (accessed, 2014)

Statkraft Press Centre (2009). http://www.statkraft.com/presscentre/news/2009/the-worlds-first-osmotic-power-plant-opened.aspx (accessed, 2012)

Su J., Yang Q., Teo J. F. and Chung T.-S. (2010). Cellulose acetate nanofiltration hollow fiber membranes for forward osmosis processes. *Journal of Membrane Science*, **355**(1–2), 36–44.

Surrey-University (Surrey University wins IChemE 2011 top water award) www.surrey.ac.uk/cce/research/water_chemical/cora/ (accessed, 2012)

Tang W. and Ng H. Y. (2008). Concentration of brine by forward osmosis: performance and influence of membrane structure. *Desalination*, **224**(1–3), 143–153.

Thomson M. and Infield D. (2002). A photovoltaic-powered seawater reverse-osmosis system without batteries. *Desalination*, **153**(1–3), 1–8.

Tomaszewska M. (2000). Membrane distillation-examples of applications in technology and environmental protection. *Polish Journal of Environmental Studies*, **9**(1), 27–36.

Tun C. M., Fane A. G., Matheickal J. T. and Sheikholeslami R. (2005). Membrane distillation crystallization of concentrated salts–flux and crystal formation. *Journal of Membrane Science*, **257**(1–2), 144–155.

Unold T. and Schock H. W. (2011). Nonconventional (non-silicon-based) photovoltaic materials. *The Annual Review of Materials Research*, **41**, 297–321.

Van der Bruggen B. (2003). Desalination by distillation and by reverse osmosis – trends towards the future. *Membrane Technology*, **2003**(2), 6–9.

Walton J., Lu H., Turner C., Solis S. and Hein H. (2004). Solar and waste heat desalination by membrane distillation, in desalination and water purification research and development program report (DWPR) no. 81. Bureau of Reclamation, Denver Federal Center, Denver, USA.

Wang K. Y., Chung T.-S. and Qin J.-J. (2007). Polybenzimidazole (PBI) nanofiltration hollow fiber membranes applied in forward osmosis process. *Journal of Membrane Science*, **300**(1–2), 6–12.

Wang K. Y., Yang Q., Chung T.-S. and Rajagopalan R. (2009). Enhanced forward osmosis from chemically modified polybenzimidazole (PBI) nanofiltration hollow fiber membranes with a thin wall. *Chemical Engineering Science*, **64**(7), 1577–1584.

Wang R., Shi L., Tang C. Y., Chou S., Qiu C. and Fane A. G. (2010). Characterization of novel forward osmosis hollow fiber membranes. *Journal of Membrane Science*, **355**(1–2), 158–167.

Wei J., Qiu C., Tang C. Y., Wang R. and Fane A. G. (2011). Synthesis and characterization of flat-sheet thin film composite forward osmosis membranes. *Journal of Membrane Science*, **372**(1–2), 292–302.

Weyl P. K. (1967). US Patent 3,340,186. Recovery of demineralized water from saline waters.

Wright J. C., Johnson R. M. and Yum S. I. (2003). DUROS osmotic pharmacuetical systems for parenteral & site-directed therapy. *Drug Delivery Technology*, **3**(1).

Xu S. C., Wang Y. X. and Wang S. C. (1994). Transport mode of membrane distillation process and the VMD method to determine the membrane configurational number. *Journal of Membrane Science*, **97**, 1–6.

Xu Y., Peng X., Tang C. Y., Fu Q. S. and Nie S. (2010). Effect of draw solution concentration and operating conditions on forward osmosis and pressure retarded osmosis performance in a spiral wound module. *Journal of Membrane Science*, **348**(1–2), 298–309.

Yang Q., Wang K. Y. and Chung T. S. (2009). Dual-layer hollow fibers with enhanced flux as novel forward osmosis membranes for water production. *Environmental Science and Technology*, **43**(8), 2800–2805.

Yang X., Wang R., Shi L., Fane A. G. and Debowski M. (2011). Performance improvement of PVDF hollow fiber-based membrane distillation process. *Journal of Membrane Science*, **369**(1–2), 437–447.

Yip N. Y. and Elimelech M. (2012). Thermodynamic and energy efficiency analysis of power generation from natural salinity gradients by pressure retarded osmosis. *Environmental Science and Technology*, **46**(9), 5230–5239.

Yu Y., Seo S., Kim I.-C. and Lee S. (2011). Nanoporous polyethersulfone (PES) membrane with enhanced flux applied in forward osmosis process. *Journal of Membrane Science*, **375**(1–2), 63–68.

Yun Y., Ma R., Zhang W., Fane A. G. and Li J. (2006). Direct contact membrane distillation mechanism for high concentration NaCl solutions. *Desalination*, **188**(1–3), 251–262.

Zhao K., Heinzl W., Wenzel M., Büttner S., Bollen F., Lange G., Heinzl S. and Sarda N. (2013). Experimental study of the memsys vacuum-multi-effect-membrane-distillation (V-MEMD) module. *Desalination*, **323**, 150–160.

Chapter 7

Brine management

Peter Sanciolo and Stephen Gray*

Institute for Sustainability and Innovation, Victoria University, P.O. Box 14428, Melbourne, 8001, Australia

**Corresponding author: Peter.Sanciolo@vu.edu.au*

7.1 INTRODUCTION

In response to increasing water needs, the application of desalination technology has grown throughout the twentieth century. In the mid to late twentieth century, this was largely to meet water needs in the Middle East and on islands with low water resources. During the end of the twentieth and start of the twenty-first century, however, the number and size of plants has rapidly increased. Figure 7.1 shows the rapid growth in installed global desalination capacity between 1980 and 2007, followed by a 5 year period of relative stagnation after the 2008 financial crisis. The growth forecasts from 2012 to 2018 including the contracted capacity (red line), and the projected growth in capacity to 2030 is also shown. Despite the period of stagnation after the financial 2008 crisis, the desalination market is growing. The drought in California and Texas, new markets opening up in Africa and Asia, strong demand in established markets such as Saudi Arabia, Morocco, Tunisia, Chile, and Jordan, and continuing opportunities in both the onshore and offshore oil and gas sector should ensure future growth in the desalination market (Global Water Intelligence, 2014).

Worldwide, across all available desalination technologies, about 63% of the desalination capacity is produced from seawater, 19% from brackish water, 5% from waste water sources, 8% from low-salt river water, and 5% from pure water for specialty applications. Much of the expected future growth will be from seawater desalination due to continued development of the market in coastal-based water poor regions such as the Arabian Gulf, the Mediterranean Sea, and the Red

Sea (Lattemann, 2010). These figures, however, vary considerably from region to region as shown in Figure 7.2.

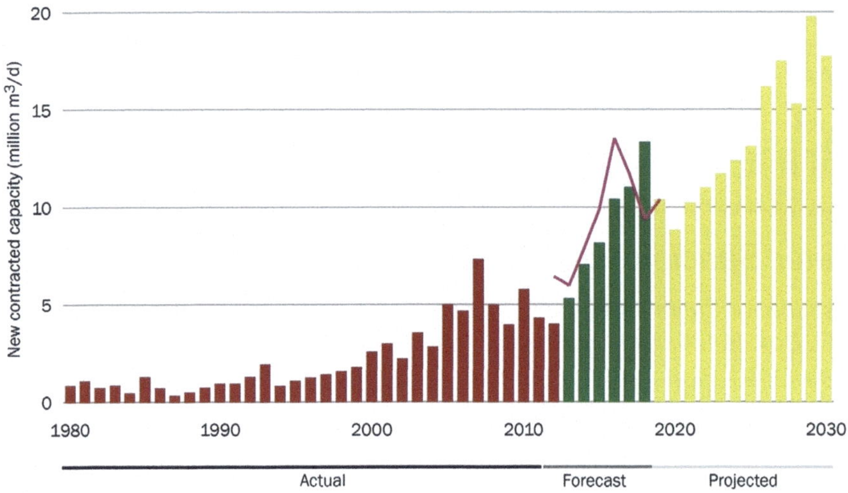

Figure 7.1 Trend in global desalination capacity. From Global Water Intelligence DesalData.

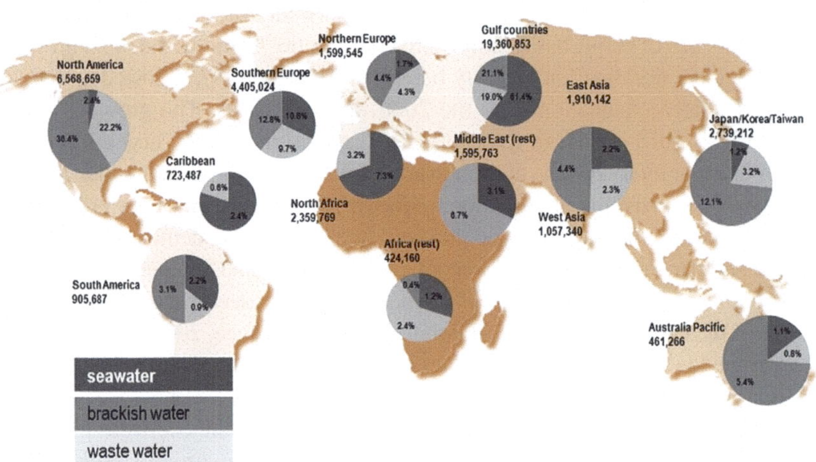

Figure 7.2 Global desalination capacities (cubic meters per day) and feedwater types. Source Sustainability Science and Engineering, Volume 2, 2010. Adapted from Buonomenna, 2013.

In some regions such as Australia, North America and Asia, brackish water and wastewater desalination constitutes a large proportion of the region's desalination capacity. This poses major challenges as these water resources are largely situated inland, and environmentally sound brine disposal practices are more difficult for inland regions.

The desalination process removes low salt water, leaving behind a waste concentrated salt solution that is commonly referred to as waste brine. Although it has its challenges, waste brine disposal from seawater desalination is generally accepted as having low adverse environmental impact provided that good dispersal of the brine is achieved, and that care is taken with chemical additives and the management of other waste streams from the process. Adverse environmental impact of brine disposal from the desalination of brackish groundwater or wastewater in inland regions, however, are much more likely due to the lower opportunity for dilution of the brine to decrease the impact of salinity and other brine constituents. Some plants have received permission by regulatory authorities to discharge to rivers, but this stands to make already salt stressed rivers more saline, compromising their use for irrigation and their ecological biodiversity. Regulatory authorities are therefore less likely to allow this mode of discharge in the future. Some plants dispose their brine to nearby sewer systems. This does not have major regulatory limitations for the desalination plant but shifts these limitations to the sewerage plant if the quantity of waste brine is such that it increases the salt and nutrient concentrations in the outfall of the sewerage treatment plant. This may also limit direct use of the wastewater for irrigation. Inland desalination plants that do not have access or permission to discharge to rivers or sewer systems can consider injection of the waste brine underground. This, however, has the potential for contamination of valuable water resources, and requires a good knowledge of the geology and availability of suitable geology. Understandably, this is highly regulated and is actively discouraged in some parts of the world such as Australia. For the increasing number of inland plants that do not have access to surface water, sewage systems or underground disposal, the only environmentally responsible option is to contain the waste on site in evaporation ponds. The cost of evaporation ponds is high so it is often beneficial to reduce the volume of waste brine. This requires investment in extra equipment which can have high operating costs, but this extra cost can be more than compensated by the savings resulting from reduced evaporation pond requirements. These reduced evaporation pond requirements, in turn, can allow a desalination plant to become viable in situations where the evaporation pond requirements would otherwise have been prohibitively expensive and/or environmentally unsustainable. Furthermore, a reduction in brine volume also reduces the capital and running cost of thermally based brine treatment processes, making zero discharge to the environment a viable prospect.

7.2 CHALLENGES IN BRINES MANAGEMENT

7.2.1 Location limitation of available options

Location poses significant limitations on the available options for brine management, and the size of the desalination operation may further reduce the available options.

In the case of seawater desalination, the choice of brine management options is relatively simple. Discharge to the sea is the most logical choice for coastal desalination plants, but this is not without challenges. It is simplistic to argue that the concentrate from desalination is concentrated seawater and all that is required to avoid environmental damage during discharge is to minimise the effect of higher salt concentration. Waste streams from desalination processes include the concentrated brine, backwash liquids and sludges from pre-treatment and cleaning processes containing suspended solids and coagulant salts as well as anti-scaling, anti-fouling, and anti-corrosion chemicals (Miri & Chouikhi, 2005; Lattemann & Hopper, 2008). As seawater is a highly corrosive liquid, wastewater may also include trace metals from corrosion processes in the desalination system. The impact of these waste streams on the receiving environment can, however, be minimised by good management. Good dispersal of the brine prevents osmotic shock that occurs when large volumes of the denser than seawater brine are discharged into one location, sinking to the seafloor and killing seagrass and other organisms that depend on this habitat (Dawoud & Mulla, 2012).

This requires location of the desalination plant such that the outfall can take advantage of ocean currents. This can require expensive pumping of seawater or waste brine if the location where the water is to be used is far from these currents. Segregation of the other streams such as backwash from pre-treatment coagulation and filtration processes allows removal of suspended solids that could decrease light penetration and affect photosynthetic organisms such as algae and seagrasses on the ocean floor. Judicious blending of clean-in-place streams with concentrate before discharge prevents localized concentrations of the cleaning chemicals that may be toxic or detrimental to sea-life in the area (Bleninger & Jirka, 2010).

Many of the good waste management practices that have allowed the widespread and growing use of seawater desalination are not available for many inland desalination plants. Unless expensive pipelines to the ocean are constructed, the available brine management options are discharge to rivers and lakes, deep well injection, evaporation ponds and zero liquid discharge (ZLD) with or without salt recovery. Discharge of waste brine to rivers and lakes is problematic since the required dispersal of the brine to prevent osmotic shock and exceedance of regulatory limits of certain constituents is not possible. In the US, water discharge is rarely allowed if the salinity difference between the discharged water and the receiving water is more than 10% (Mickley, 2004). Concentrates produced from groundwaters can also have high concentrations of gases, such as carbon dioxide, ammonia, and hydrogen sulfide, which can be toxic to aquatic animals (Mickley,

2006). The disposal of these to surface waters would require concentrate treatment such as pH adjustment, chlorination, dechlorination, degasification, and aeration. High levels of calcium and fluoride in RO concentrate have also been shown to be toxic to test organisms (Mickley, 2006). Some rivers may have the required flow for adequate dispersal of contaminants at some times of the year but rarely all year round. Despite this, however, the option for discharge to rivers does exist, even in countries with high levels of environmental protection such as the USA (Mickley, 2010). The level of protection rivers and streams receive in the US and many other parts of the world depends on the river's current state of degradation and their intended use. In the US state of California, for example, Antidegradation Policy aims to protect bodies of water with high water quality for beneficial uses and from any adverse impacts to water quality. Surface waters are classified on the basis of their beneficial use and their composition is regulated according to this intended use (U.S. Department of the Interior, 2009). For inland surface waters, these objectives are often intended to protect agricultural uses, which can result in stringent limits on salinity-related constituents. This has resulted in effluent limits for TDS for some water bodies that are considered to be unachievable and have been included in wastewater discharge permits to protect salt sensitive avocado and strawberry crops. The total maximum daily load in the Santa Clara River for chloride adopted in 2002, for example, called for a waste-load allocation of 100 mg/L as an instantaneous maximum—a number that Sanitation Districts of Los Angeles County have difficulty meeting in their effluent discharges.

It is noteworthy that the distinction between discharge to the sea and discharge to a river becomes blurred when discharge is to estuarine environments in the regions where rivers meet the sea. The Queensland EPA, for example, permits the discharge of RO concentrate into the Brisbane River at Luggage Point, Gibson Island and Bundamba discharging at 1.5 km, 9.7 km and 67 km upstream from the mouth of the Brisbane River, respectively. The rationale for this is that the RO concentrate is not hypersaline and therefore salinity was not identified as the predominant environmental concern. The environmental concerns, however, were focussed around the presence of toxicants in the form of chloramines and ammonia being injected into the influent for the purpose of protecting the RO membranes. Furthermore, the RO treatment process concentrates metals and metalloids to potentially environmentally harmful concentrations. Therefore, the focus of the Queensland EPA assessment revolved around ensuring that the dilution dynamics in the near-field mixing zone was capable of controlling the extent of a potential toxicity zone to an acceptable distance. The toxicant concentrations initially presented to the Queensland EPA in the Environmental Impact Statement (EIS) were later revised when data from pilot Reverse Osmosis (RO) plant results became available. Comparison of the toxicant concentrations with ANZECC/ARMCANZ (2000) toxicant trigger values (TTVs) indicated that a standard Sewage Treatment Plant (STP) outfall would be insufficient to maintain a chronic toxicity zone within an acceptable distance. The diffuser designers were then changed to achieve the required number of dilutions to negate observable chronic toxicological effects beyond the authorised 5 m chronic toxicity zone (Tripodi & Ramsay, 2009).

Another option for discharge of waste brine is to a nearby sewer system (Greenlee *et al.* 2009). This is a particularly attractive option for both coastal and inland desalination operations that use the sewage system treated wastewater as a feedwater for the desalination process. Since this is not an environmental discharge, there are no regulatory requirements other than those imposed by the sewerage treatment plant authorities in licencing agreements that are established between the sewage treatment plant and the desalination proponents. The removal of low salinity water from the wastewater by the desalination plant, however, has the potential of significantly increasing the salinity of the wastewater, thereby making it less attractive as a recycled water for irrigation. The desalination plant waste brine can also increase the concentration of EPA regulated toxicants and nutrients in the outfall of the sewage treatment plant to levels above critical limits if the volume treated by the desalination is a considerable proportion of the whole, so this option is usually limited to small desalination facilities taking water from large sewage treatment plants.

Injection of the brine underground is another option that is available for inland desalination plants. Injection into a constructed well is referred as Deep Well Injection, and this is a popular brine management option in some parts of the USA. According to the survey results of Mickley (Mickley, 2009) this is the third most widely used method of brine discharge in the USA (13% of plants use this method), after surface water discharge (45%) and sanitary sewer discharge (27%), but it is only used in two states. However, this option has been less popular in Australia. This may be attributed to the availability of suitable geology and/or the very stringent regulatory requirements to safeguard the integrity and quality of Australian groundwater resources. In Western Australia and all other Australian states and territories, for example, deep well injection of waste has been discouraged for fear of contaminating valuable and vital groundwater resources (WA EPA, 2003) Nevertheless, guidance on deep and shallow liquid injection of industrial waste is available if the stringent requirements can be met. Underground disposal of the RO concentrate is also used in in-situ desalination (ISD) where the RO membranes are inserted directly into a borehole drilled into a brackish groundwater aquifer. Desalinated water is pumped from the aquifer through the membrane, leaving the concentrate behind in the groundwater. Despite the benefit associated with decreased cost of above-ground infrastructure, ISD requires a thorough understanding of the aquifer's hydrology and water quality. The concentrate remains in the vicinity of the RO membranes and the desalination performance depends on the groundwater movement and aquifer recharge (National Water Commission, 2011).

If discharge to surface waters or nearby sewer system or injection of the waste brine underground are not available, the only remaining options for inland desalination are to contain the waste brine on site or pay to have it treated elsewhere. The option of land application of the waste brine is also sometimes cited in the literature on brine management, but this will not be discussed here as it is rarely possible or desirable due to environmental concerns. If the brine is to be contained on site or transported to a suitably approved disposal facility, it is advantageous to reduce the volume of brine. This can be done via the use of evaporation ponds. Evaporation ponds have been a popular option in the Middle East and Australia, and in some USA locations such as Texas (Greenlee *et al.* 2009). The cost of this management option is, however, usually high and increases dramatically with increased waste brine flow rate (Mickley, 2009). Apart from being capital intensive, this option requires a suitably hot and dry climate. Evaporation ponds must also be suitably constructed and monitored for leaks and must not be situated above valuable freshwater aquifers in order to avoid the possibility of contamination of these aquifers in the event of a leak (Greenlee *et al.* 2009). In order to minimise the cost of evaporation ponds and, in some cases, to make the desalination operation economically viable, it is important to maximise the water recovery achieved during desalination in order to minimise the volume of brine to be contained in the evaporation ponds. This can be done using a variety of technologies that have now become available in response to the need for low salt water in inland regions and the need to desalinate coal seam gas (CSG) or produced water for environmental discharge and/or beneficial uses. For many of the growing number of inland desalination operations that have resulted from this need, the only practical option is one of containment of the brine on site using evaporation ponds. There are also technologies available, such as technologies for valuable salt recovery that can offset the cost involved in containing the waste on site. These will be discussed in Section 7.3.

7.2.2 Cost of options

The costs associated with brine management are feedwater specific as they depend on the water recovery that can be achieved and the resulting volume of waste brine to be managed.

Water recovery limits are set by feedwater characteristics. This applies to whatever brine management option is to be implemented. For seawater, the RO recovery limit is set by osmotic pressure. For less salt laden feedwaters, the recovery limits are set by scale formation. The formation of insoluble calcium, magnesium and silica deposits can occur as the desalination process removes very low salt water, leaving behind the precursors of these deposits. When the concentration of these scale precursors exceeds the solubility limit of scale compounds, precipitation can occur. For calcium carbonate scale formation, this precipitation can be mitigated by pH control as the solubility of this scale compound is higher at lower pH values. Scale formation can also be mitigated by addition of antiscalants which shift the precipitation to higher supersaturation values by adsorbing on the surface of newly formed precipitate nuclei (Gloede & Melin, 2006).

As shown in Figure 7.3, there is an optimum water recovery (Y) that is a function of the specific energy cost (SEC) of generating desalted water, the specific brine management costs (SBC) per unit volume of produced permeate and the normalised brine management cost per unit volume of brine (b norm) (Zhu *et al.* 2010). The higher the cost of brine management (b norm), the higher the optimum water recovery. For waters with high scaling potential, the water recovery is below this optimum, and scaling mitigation that allows an increase in water recovery towards the optimum can result in cost savings if the additional cost of scaling mitigation is not excessive.

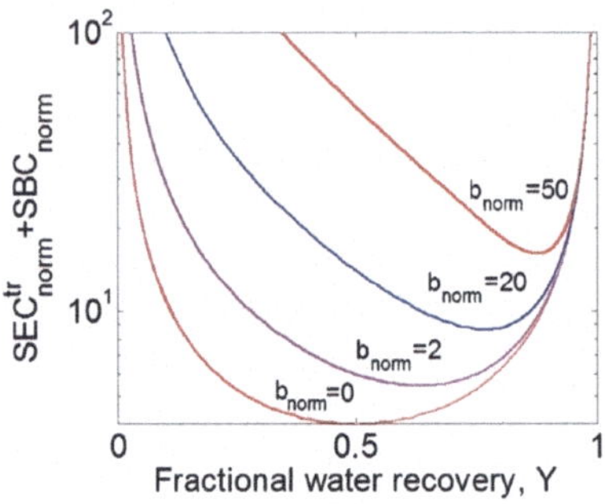

Figure 7.3 Variation of the combined energy and brine management costs with product water recovery for a single-stage RO (with an ideal pump but without energy recovery), operated at the limit of the thermodynamic restriction. From Zhu *et al.* 2010.

Decreasing the scaling potential of the water by removing the scale precursor ions is one way to achieve higher water recoveries, and this can considerably decrease running costs. For agricultural drainage water, for example, the operating costs can be reduced from \$0.71/m^{-3} to \$0.45/m^{-3} by the use of intermediate concentrate demineralization (ICD) to allow an increase in water recovery from 58% to 92%, as shown in Figure 7.4. (Gray *et al.* 2011).

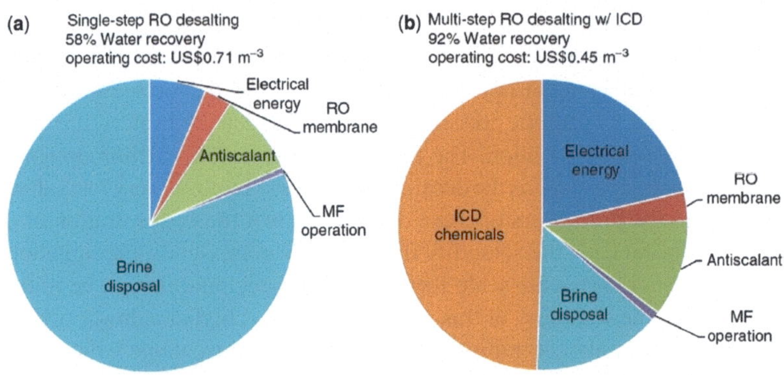

Figure 7.4 Operating cost estimates for a single step RO operated at 58% water recovery (a), compared to multistage RO operation with interstage chemical demineralization operated at 92% water recovery (b). Assuming deep well injection brine disposal option. From Gray *et al.* 2011.

The mitigation of scale formation allows high recovery to be achieved, but this can result in high energy consumption due to the high osmotic pressure at these high salt concentrations. Some studies have found that the use of more stages reduces specific energy consumption (SEC), allowing improved water recovery to be achieved. This improvement, however, is less pronounced when the number of stages exceeds four. The Brackish Water Reverse Osmosis (BWRO) system can also be operated in Closed Circuit Desalination (CCD) mode. A comparison of the energy requirements of these different configurations is shown in Figure 7.5 (Qiu & Davies, 2012).

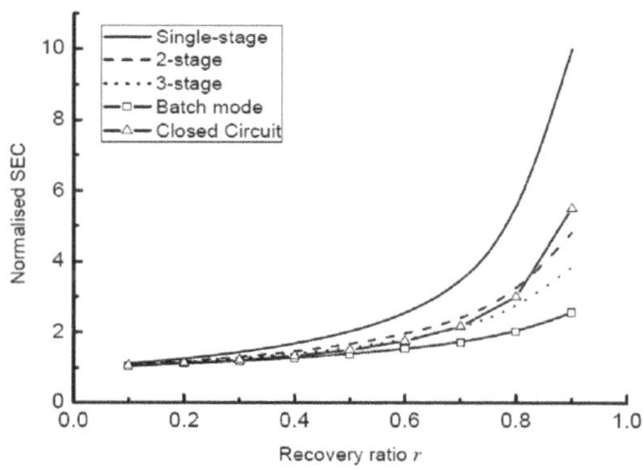

Figure 7.5 Theoretical limit of normalized SEC in batch-mode RO and multi-stage ROs with ERD, From Qiu and Davies, 2012.

A high water recovery process using 3 stage RO with interstage demineralization has been used to achieve water recoveries of 97% for wastewater with severe fouling potential (Rautenbach *et al.* 2000). The process train is shown Figure 7.6. This process uses disk-tube nanofiltration (NF) modules to further concentrate the RO concentrate from two previous RO stages and enhance the performance of a crystalliser which removes the scaling compounds before the concentrate is further treated by another RO stage to remove more water and further reduce the brine volume. The high overall RO recovery was achieved with periodic NF feed-channel flushing (every 30 s) and alkali cleaning (every 250–300 h) as preventive actions against fouling and/or scaling.

Constituents other than scale precursors can also be a determining factor in how the waste brine is managed, and the cost of brine management. The treatment of wastewater by RO will naturally give rise to a concentrate with higher concentrations of various pollutants in the wastewater. For municipal wastewater desalting operations this can be nutrients and salts, as well as environmentally persistent pharmaceutical pollutants such as beta blockers (Perez-Gonzalez *et al.* 2012).

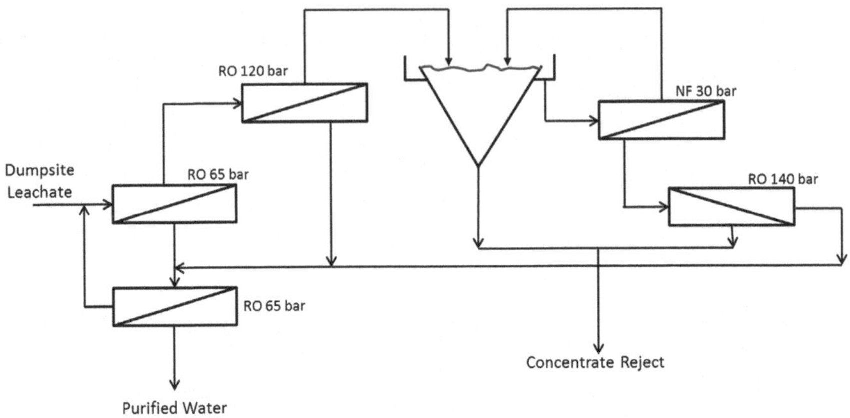

Figure 7.6 Three stage RO with interstage crystallisation. Adapted from Rautenbach *et al.* 2000.

Concerns regarding the quality of the waste brine can also occur for some groundwaters that have high levels of soluble arsenates. Although these can be effectively treated by ferric coagulant addition prior to disposal, this adds another waste process stream which has a cost associated with its management. Xu *et al.* (Xu *et al.* 2013) for example, found that 10 mg/L Fe^{3+} (~20 mg/L $Fe(OH)_3$) was required to reduce the levels of arsenate from 29 ug/L to 6 ug/L. The resulting sludge would require disposal in landfill at high cost due to the presence of arsenate. These authors, however, indicated that the use of electrodialysis (ED) can lead to the recovery of valuable quantities of calcium and magnesium for agricultural applications to partly offset these costs.

The relative cost of the different disposal options for most situations (Mickley, 2009) are shown in Figure 7.7. This is a generalised assessment as there are many complicating factors and assumptions involved, and the highly variable and site-specific costs of conveyance of concentrate to the site of disposal are not included. It can, however, be seen that the capital cost of the use of evaporation ponds is comparatively very high at high concentrate flow rates, making it imperative for inland desalting operations that have no other practical option other than containment of the concentrate on site in evaporation ponds to minimise the volume of brine. This is a particularly pressing need for the coal seam gas industry which depends on desalting technology to treat very large volumes of produced water for environmental discharge or for other beneficial uses [QLD Department of Environment and Resource Management]. In most of the current coal seam gas projects Environmental Impact Statements, it is envisaged that desalination plants will be set up throughout the gas fields to treat the coal seam gas production water by reverse osmosis (RO) (Hunter & Taylor, 2013). The large volume of CSG water being extracted is expected to increase as the development of coal seam gas fields continues (Klohn Crippen Berger, 2012). Figure 7.7 also shows that deep well injection has higher capital cost than disposal to sewer and disposal to surface water, but that it has lower capital costs than disposal to evaporation ponds at higher concentrate flows.

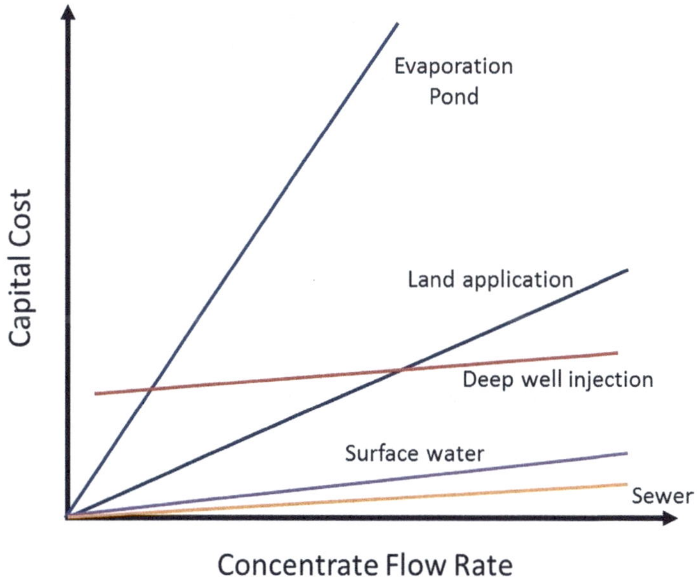

Figure 7.7 Relative capital costs of conventional disposal options. Adapted from Mickley, 2009.

7.2.3 Regulatory requirements and limitations

The regulatory framework in which desalination takes place has a major bearing on the choice of brine management practice. This framework is in a state of development for most countries as the use of desalination for water supply increases and regulatory authorities seek to prevent harm to the environment and health from the growing number of desalination activities. This regulation will drive change in management practices and in new technology to allow new regulatory limits to be achieved.

A relatively recent (2009) US Bureau of Reclamation report, for example, states that, with worsening drought conditions in California and increasing competition for water from traditional supply sources, the trend to identify and develop additional water supply sources will continue. This expansion will result in increased concentrations and quantities of brine-concentrate in southern California.

The report states that with increased production of brine-concentrate, as well as a better understanding of constituents' impact on human and environmental health, it is likely that regulation will continue to become more stringent (U.S. Department of the Interior, 2009; Greenlee *et al.* 2009).

In Australia, primary responsibility lies with the States & Territories (environmental protection or water resource protection agencies) and legislation, regulations, guidelines, and policies, vary between States and vary depending on the source of brine. The focus, however, is one – ensuring that the management of storage and discharge of water are safe, stable and non-polluting. In Victoria, for example, desalination needs to comply with the state environment protection policies (SEPPs): SEPP (Waters of Victoria), SEPP (Groundwaters of Victoria), and state Industrial Water Management Policy (Victorian Government, 2009). SEPPS don't specifically address desalination, and waste brine is generally treated as industrial waste. As world demand for fresh water increases and desalination is increasingly used to provide this water from new wastewater and groundwater sources, regulatory authorities are increasingly being called upon to specifically regulate this activity. In a paradigm of high environmental and health concern, these new regulations are likely to be increasingly more stringent.

Similarly, a 2009 European based study states that, there is a lack of clear European concentrate disposal regulation and that this has led to uncontrolled discharge or dumping. (PRODES, 2010). The European Water Framework Directive (WFD) aims to achieve good qualitative and quantitative status for all water bodies including near-shore marine waters by 2015. It lists desalination as one of many supplementary measures to achieve the objectives of improved water management and protection. The discharge of the concentrate from desalination processes are not explicitly listed in the Water Framework Directive and new standards are expected. These new regulations stand to shape the choice of brine management options and the economics of desalination. The situation in the Middle East is similar, with increasing regulations, plant size and public concerns being cited as limiting the disposal options and challenging the search for technically, environmentally and financially feasible methods of waste brine management (Bleninger & Jirka, 2010).

Paradoxically, high water recovery operations makes brine management easier by decreasing the volume of brine to be managed, but also leads to the production of brine that is more difficult to dispose of from a regulatory perspective. Concentrated brine can contain considerably higher concentrations of potentially toxic constituents, some of which may be below the detection limit in the less concentrated brine, but which may trigger classification of the concentrate as a more hazardous waste. In concert with increasingly stringent regulation, this is driving an evolution of brine management towards containment of the brine on site and disposal of a dry product in landfill and/or to salt recovery, i.e. toward zero liquid discharge (ZLD).

7.3 TECHNOLOGICAL APPROACHES TO BRINES MANAGEMENT AMELIORATION

For many inland utilities with desalination needs, location and regulatory constraints prevent discharge of the waste brine to the environment. In such situations, the only option is to contain the brine on site. In its simplest form, this requires the construction of sufficiently large evaporation ponds such that the loss of volume through evaporation is greater than the waste volume addition to the evaporation ponds. This is only feasible for small desalination plants situated in hot arid regions with low land value. For inland desalination plants that don't fit this description, reduction of the brine volume is required for the use evaporation ponds to be a feasible brine management option.

The containment of the waste brine on site with no discharge to the environment is often referred to as zero liquid discharge (ZLD). Most current applications of ZLD treat wastewater or power plant cooling water using thermal desalination and/or evaporation ponds. This or similar technologies can also be used for the treatment of waste RO brines. Thermally based processes, however, are energy-intensive and expensive to construct and operate, and evaporation ponds require large land areas and, in many cases, expensive lining and leak detection systems. (Bond & Veerapaneni, 2007). There are, however, now a range of technologies that promise to make ZLD desalination a feasible solution to brine management needs. These technologies do this by decreasing the brine's scaling potential, which allows further treatment of the brine to recover more usable water and produce lower waste brine volume, and/or by using novel membrane based processes that can tolerate the high scaling potential of the waste brine, and/or by using energy efficient thermal based process.

7.3.1 Brine volume reduction by decreasing brine scaling potential

It has been shown that the level of calcium, barium and silica are the key water quality parameters that determine the scaling potential of the water and hence the achievable water recovery in reverse osmosis desalination, and that the removal of these scale precursor substances can markedly decrease the scaling potential, increase the water recovery and thus decrease the brine volume. A relatively recent US Bureau of Reclamation report (Bond & Veerapaneni, 2007) showed that it may be possible to achieve zero liquid discharge at 50 to 70 percent less cost and 70 to 75 percent less energy than the traditional ZLD approach by treating the concentrate to decrease its scaling potential and then putting the resulting treated concentrate through another desalination stage to maximise the water recovery and minimise the brine volume prior to thermal concentration and evaporation pond treatment. Furthermore, the cost analysis indicated that ZLD treatment costs decreased in direct proportion to increasing water recovery in the primary desalination stage.

In locations where the use of evaporation ponds is possible, if a sufficiently high overall desalination water recovery can be achieved, costs may be further reduced by removing the thermal concentration process altogether (Poulson, 2010).

The scaling potential of the brine can also be reduced by control of pH. It has long been known that if the major scale precursor ions in the water are Ca and carbonate and phosphate, scale formation can be avoided and higher water recoveries can be achieved by lowering the pH [Dow Water & Process Solutions]. This method of scale mitigation is not effective for other calcium scale compounds such as calcium sulphate or calcium fluoride because these have a solubility that is not pH sensitive. The control of pH is also effective for silica scale mitigation. It has been shown that silica scale formation on RO membranes can be delayed by operating the desalination process at very low pH values (Tarquin *et al.* 2010) and avoided at high pH values in the absence of calcium and magnesium (Subramani & Jacangelo, 2014).

7.3.1.1 Chemical precipitation

A precipitation process that is analogous to the classic lime soda or caustic softening process has been used to decrease the concentration of calcium in primary RO concentrate, thus allowing further RO treatment of this concentrate by RO to remove more fresh water and reduce the volume of brine. An overall water recovery of 95% was demonstrated with this process (Zhu *et al.* 2010). A typical process train is shown in Figure 7.8.

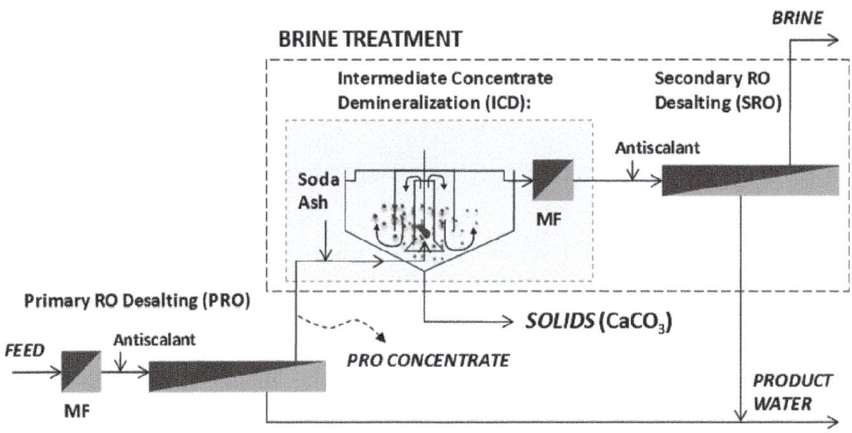

Figure 7.8 Brine treatment process for RO concentrate volume reduction using intermediate concentrate demineralization (ICD) via chemical precipitation and microfiltration (MF), followed by secondary RO desalting. From Zhu *et al.* 2010.

Similar results are projected for pellet reactor softening using a fluidised bed reactor (Bond, R.G. & Veerapaneni, 2007). These precipitation processes are effective for calcium removal but leave behind soluble silica which is also a major scale precursor that usually manifests itself as silica and/or aluminosilicate scale at higher water recoveries once other scale precursor ions such as calcium and magnesium that cause scale at lower water recoveries are removed (Cob *et al.* 2012).

Electrocoagulation is another process to remove divalent ions and metals. It has been used as a pre-treatment for RO, but may also be used for treatment of RO concentrate and thus allow further volume reduction of the waste brine and generation of more usable water. This process involves the use of an electrolytic reactor with sacrificial aluminium or iron electrodes to generate iron or aluminium hydroxides at the anode and hydroxyl ions at the cathode, leading to the formation of calcium carbonate and calcium hydroxide solid at the cathode which is coagulated and captured by the iron or aluminium hydroxide from the anode. It has been used for calcium, chemical oxygen demand (COD) and turbidity removal from coal seam gas product water (Zhao *et al.* 2014). Similarly, electocoagulation has also been used as a pre-treatment to remove silica form coal seam gas produced water, (Den & Wang, 2008) but can also be used to treat RO concentrate to allow higher water recoveries to be achieved. This process generates a metal based sludge that also needs to be managed.

7.3.2 Seeded precipitation

Calcium has also been removed from RO concentrate by seeding with calcium sulphate seed to accelerate precipitation of the calcium as calcium sulphate and/or calcium carbonate from a concentrate that is supersaturated with respect to these calcium compounds (Rahardianto *et al.* 2007). The process train is similar to that used for precipitation (see Figure 7.8), the only difference being that a seed material such as calcium sulphate or calcium carbonate is added to the precipitation reactor and is recycled to reduce operating costs.

This concept was further developed in subsequent work by the same researchers to overcome the adverse effect of antiscalant present in the RO concentrate on seeded precipitation (Rahardianto *et al.* 2010). Seeded precipitation was used in conjunction with chemical lime pre-treatment to generate calcium carbonate precipitate from the available carbonate for scavenging of antiscalant, thus allowing seeded precipitation of gypsum using gypsum seed to take place faster. A lime dose of 2.7 mM was found to double the extent of precipitation of calcium sulphate to close to the solubility limit (see Figure 7.9).

It has been reported that high recovery processing of municipal desalination brine is not usually practiced. This has been attributed to the high capital and operating costs for most municipal settings (Mickley, 2012). This may also be attributed to the high levels of inhibitors to precipitation of calcium carbonate, rendering chemical and seeded precipitation processes and high recovery RO operation less efficient (Sanciolo *et al.* 2012; Sanciolo *et al.* 2008).

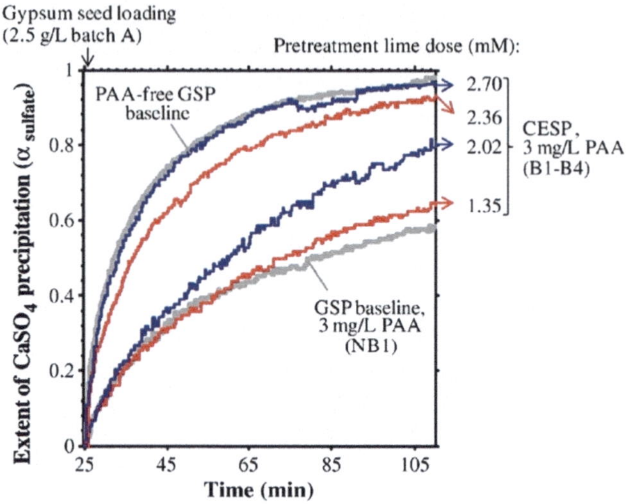

Figure 7.9 Effect of the initial antiscalant (PAA) concentration on the extent of $CaSO_4$ precipitation in the gypsum seeded precipitation stage. From Rahardianto *et al.* 2010.

7.3.3 Adsorption

A USA based pilot study of the use of activated alumina as part of a ZLD process for waste brine management (Bond, R.G. & Veerapaneni, 2007) evaluated adsorption of silica at four inland desalination sites using activated alumina. Virgin activated alumina was effective in reducing the silica concentrations of more than 150 mg/L to less than 5 mg/L in 10 minutes. Calcium and barium were also removed during this treatment. The silica adsorption capacity, however, was consumed quickly and bed service times were expected to be short (one day). Testing to optimize regeneration and evaluation of its silica removal performance after several regeneration cycles was identified as an additional research need. Of all the processes tested in this study (chemical precipitation with NaOH or lime, fluidized bed crystallization, ion exchange, chemical precipitation with alum, chemical precipitation with sodium aluminate, and adsorption with activated alumina), adsorption with activated alumina was found to achieve the greatest percentage of silica removal and was the only test that also removed significant quantities of calcium and barium in the pH range of 8 to 9. Adsorption onto activated alumina, therefore, appears to hold potential for concentrate treatment from the RO treatment of highly buffered coal seam gas process waters that are not amenable to the pH adjustment required for chemical and seeded precipitation processes.

A more recent Australian study investigated the use of activated alumina for silica removal from groundwater and from coal seam gas water (Sanciolo *et al.* 2013). It was found that the coal seam gas water from Queensland required an alumina dose of 2 g/L to achieved silica reductions that would allow water recoveries above 90%. Another groundwater from Western Australia, however, required 10 g/L to achieve the same projected water recovery. Regeneration of the adsorbent using 2% NaOH was found to be effective for at least three regeneration cycles (Sanciolo *et al.* 2014). The regenerated alumina was found to be less soluble and to have a higher adsorption capacity than the virgin alumina. The quantity of adsorbed silica liberated during the regeneration process, however, was also lower for the regenerated alumina than for the virgin alumina, suggesting the formation of aluminium silicate on the alumina surface during regeneration. Regeneration was also found to release calcium, suggesting the possibility of the formation of calcium silicate and/or calcium aluminium silicate on the adsorbent surface. A better understanding of the complex chemical interactions involved in silica removal by adsorption and the feedwater specificity of the outcome of adsorption processes was identified as a future research focus that may allow better control over the process.

7.3.4 Ion exchange

Ion exchange is a key feature of the High Efficiency Reverse Osmosis (HERO) process discussed in Section. 7.3.7. It has been found to be effective for brines with water recovery limits that are determined by barium sulphate precipitation (Bond, R.G. & Veerapaneni, 2007). It was found that barium removal from concentrate is possible and this removal is sustainable for at least 4,000 bed volumes. It was, however, found that efficient removal of barium can also be achieved with fluidized bed crystallization or with chemical precipitation. Ion exchange has also been found to be effective for heavy metal and bicarbonate reduction from coal seam gas water [QLD Department of Environment and Resource Management].

7.3.5 Biological sulphate reduction

The removal of sulphate by conversion to sulphide by sulphate reducing bacteria has been used to achieve improved water recovery (Subramani & Jacangelo, 2014). The resulting decrease in sulphate concentration decreases the scaling potential of waters that have calcium, barium or strontium, thereby allowing higher water recoveries and lower waste brine volume. This process would be expected to be expensive, as the reduction reaction requires the addition of an electron donor such as acetate, glucose or ethanol, and removal of the resulting sulphide requires acid stripping under acidic conditions. It may, however, find an application if sufficiently high water recoveries can be achieved such that small volumes of waste brine need to be treated. This would require halophilic sulphate reducing bacteria (Rees *et al.* 2010).

7.3.6 Low pH operation

In acid solution, attainment of the precipitation equilibrium for silica is very slow, sometimes taking months or years (Krauskopf, 1965). This low silica precipitation rate at low pH values has been exploited to achieve high water recovery (above

90%) in batch mode RO processing (Tarquin *et al.* 2010). A concentrate from an RO process that was run at approximately 80% water recovery and containing an average silica concentration of 131 mg/L was treated at low pH with an antiscalant against calcium sulphate precipitation, to approximately 80% water recovery without scale formation, resulting in an overall water recovery of 96%.

Figure 7.10 shows the silica concentration follows the expected concentration enrichment for water recoveries up to ~80%, beyond which the measured silica concentration is below the expected concentration, indicating precipitation of the silica has taken place.

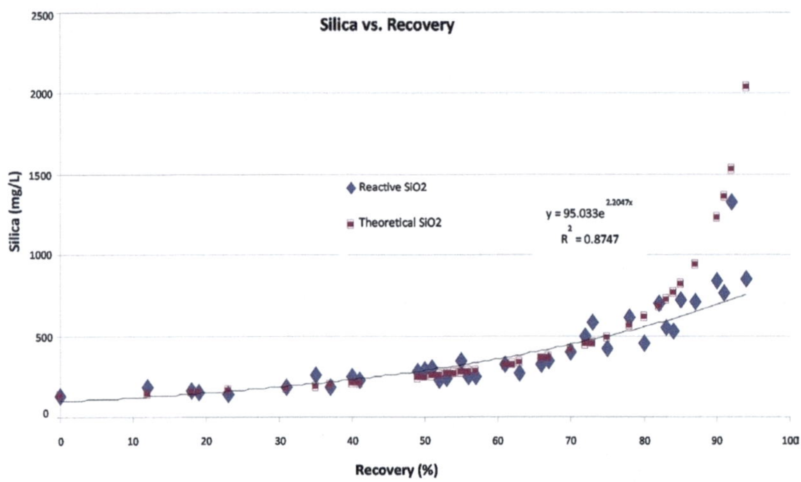

Figure 7.10 Low pH (3.6) treatment of RO concentrate. Adapted from Tarquin *et al.* 2010.

7.3.7 Lime softening pre-treatment with interstage ion exchange removal of multivalent cations and high pH RO operation (HERO process)

Silica has a high solubility at high pH values (Krauskopf, 1965) so removal of multivalent cations which can lead to the formation of metal silicates (Milne *et al.* 2014) and operation of the RO process at high pH decreases the silica scaling potential of the water, allowing greater water recoveries to be achieved. This is essentially the mode of operation of the High Efficiency Reverse Osmosis (HERO) process (Figure 7.11). The process train consists of lime softening and ion exchange to remove multivalent scale precursors, a degasification step to remove carbon dioxide, and a caustic addition stage to increase the pH of the RO feed water. This process has been used to achieve high feed water recovery. The combined recovery of the process was estimated to be greater than 90% for brackish waters with typical target recovery rates of approximately 95% (Subramani & Jacangelo, 2014).

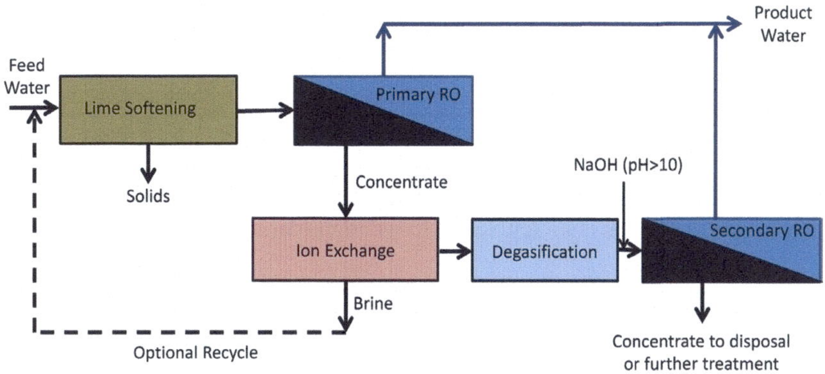

Figure 7.11 HERO process. Adapted from Subramani and Jacangelo, 2014.

7.3.8 Two stage RO concentrate lime softening with interstage recycling (ARROW process)

The size and cost of the scale mitigation stage is dictated by the volume to be treated. It is, therefore, advantageous to precipitate the scale precursors from the waste brine after it has been through the second RO stage and then feed the treated brine back through the second RO in order to remove more water from the treated concentrate, thus achieving high water recovery. This is essentially the configuration of the Advanced Reject Recovery of Water (ARROW) process (See Figure 7.12). The softening process uses chemical precipitation to reduce calcium, magnesium and silica scale precursors, or ion exchange if silica is not a concern. Feed water recoveries greater than 95% have been reported for this type of configuration (Mickley, 2009). This is a new technology that in 2009 had only one full scale application (33 gpm, New Jersey, USA).

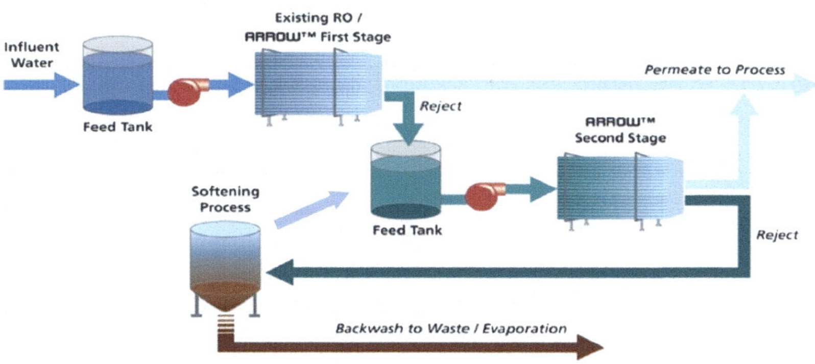

Figure 7.12 Process configuration of the ARROW process. Adapted/modified from US Bureau of Reclamation, 2009.

7.4 BRINE VOLUME REDUCTION THROUGH MEMBRANE BASED PROCESSES THAT CAN TOLERATE HIGH SCALING POTENTIAL

Chemical treatment to reduce the scaling potential of the waste brine and the disposal of the resulting sludge (if any) can be expensive due to the additional chemical and energy requirements. An alternative to this is to further treat the brine using a technology that is more tolerant to the high scaling potential of the water.

7.4.1 Electodialysis reversal (EDR)

Electrodialysis reversal (EDR) is a desalination processes in which an electric field is the driving force for separation of water from salt. Like reverse osmosis, this too is subject to a water recovery limit that is governed by scale formation. The advantage of EDR, however, is that the electric field can be readily and frequently reversed, converting product water streams into waste streams and allowing the newly formed scale to be re-dissolved and rinsed out. Treatment of RO concentrate using EDR has achieved water recoveries as high as 97% (Mickley, 2009). The combination of RO and EDR (See Figure 7.13) is particularly effective when silica is the primary mineral scalant of concern since EDR is not constrained by uncharged species (Gray *et al.* 2001).

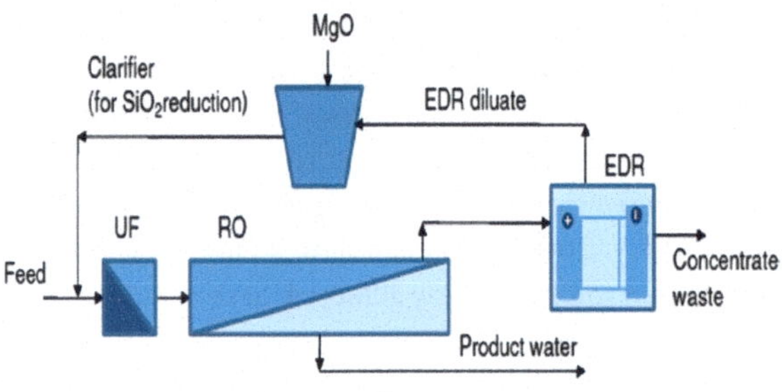

Figure 7.13 RO-EDR process with silica precipitation. From Gray *et al.* 2001.

7.4.2 Electrodialysis metathesis (EDM)

Like EDR, EDM uses an electric field to drive separation of ions from water. The ions are made to migrate through a suitable membrane under the influence of an applied electric field. In EDM, however, sodium chloride is supplied in an adjacent cell, resulting in the formation of separate concentrated calcium chloride and sodium sulfate streams. These salts are highly soluble and do not form scale, allowing operation to high water recoveries. This configuration is applicable for treating RO concentrate with high scaling potential due to $CaSO_4$ and $CaCO_3$ (Figure 7.14). Bond *et al.* (Bond *et al.* 2011) used an EDM system to treat RO concentrate (TDS of 16,000 mg/L), separating into two streams of highly soluble salts. The recovery obtained from the EDM was more than 99% and the concentrate from the EDM was further treated with a crystallizer to achieve ZLD. Feed water recovery was found to decrease with increasing TDS of the RO concentrate and the energy consumption of the EDM process increased with TDS.

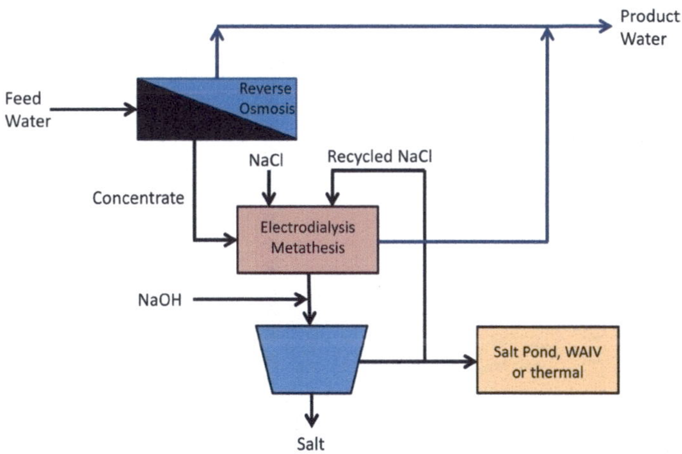

Figure 7.14 Process schematic of EDM process for treatment of RO concentrate. Adapted from Subramani and Jacangelo, 2014.

7.4.3 Vibratory shear enhanced processing (VSEP)

The concentration of salt near the surface of conventional RO membranes is higher than in the bulk due to the phenomenon of concentration polarization (Nam *et al.* 2014). The process of membrane fouling by precipitation of scale compounds is, therefore, likely to begin at the membrane surface. Modern membrane modules attempt to minimise this polarization by the use of spacers that create turbulence which disrupts the establishment of the concentration polarization layer, but there is only so much of this disruption that can take place in conventional spiral wound membrane modules. Greater disruption of the concentration polarization layer can be achieved by vibrating the membrane. This, however, requires a different module configuration as shown in Figure 7.15.

The shear waves produced using Vibratory Shear-Enhanced Processing (VSEP) decrease precipitation due to concentration polarization and prevent the caking of any colloidal material that may have formed in the bulk solution, thereby minimizing fouling and allowing higher water recoveries than can be achieved with conventional membrane system. VSEP employs torsional oscillation at a rate of 50 times per second (50 hertz). Unlike the conventional RO membranes, VSEP performance does not require pretreatment to remove colloidal material. The VSEP system can use RO or NF membranes in a single-stage or multiple-stage arrangement. VSEP has not been used in a full-scale concentrate application. The process has, however, been used in agricultural and industrial applications (US Bureau of reclamation, 2009).

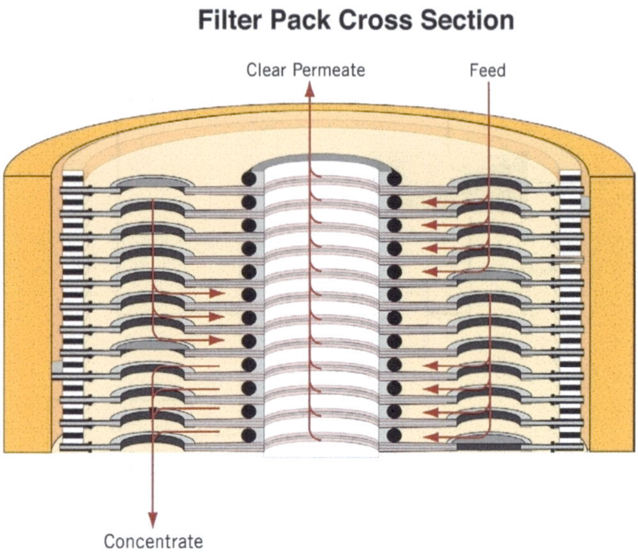

Figure 7.15 Vibratory Shear Enhanced Processing (VSEP) filter pack arrangement on resonating drive system. From VSEP.

7.4.4 Slurry precipitation and recycle reverse osmosis (SPARRO)

Seeded precipitation has been found to be effective in removing scale precursors from RO brine, thus allowing further treatment by RO and higher water recoveries. The precipitation takes place in a large clarifier and microfiltration is used for removal residual precipitated solids from the clarifier prior to introduction of the treated RO concentrate to the second RO stage (see Section 7.3.2). The Slurry Precipitation and Recycle Reverse Osmosis process (SPARRO) avoids the need to remove the seed after precipitation by using a tubular RO membrane rather than a conventional spiral wound membrane. This membrane configuration is not as prone to fouling and blockage in the presence of solids as conventional spiral wound membrane modules, thus allowing the seeds to be present as the solubility products of the scale material is exceeded during the concentration process within the membrane tubes and preventing scale formation on the membrane surface. A conceptual process train of SPARRO is presented in Figure 7.16a. The tubular RO membranes used in SPARRO is shown in Figure 7.16b (US Bureau of reclamation, 2009). The expected water recovery for the SPARRO process varies between 90 and 95%. (Burbano & Bradhuber, 2012).

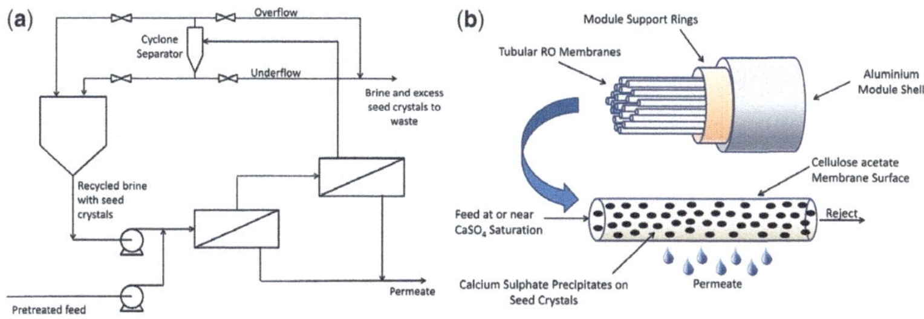

Figure 7.16 (a) SPARRO process train adapted from the US Bureau of Reclamation, 2009 and (b) tubular RO membrane, adapted from KAUST, 2010.

7.4.5 Membrane distillation (MD)

Membrane distillation is a desalination process that relies on a temperature difference across a hydrophobic membrane to drive the movement of water, in the form of vapour, from the hotter side to the cooler side of the membrane (see Figure 7.17). There are a number of variants of the MD process, the most simple being direct contact membrane distillation (DCMC). MD has been applied to the treatment of RO concentrate giving water recoveries from the RO brine of close to 90% as MD performance is only weakly influenced by concentration polarization phenomena, so high concentration values (up to saturation) can be fed into the process (Burbano & Bradhuber, 2012). Although a large amount of energy is required to heat the water or brine to be treated, the process can use low grade waste heat, thus greatly improving the economic viability of the process. It has been estimated that 15 kWh/m^3 of product water would be required to operate an RO-MD plant without the use of waste heat, but that the use of waste heat would reduce this to 2.25 kWh/m^3. (Criscuoli & Drioli, 1999).

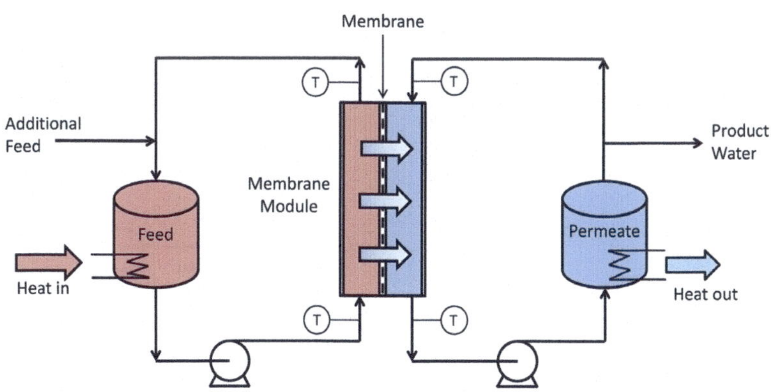

Figure 7.17 Schematic illustration of direct contact membrane distillation.

7.4.6 Forward osmosis (FO)

Forward osmosis technology exploits the natural tendency for water to move from a low solute concentration environment to a higher solute concentration environment. If a highly concentrated solution is separated from a salt solution or waste brine by a suitable membrane, water will be drawn from the salt solution or brine to this concentrated (draw) solution. The draw solution can then be processed to regenerate the draw solution and recover clean water. This regeneration can be done by choosing a draw solute that can be readily separated from the drawn water and recycled back to the process. Ammonium bicarbonate, for example, has been used as a draw solution and recycled by low temperature distillation of the ammonia and carbon dioxide decomposition products (McCutcheon *et al.* 2005). The draw solution can also be regenerated by reverse osmosis (see Figure 7.18).

The use of the configuration shown in Figure 7.18 was found to give water recovery of approximately 90% from the RO brine used as feed (Martinetti *et al.* 2009). FO is still an emerging technology but it promises to be less prone to membrane fouling than pressure-driven membrane processes. Membrane fouling in FO is relatively lower and more reversible than RO (Mi & Elimelech, 2010). This is mostly due to its ability to operate the feed solution at a much lower pressure, thus avoiding cake layer compaction. Further developments in membranes and in draw solutions are required.

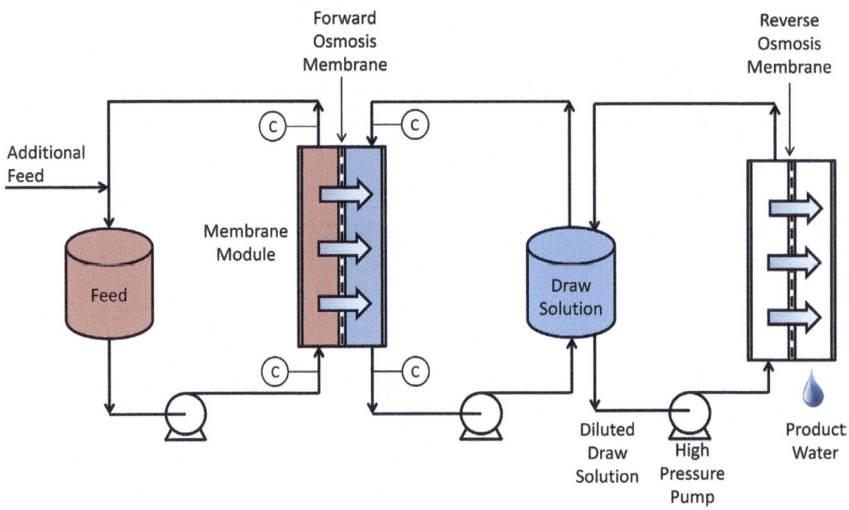

Figure 7.18 Forward osmosis with RO regeneration of draw solution. Adapted from Martinetti *et al.* 2009.

7.5 BRINE VOLUME REDUCTION VIA THERMALLY BASED PROCESSES

From an economic, and arguably short sighted perspective, disposal (e.g. to surface waters, deep well injection) of the RO brine without further processing is the best option. If due to location and regulatory environment these disposal options are not available, reduction of the volume of brine is advantageous as it allows containment of the brine on site and management of the fate of the resulting dry salt product. This is commonly termed zero liquid discharge (ZLD) and can involve thermal based processes if the brine volume reduction achieved in high recovery RO operations is insufficient. The last stage of the ZLD process is often an evaporation pond, but this is not sometimes possible due to the site climate characteristics and land values.

If the water recovery is sufficiently high and the evaporation pond construction and operation option is available (i.e. if the location is sufficiently dry and the land costs are sufficiently low), the use of a thermal brine concentrator can be avoided altogether. In such cases the added expense of chemicals and sludge disposal associated with high recovery operation can be offset by the lower cost of evaporation pond construction and the recovery of a significantly greater amount of fresh water from the original waste brine. The cost of evaporation pond construction can be further reduced by enhancing the evaporation rate via the use of wind-aided intensification of vaporization (WAIV). Reduced evaporation pond area requirements by a factor of 10 have been reported when using WAIV (Gilron *et al.* 2003).

The most widely used thermal process in conventional ZLD plants is thermal/mechanical vapor recompression (MVR). The MVR equipment is commonly referred to as brine concentrator equipment and is often operated in a seeded slurry mode, in which calcium sulfate is circulated to provide a site for coprecipitation and/or adsorption of salts during the evaporation process (Mickley, 2009). Brine concentrator recovery depends on the feed water quality, but typically ranges from 90% to 98% of the feed concentrate stream. Brine concentrators are, however, energy-intensive, requiring approximately 18–26 kWh/m^3 energy consumption (Mickley, 2008). Other thermal processes that can be used in conjunction with brine concentrators include crystallizers and spray driers. These are also energy intensive and the advantage of their use would need to be weighed up against the expected savings in evaporation pond requirements that results from their use.

The cost associated with the use of thermal processes in ZLD operations may be offset by recovery of valuable salts from the brine. Technologies such as the SAL-PROC selective precipitation based process and the Zero Discharge Desalination (ZDD) process based on electrodialysis metathesis (see Section 11.3.2.2) have been used individually and in combination with RO and other concentration processes to yield valuable salt from waste RO brine concentrate (Morillo *et al.* 2014). Despite the high capital, chemical and energy costs, such salt recovery can be a viable option (Davis, 2006). It is important to emphasise, however, that the benefit of salt recovery can go beyond consideration of the value of the salt product versus the cost of recovery of the salt and the cost of disposal, because salt recovery can be a sustainable waste management option in situations where none otherwise would exist. In coal seam gas product water desalination, for example, the high volume and salt content of the feedwater leads to very large quantities of salt for disposal (Ford *et al.* 2012) and reduction of the volume of salt produced through selective recovery of salts may be crucial to the long term environmental and economic sustainability of the waste management measures.

Analysis of ZLD processing options has shown that inclusion of high recovery RO before a brine concentrator dramatically reduces the size of the thermal volume reduction requirements and therefore reduces the capital cost and the energy costs. The decreased energy costs, however, are partly offset by the increased costs of chemicals and sludge disposal (Mickley, 2009). So from a purely cost perspective there is often little to be gained by inclusion of high recovery RO in a ZLD based desalination scheme including brine concentrators. There are, however, some real advantages to high recovery RO operation in the ZLD scheme with brine concentrators – a much smaller environmental footprint and a reduced liability risk. Environmental regulations have always been a driver for innovation and the current trend towards more stringent regulation of brine management (U.S. Department of the Interior, 2009) will undoubtedly drive further development in this field.

7.6 CONCLUSIONS

As human population increases and the per capita consumption of water increases, our fresh water resources from rivers, streams and groundwater sources are becoming increasingly stressed. We are, therefore, driven to use ever greater quantities of saline waters such as seawater and brackish groundwater and wastewaters that require desalination. Where deep well injection is not available, the waste product from inland desalination has in the past been responsibly discharged to the environment, either directly or through municipal wastewater treatment plants. As more and more desalination plants are planned and built to meet our growing water needs, however, we are faced with the prospect of degradation of environmental ecosystems and the very waters we rely on for food production. Our regulatory authorities have responded to this prospect by implementing increasingly stringent environmental laws and regulations that make disposal increasingly more difficult and encourage management of waste brine in a more responsible manner. These regulatory forces, together with sometimes opposing economic forces, are driving an evolution of brine management towards containment of the brine on site and disposal of a dry product in landfill, i.e. toward zero liquid discharge (ZLD).

The inclusion of additional process stages to reduce brine volume increases to the cost of desalination when compared to desalination with discharge of the waste brine to the environment. When compared to desalination which, by necessity, uses more expensive brine management options such as deep well injection and evaporation ponds, however, brine volume reduction can be cost neutral or sometimes economically favourable. The expense associated with volume reduction is offset by the savings to be made from managing smaller brine volumes by deep well injection or evaporation ponds. In cases where deep well injection is used as the management option, this lower waste brine volume can extend the useful life of the well. The reduced evaporation pond area requirements can also allow this option to be used in high flow applications such as coal seam gas product water desalination where the evaporation pond requirements would otherwise have been prohibitively expensive and/or environmentally unsustainable. The quest is now to make these expensive brine management options less expensive by optimization and innovation. We have made considerable progress towards this end in recent times and cost effective environmentally responsible brine management appears to be within reach.

7.7 REFERENCES

ANZECC/ARMCANZ, Australian and New Zealand Guidelines for Fresh and Marine Water Quality (2000). http://www.environment.gov.au/water/quality/publications/australian-and-new-zealand-guidelines-fresh-marine-water-quality-volume-1 (accessed 24 June 2015).

Bleninger T. and Jirka G. H. (2010). Middle East Desalination Research Center, Environmental planning, prediction and management of brine discharges from desalination plants, Final report, MEDRC Series of R&D Reports, MEDRC Project: 07-AS-003.

Bond R., Batchelor B., Davis T. and Klayman B. (2011). Zero liquid discharge desalination of brackish water with an innovative form of electrodialysis: electrodialysis metathesis, *Florida Water Resource Journal*, 38–42.

Bond R. G. and Veerapaneni S. (2007). Zero Liquid Discharge for Inland Desalination, American Water Works Association Research Foundation, Final Report 3010, Denver, Colorado, USA.

Buonomenna M. G. (2013). Nano-enhanced reverse osmosis membranes. *Desalination*, **314**, 73–88.

Burbano A. and Brandhuber P. (2012). Water Environment Research Foundation, Demonstration of Membrane Zero Liquid Discharge for Drinking Water Systems. A literature review.

Cob S. S., Beaupin C., Hofs B., Nederlof M. M., Harmsen D. J. H., Cornelissen E. R., Zwijnenburg A., Guner F. E. G. and Witkamp G. J. (2012). Silica and silicate precipitation as limiting factors in high-recovery reverse osmosis operations. *Journal of Membrane Science*, **423–424**, 1–10.

Criscuoli A. and Drioli E. (1999). Energetic and exergetic analysis of an integrated membrane desalination system. *Desalination*, **124**(1–3), 243–249.

Davis T. A. (2006). Zero Discharge Seawater Desalination: Integrating the Production of Freshwater, Salt, Magnesium, and Bromine, Desalination and Water Purification Research and Development Program Report No. 111, United States Department of the Interior, Bureau of Reclamation, Denver, Colorado, USA.

Dawoud M. A. and Al Mulla M. M. (2012). Environmental impacts of seawater desalination: arabian gulf case study. *International Journal of Environment and Sustainability*, **1**(3), 22–37. ISSN 1927–9566.

Den W. and Wang C. (2008). Removal of silica from brackish water by electrocoagulation pretreatment to prevent fouling of reverse osmosis membranes. *Separation and Purification Technology*, **59**(3), 318–325.

Dow Water & Process Solutions. FILMTEC™ Reverse Osmosis Membranes Technical Manual. http://www.dowwaterandprocess.com/en/resources/reverse-osmosis-technical-manual) (accessed June 2014).

Ford J. A., Steen J. and Verreynne, M. (2014). How environmental regulations affect innovation in the Australian oil and gas industry: going beyond the porter hypothesis. *Journal of Cleaner Production*, **84**, 204–213.

Gilron J., Folkman Y., Savliev R., Waisman M. and Kedem O. (2003). WAIV – wind aided intensified evaporation for reduction of desalination brine volume. *Desalination*, **158**(1–3), 172, 205–214.

Gloede M. and Melin T. (2006). Potentials and limitations of molecular modelling approaches for scaling and scale inhibiting mechanisms. *Desalination*, **199**(1–3), 26–28.

Global Water Intelligence (2014). The 27th Worldwide Desalting Plant Inventory: Annual Capacity Increments. http://www.globalwaterintel.com/archive/15/10/chart-of-the-month/chart-month-27th-worldwide-desalting-plant-inventory-annual-capacity-increments.html#sthash.RkMZkNZr.dpuf (accessed 24 October 2015).

Gray Semiat R., Duke M., Rahardianto A. and Cohen Y. (2011). Seawater use and desalination technology. In: Treatise on Water Science, K. H. Peter Wilderer, Tom Vereijken (eds), Elsevier, pp. 73–105.

Greenlee L. F., Lawler D. F., Freeman B. D., Marrot B. and Moulin P. (2009). Reverse osmosis desalination: water sources, technology, and today's challenges. *Water Research*, **43**(9), 2317–2348.

Hunter T. and Taylor M. (2013). Coal Seam Gas: An Annotated Bibliography. Centre for international minerals and energy law.

KAUST (2010). Industry Collaboration Program (KICP). The KICP Annual Strategic Study, Promoting Wastewater Reclamation and Reuse in the Kingdom of Saudi Arabia: Technology Trends, Innovation Needs, and Business Opportunities, 2010–2011.

Klohn Crippen Berger (2012). Forecasting coal seam gas water production in Queensland's Surat and southern Bowen basins, Summary. Prepared for the Department of Natural Resources and Mines. Healthy HeadWaters Coal Seam Gas Water Feasibility Study.

Krauskopf K. B. (1965). Disolution and precipitation of silica at low temperatures. *Geochimica et Cosmochimica Acta*, **10**(1–2), 1–26.

Lattemann S. (2010). Development of an Environmental Impact Assessment and Decision Support System for Seawater Desalination Plants. PhD thesis, Delft University of Technology, Deft, The Netherlands.

Lattemann S. and Hopner T. (2008). Environmental impact and impact assessment of seawater desalination. *Desalination*, **220**(1–3), 1–15.

Martinetti C. R., Childress A. E. and Cath T. Y. (2009). High recovery of concentrated RO brines using forward osmosis and membrane distillation. *Journal of Membrane Science*, **331**(1–2), 31–39.

McCutcheon J. R., McGinnis R. L. and Elimelech M. (2005). A novel ammonia–carbon dioxide forward (direct) osmosis desalination process. *Desalination*, **174**(1), 1–11.

Mi B. and Elimelech M. (2010). Gypsum scaling and cleaning in forward osmosis: measurements and mechanisms. *Environmental Science and Technology*, **44**, 2022–2028.

Mickley M. C. (2004). Review of Concentrate Management Options, Groundwater Report 363. Technical Papers, Case Studies and Desalination Technology Resources. Future of Desalination in Texas, Vol.II. Texas Water Development Board. (Available from website: ⟨www.twdb.state.tx.us/iwt/desal/docs/Volume2Main.asp⟩) (accessed July 2014).

Mickley M. (2006). Membrane Concentrate Disposal: Practices and Regulation. Mickley and Associates, Desalination and Water Purification Research and Development Report No. 123, Second Edition.

Mickley M. (2008). Survey of High-Recovery and Zero Liquid Discharge Technologies for Water Utilities, Final Report, Water Reuse Foundation.

Mickley M. (2009). Treatment of Concentrate, Desalination and Water Purification Research and Development Program Report No. 155, United States Department of the Interior, Bureau of Reclamation, Denver, Colorado, USA. (Available from https://www.usbr.gov/research/AWT/reportpdfs/report155.pdf (accessed 14 July 2014).

Mickley M. (2010). State-of-the-Science White Paper: Brackish Groundwater Concentrate Management, Prepared for New Mexico State University (NMSU) and Consortium for High Technology Investigations on Water and Wastewater, (CHIWAWA).

Mickley M. (2012). US municipal desalination plants: number, types, locations, sizes and concentrate management practices, *IDA Journal*, First Quarter, 44–50.

Milne N. A., O'Reilly T., Sanciolo P., Ostarcevic E., Beighton M., Taylor K., Mullett M., Tarquin A. J. and Gray S. R. (2014). Chemistry of silica scale mitigation for RO desalination with particular reference to remote operations. *Water Research*, **65**, 107–133.

Miri R. and Chouikhi A. (2005). Ecotoxicological marine impacts from seawater desalination plants. *Desalination*, **182**(1–3), 403–410.

Morillo J., Usero J., Rosado D., El Bakouri H., Riaza A. and Bernaola F. J. (2014). Comparative study of brine management technologies for desalination plants. *Desalination*, **336**, 32–49.

Nam J., Park J., Kim J., Kwon S., Chon K., Lee E., Kim H. and Jang A. (2014). The evaluation on concentration polarization for effective monitoring of membrane fouling in seawater reverse osmosis membrane system. *Journal of Industrial and Engineering Chemistry*, **20**(4), 2354–2358.

National Water Commission (2011). Brackish Groundwater: A Viable Community Water Supply Option? Waterlines Report Series No 66.

Perez-Gonzalez A., Urtiagam A. M., Ibanez R. and Ortiz I. (2012). State of the art and review on the treatment technologies of water reverse osmosis concentrates. *Water Research*, **46**(2), 267–283.

Poulson T. K. (2010). Strategic Alternatives for Brine Management in the Valley of the Sun, Central Arizona Salinity Study, Concentrate Management Sub-Committee, Phoenix, USA.

PRODES (2010). Guidelines for the Regulation of Desalination, Deliverable 6.2 Report.

Qiu T. and Davies P. A. (2012). Comparison of configurations for high-recovery inland desalination systems. *Water*, **4**, 690–706. Open Access, ISSN 2073–4441, www.mdpi.com/journal/water (accessed 24 June 2015).

QLD Department of Environment and Resource Management, Guideline; Environmental Protection (Waste Management) Regulation 2000. Approval of coal seam gas water for beneficial use, www.derm.qld.gov.au ABN 46 640 294 485.

Rahardianto A., Gao J., Gabelich C. J., Williams M. D. and Cohen Y. (2007). High recovery membrane desalting of low-salinity brackish water: integration of accelerated precipitation softening with membrane RO. *Journal of Membrane Science*, **289**, 123–137.

Rahardianto A., McCool B. C. and Cohen Y. (2010). Accelerated desupersaturation of reverse osmosis concentrate by chemically-enhanced seeded precipitation. *Desalination*, **264**(3), 256–267.

Rautenbach R., Linn T. and Eilers L. (2000). Treatment of severely contaminated waste water by a combination of RO, high-pressure RO and NF - potential and limits of the process. *Journal of Membrane Science*, **174**(2), 231–241.

Rees G. N., Baldwin D. S., Watson G. O. and Hall K. C. (2010). Sulfide formation in freshwater sediments, by sulfate-reducing microorganisms with diverse tolerance to salt. *Science of the Total Environment*, **409**(1), 134–139.

Sanciolo P., Zou. L., Gray S., Leslie G. and Stevens D. (2008). Accelerated seeded precipitation pre-treatment of municipal, wastewater to reduce scaling. *Chemosphere*, **72**(2), 243–249.

Sanciolo P., Ostarcevic E., Atherton P., Leslie G., Fane T., Cohen Y., Payne M. and Gray S. (2012). Enhancement of reverse osmosis water recovery using interstage calcium precipitation, *Desalination*, **295**, 43–52.

Sanciolo P., Milne N., Ostarcevic E., Beighton M., Peril G., Taylor K., Mullet M., Tarquin A. and Davis T. (2013). Silica Removal from Groundwater for RO Water Recovery and Waste Brine Volume Reduction, NCEDA Hosted International Desalination Workshop No.6, Woodard Conference Centre, Melbourne University, 28–29 November 2013, (http://shop.desalination.edu.au/dr-peter-sanciolo-silica-removal-from-groundwater-for-reverse-osmosis-water-recovery-enhancement-and-waste-brine-volume-reduction.html) (accessed July 2014).

Sanciolo P., Milne N., Taylor K., Mullet M. and Gray S. (2014). Silica scale mitigation for high recovery reverse osmosis of groundwater for a mining process. *Desalination*, **340**, 49–58.

Subramani A., Jacangelo J. G. (2014). Treatment technologies for reverse osmosis concentrate volume minimization: a review. *Separation and Purification Technology*, **122**, 472–489.

Tarquin A. J., Fahy M. P. and Balliew J. E. (2010). Concentrate Volume Reduction Research in El Paso, Texas, World Environmental and Water Resources Congress 2010: Challenges of Change, 3507–3518.

Tripodi N., Ramsay I. (2009). Managing Increasing Brine Concentrate Discharges in Drought Conditions. Freshwater and Marine Science, Queensland EPA.

U.S. Bureau of reclamation (2009). Brine-Concentrate Treatment and Disposal Options Report, Southern California Regional Brine-Concentrate Management, Study – Phase I Lower Colorado Region.

U.S. Department of the Interior, Bureau of Reclamation (2009). Regulatory Issues and Trends Report Southern California Regional Brine-Concentrate Management Study – Phase I Lower Colorado Region.

Victorian Government (2009). Minister for Planning, Victorian Desalination Project Assessment Under Environment Effects Act 1978.

WA EPA (2003). Guidance for the assessment of environmental factors (in accordance with the Environment Protection Act 1986). Deep and shallow well injection of liquid industrial waste, No.4, March.

Xu P., Capito M. and Cath T. Y. (2013). Selective removal of arsenic and monovalent ions from brackish water reverse osmosis concentrate. *Journal of Hazardous Materials*, **260**, 885–891.

Zhao S., Huang G., Cheng G., Wang Y. and Fu H. (2014). Hardness, COD and turbidity removals from produced water by electrocoagulation pretreatment prior to Reverse Osmosis membranes. *Desalination*, **344**, 454–462.

Zhu A., Rahardianto A., Christofides P. D. and Cohen Y. (2010). Reverse osmosis desalination with high permeability membranes — Cost optimization and research needs. *Desalination and Water Treatment*, **15**, 256–266.

Chapter 8

Impact of seawater desalination by reverse osmosis on the marine environment

Nurit Kress and Bella Galil*

Israel Oceanographic and Limnological Res., The National Institute of Oceanography, P.O. Box 8030, Tel Shikmona, Haifa, Israel 31080

**Corresponding author: nurit@ocean.org.il*

8.1 INTRODUCTION

Freshwater is a rare commodity and increasingly scarce. Less than 1% of the water in the hydrosphere is easily accessible freshwater with a heterogeneous global distribution. More than a third of the world's population lives in areas with water shortages, and their number are expected to increase to two thirds by 2025. Growing populations and higher standards of living increase demand on dwindling water resources and contribute to vulnerability to drought. The increasing need for potable water in conjunction with technological advances has transformed desalination into a fast growing industry (WWAP, 2012) with the global production capacity of desalinated water increasing from 17.3 to 68 million m^3day^{-1} (Mm^3d^{-1}) between 1994 to 2009, and is projected to reach 130 Mm^3 d^{-1} by 2016 (Water-world 22nd-GWI/IDA; Yermiyahu *et al.* 2007). The desalination market comprises different source waters: wastewater was used to produce ca. 5% of the total volume of desalinated water, 19% was produced from brackish water and 63% from seawater (Lattemann, 2010). The leading desalination processes are thermal Multi Stage Flash – MSF and membrane-based Reverse Osmosis (RO). Other processes such as Multi Effect Distillation (MED), Vapor Compression (VC), nanofiltration (NF), electro dialysis (ED), are utilized on smaller scales. Newer technologies such as nanoporous materials for membranes are under development, and forward osmosis and integrated systems are being examined for possible upscaling (Macedonio *et al.* 2012; Perez-Gonzalez *et al.* 2012).

Seawater desalination production is concentrated in the Persian Gulf (45% of global production), Red Sea (14%, mainly Saudi Arabia) and Mediterranean Sea (14%, mainly Spain). The largest seawater desalination plant is Jebel Ali (Phase 2) in the United Arab Emirates (MSF process, 300 $Mm^3\ y^{-1}$), and the largest operating seawater RO (SWRO) plant is in Soreq, Israel (150 $Mm^3\ y^{-1}$). The Victorian desalination project (Wonthaggi, Australia) has the same production capacity but has not produced water since the completion of the plant in December 2012. Seawater desalination has been developing rapidly in Australia, China, United States and Latin America. A combination of larger plants, modern membrane technology and improved energy systems have reduced the cost of SWRO desalinated water by ca. 3 times over the past 15 years (Campbell & Jones, 2005; Service, 2006; Tal, 2006; Fritzmann *et al.* 2007; Elimelech & Phillip, 2011). SWRO is expanding even in the Gulf States which have been relying on thermal processes, because it minimizes seawater intake and brine discharge, cuts back on thermal pollution and reduces energy usage (Darwish *et al.* 2013).

In this chapter we discuss the impact of SWRO desalination on the marine environment. We describe the process and its possible effects and review the development of environmental research through case studies. We appraise yet unaddressed or not fully answered concerns and end with recommendations for an integrated approach towards environmentally friendlier practices.

8.2 THE SWRO PROCESS

The SWRO process uses high pressure to force water molecules through a semi-permeable membrane that retains the salts, producing fresh water and brine (up to 50% conversion factor). At various stages of the process chemicals are added which may be subsequently disposed with the brine: coagulants in the pre-treatment stage (iron or aluminum salts, polymers); biocides (such as chlorine) and neutralizers (sodium sulfite); antiscalants to prevent fouling of the membranes (polyphosphates, polyphosphonates, polyacrylic acid, polymaleic acid); cleaning solutions for RO membranes (acidic and alkaline solutions and detergents); and pH and hardness adjustors for the product water (lime). The successive steps, usage of chemicals, energy recovery and improved efficiency have been extensively described (Fritzmann *et al.* 2007; NRC, 2008; NWC, 2008; UNEP, 2008; Greenlee *et al.* 2009; Elimelech & Phillip, 2011; Ghaffour *et al.* 2013 among others). Both, seawater intake and brine disposal into the sea can impact the marine environment.

Seawater intake may be sited near the shore line, as a submerged intake, typically at 10–15 m water depth and ca. 4 m above the seafloor, or utilize other sources, such as seawater used for cooling in power plants. The massive volume of seawater pumped into the plants (at least twice the production) entail the entrainment and impingement of their biota (NRC, 2008; UNEP, 2008). The few quantitative studies of this phenomenon have been carried out in power plant seawater intake (Mayhew *et al.* 2000; Barnthouse, 2013). The biota that enters the desalination plant does not survive the filtration at the pre-treatment stage, whereas some biota may survive the cycle at the power plants. Entrainment and impingement can be reduced by controlling intake rates, using subsurface intakes (Missimer *et al.* 2013) or, where appropriate, placing the intake offshore. The global increase in gelatinous blooms (scyphozoan and ctenophorans) and algal blooms has already impacted plant operations dependent on seawater intake: jellyfish blocked seawater cooling intakes at the Birka Power and Desalination Plant and Ghubra Desalination Plant, Oman, where 300 tonnes of jellyfish damaged intake screens causing a 50 per cent reduction in output. A year earlier jellyfish blocked Oman LNG's seawater cooling system intake[1]. The outbreak of an alien invasive dinoflagellate in 2008–9 caused the closure of plants in the Gulf of Oman (Richlen *et al.* 2010) and the potential impact of harmful algae was evaluated for the southern Californian coast (Caron *et al.* 2010). Re-positioning intakes to reduce impingement may avoid such crises.

Brine discharge. SWRO brine, referred to also as concentrate, retentate, residual or reject stream, has twice the salinity of the seawater at the intake (assuming an optimal 50% conversion). The brine, chemicals (i.e coagulants, neutralized biocides, anti-scalants, excess of pH and hardness adjustors and cleaning solutions, depending on the plant) and the entrained biotic debris are discharged at the shoreline, alone or mixed with other discharges, often cooling waters from power plants, or discharged further off shore through outfalls. Diffusers are usually installed to increase dilution. Brine disposal increases salinity in the immediate surroundings and affects the biota nearby. The challenge involved in locating a brine discharge site is finding an area of low ecological sensitivity, with strong ocean currents to disperse the brine or co-discharge and dilute it; and at the same time optimize construction costs. This is usually unrealistic and the end result is a compromise among the prevailing constraints.

Advances in desalination technology have already reduced the amount of chemicals used and the quantity of brine. Micro filtration (0.1–10 μm) and ultra-filtration (0.1–0.01 μm) may replace the pre-treatment through chemical coagulation (Fritzmann *et al.* 2007), biological pre-treatment, such as bioflocculation, may replace chemical coagulants (Bar-Zeev *et al.* 2013), and back wash from sand filters is dried and residuals disposed on land, where commercially important brine components can be isolated by new techniques (Macedonio *et al.* 2012; Pérez-González *et al.* 2012).

[1]http://gulfnews.com/news/gulf/uae/general/jellyfish-choke-oman-desalination-plants-1.355525 (accessed May 6, 2003).

8.3 REVIEW OF THE DEVELOPMENT OF MARINE ENVIRONMENTAL RESEARCH PERTAINING TO SWRO

The literature on desalination is focused on plant planning, site selection and construction, operational aspects, technological improvements, and energy cost. Until recently it was assumed that, when properly engineered and constructed, effluents are environmentally safe (Ahmed *et al.* 2001; Campbell & Jones, 2005; Alpert *et al.* 2007; Lattemann & Hopner, 2008a; NWC, 2008; Safrai & Zask, 2008; Shannon *et al.* 2008). Desalination proponents deny the possibility of environmental impacts: 'seawater desalination plants temporarily remove a small portion of ocean water, produce fresh drinking water, which in turns is returned to the ocean via the ocean discharges of the wastewater treatment plants located in the vicinity of the desalination plant, thereby re-uniting the separated fresh water and salts, both of which originated from the ocean, within a period much shorter than the seasonal interval which returns the water removed from the ocean by evaporation.' (Voutchkov, 2011). Yet, an ever increasing body of work suggests that desalination effluents do indeed impact the marine biota at the vicinity of the outfall. Most of the references concern theoretical analysis of entrainment and impingement of organisms at the intake; increased salinity and stratification, reduced vertical mixing, decrease in oxygen concentration, increased turbidity, eutrophication, decreased or increased production and toxicity at the outfall (UNEP/MAP/MEDPOL, 2003; Fritzmann *et al.* 2007; NRC, 2008; UNEP, 2008; Khan *et al.* 2009; Lattemann, 2010; Ahmad & Baddour, 2014). Other studies are descriptive and provide little quantitative data. The number of articles publishing quantitative effects *in situ* or in lab experiments is small and limited in scope (Roberts *et al.* 2010). Most of the publications emphasize the effects of salinity on the benthic communities and those are site- and organism specific (Walker *et al.* 1988; Walker & McComb, 1990; Fernández-Torquemada *et al.* 2005; Raventos *et al.* 2006; Gacia *et al.* 2007; Koch *et al.* 2007; Sánchez-Lizaso *et al.* 2008).

Since Roberts *et al.* (2010) critical review of the studies on the environmental impacts of desalination effluents, a number of peer reviewed publications have been published. To these we may add some of the massive scientific body of work on the impacts of global change on marine temperature and salinity which is also of relevance to desalination environmental research (see below).

8.3.1 Salinity[2] and temperature of the receiving waters

One point is beyond dispute: salinity increases in the vicinity of the discharge point. Environmental Impact Assessment (EIA) studies require modeling of the dispersion of brine in the receiving environment that serve in turn to the planning of efficient dispersal strategies, such as discharge through diffusers or co-discharge

[2]Salinity is measured as a conductivity ratio and has no physical units (Millero, 1993).

and dilution. From here the narratives diverge, some see no environmental effects while others point to significant impacts on the biota.

Brine plumes from SWRO plants may vary significantly depending on site characteristics, effluent volume, mode of discharge, and the prevailing hydrographic conditions. For example, in San Pedro del Pinatar (SE Spain), two SWRO plants (total production 50 $Mm^3 y^{-1}$) discharge their brine (salinity of 70) through a 5 km long marine outfall at 38 m depth. High salinity (45, ambient 37–38) was observed up to 2 km from the outfall in winter, though in summer high salinity was confined to the surroundings of the outfall (Fernandez-Torquemada *et al.* 2009). In Tampa Bay, Florida, USA, one SWRO plant (35 $Mm^3 y^{-1}$) discharges brine (salinity of 54–62, ambient 26) into the bay after mixing (70:1 ratio) with cooling waters from a nearby power plant. Salinity at the vicinity of the discharge increased up to 2, well within the maximum threshold permitted (McConnell, 2009). In Ashqelon's SWRO plant (120 $Mm^3 y^{-1}$), Israel, brine is co-discharged with cooling waters at the shoreline, the positively buoyant mixture disperses at the surface up to a distance of 2 km from the discharge point. The mixed brine-cooling waters discharge increase salinity (up to 1.84 above ambient 39) and water temperature (up to 7.8°C in the winter) at the outfall (Drami *et al.* 2011). From 2007 to 2013 the Palmachim SWRO plant (45–90 $Mm^3 y^{-1}$) Israel, discharged brine through a 1 km long marine outfall at ca. 10 m water depth. High salinity was measured at the outfall (45, ambient average 39). The hypersaline plume was confined to the bottom 1–2 m water layer and detected even at a distance of 1,000 m from the outfall, encompassing an area of approximately 0.4 km^2 (Kress & Galil, 2012). Since 2014, the brine is discharged at 20 m depth. The Perth SWRO plant, Western Australia (53 $Mm^3 y^{-1}$) discharges brine (65, ambient 33 to 37) into Cockburn Sound through an outfall with a 40 port diffuser, 500 m offshore in 10 m of water. Salinity near the outfall at the shallow eastern shelf (ca. 350 m westwards) increased by 1 in bottom waters, but was within the range of natural variability. Slight temperature stratification in bottom waters close to the diffuser were observed (Holloway, 2009). A modeling study of the effect of a SWRO plant (92 $Mm^3 y^{-1}$) and climate change on salinity in northern Spencer Gulf, Australia, predicted an increase of 0.11 in salinity due to brine discharge and 0.09 and 0.29 due to climate effects in 2030 and 2070, respectively (Nunes-Vaz, 2012).

Studies of the effect of thermal desalination in the enclosed Persian Gulf, where much of the global production is clustered, acknowledged that there is an effect on water temperature and salinity and recommended caution (Purnama *et al.* 2005; Lattemann & Hopner, 2008b). Recently, a regional increase in salinity was reported (Uddin *et al.* 2011). The salinity at the discharge area of the Az Zour plant, Kuwait (MSF process, 183 $Mm^3 y^{-1}$ co-located with a power plant) increased from 36–41.5 to 41–43 from 1993 to 2002, while the natural seasonal variation decreased, with no concurrent change in temperature. Following the expansion of the plant, salinities increased to 50. In The effluents of the Subiya Power and desalination plant situated on the northern Kuwaiti coast, are diluted by river discharge from the Shaat Al Arab (Sheppard *et al.* 2010; Uddin *et al.* 2011). Only recently the Gulf's environment was looked at in an integrated approach (Sheppard *et al.* 2010; Sale *et al.* 2011).

8.3.2 Hypersalinity effects on biota

Salinity and temperature have long been perceived as inhibitory environmental factors for survival and growth of marine biota (Murray & Wingard, 2006; Wiltshire *et al.* 2010). Salinity thresholds determine energy consumption, absorption, expenditure, and therefore, scope for growth and ultimately, survival. But whereas temperature thresholds have been extensively investigated in the past decade due to increasing research effort into climate change effects (Tasker, 2008; Byrne, 2011; Monaco & Helmuth, 2011), effects of elevated salinity have received lesser attention. Verifying empirical salinity thresholds is important for identifying, mapping and modeling the impact of desalination brine plumes on the biota.

Most published studies targeted sea grasses. The beds of *Posidonia oceanica* (L.) Delile, 1813, a sea grass endemic to the Mediterranean Sea, constitute some of the Sea's most ecologically important shallow-water marine habitats, providing key ecological services to a rich and diverse biota, producing and exporting organic carbon, stabilizing sediment and reducing coastal erosion and supporting commercial fisheries through their role as nurseries. Its meadows have been declining due to bottom-trawling, fish farming, coastal constructions, and invasive macrophytes. This stenobiotic *P. oceanica* is notoriously susceptible to elevated salinity. Short-term (15 days) mesocosm experiments revealed significant reduction in leaf growth, increased mortality, increased necrosis, and increased senescence, with even an increase in salinity of 1 above the natural range causing significant effect and total mortality at a salinity of 50 (Fernández-Torquemada *et al.* 2005). The effects of increased salinity on water relations and osmolyte (carbohydrates and amino acids) concentrations in *P. oceanica* were studied in a mesocosm system for 47 days. Analyses of leaf-tissue osmolality under hypersaline stress indicated that osmotic adjustments interfered with leaf growth and shoot survival (Sandoval-Gil *et al.* 2012b). Water relations, amino acids, carbohydrates, ions, photosynthesis, respiration, chlorophyll a fluorescence, leaf growth and morphology, and plant mortality in P. oceanica plants exposed to 43 for one and three months followed by a month recovery were studied in a mesocosm experiment. Three months long salinity stress induced excessive ionic exclusion capacity, increased leaf cell turgor, reduced plant carbon balance, increased leaf aging and decay and increased mortality. Even a month later plants showed unbalanced leaf ionic content, impaired photosynthesis, declining internal carbon resources and decreased leaf growth (Marín-Guirao *et al.* 2013). In an *in situ* salinity manipulation experiment a dozen 3 m^2 plots, located in a bed of *P. oceanica*, were treated with desalination effluent diluted to salinity of 1 and 2.5 above mean natural salinity for a 3-month period (Ruiz *et al.* 2009). At the end of the experiment, plant survival, number of leaves, maximum leaf length, leaf growth rate and biomass were lower in the plots treated with the higher salinity values. Importantly, shoot decline continued for a further 3 months after the experiment. A field survey of a shallow *P. oceanica* meadow exposed to reverse osmosis brine discharge for more than 6 years revealed low shoot abundance, significant reduction in leaf size, and overload of epiphytes in the area nearest the outfall (salinity of 38.4–39.8) (Sánchez-Lizaso *et al.* 2008). However, the nitrate-enriched groundwater used for desalination resulted in nutrient loaded brine, and the relative contributions of high salinity and eutrophication have not been established (Garcia *et al.* 2007). Two other Mediterranean seagrasses, *Cymodocea nodosa* (Ucria) Ascherson, 1870 and *Zostera noltii* Hornem. proved sensitive to increases in salinity, with significant decrease in shoot growth rate at salinity higher than 41 and total mortality at 56, in short-term (10 days) laboratory experiments (Fernández-Torquemada & Sánchez-Lizaso, 2011; Fernández-Torquemada & Sánchez-Lizaso, 2013). However, it is difficult to narrow the threshold of salinity tolerance further from the literature. Hypersaline exposure *in situ* often coincides with additional stress factors such as increased temperature, nutrients and sedimentation, potentially altering the experimentally derived values. Moreover, seagrasse's tolerance to hypersalinity stress varies (Walker *et al.* 1988; Koch *et al.* 2007; Sandoval-Gil *et al.* 2012a; Sandoval-Gil *et al.* 2012b). Lack of long term continuous *in situ* salinity monitoring has so far precluded the derivation of empirical salinity tolerance thresholds that take into account issues such as interstitial water salinity, seed germination and seedling sensitivities, and long term responses to salinity stress mitigation. Nevertheless, guidelines were set to protect the *P. oceanica:* salinity should not exceed 38.5 (ca. 1.3% above ambient salinity) for more than 25% of the time annually and not to exceed 40 for more than 5% of the time (Sánchez-Lizaso *et al.* 2008). Salinity threshold were also set in Western Australia and southern California (see case studies below).

Biota, other than seagrass, was also affected by hypersalinity. However, the numbers of studies are small and contain conflicting evidence. Brine discharge changed the benthic community at the southeastern Mediterranean coast of Spain (Del Pilar Ruso *et al.* 2007) and possibly in the Gran Canaria (Riera *et al.* 2011; Riera *et al.* 2012) while Echinoderm disappeared near the outfall of the Dhekelia SWRO in Cyprus (Argyrou, 1999). However, no effect was found in the northwest Mediterranean (Raventos *et al.* 2006) nor in southwest Florida (Hammond *et al.* 1998). Laboratory experiments showed that stressful combinations of temperature and salinity substantially reduced larval performance and development of the barnacle *Amphibalanus improvises* (Nasrolahi *et al.* 2012), while salinity was shown to affect the silica structure of diatoms (*Thalassiosira pseudonana and Chaetoceros muelleri*) (Vars *et al.* 2013). Hypersalinity was shown in the laboratory to decrease embryos survival of the giant Australian cuttlefish *Sepia apama* and reduce mean weight and mantle length (Dupavillon & Gillanders, 2009). Changes in microbial abundance and population were detected by Drami *et al.* (2011, see Section 4.1) and by van der Merwe *et al.* (2014a) in Saudi Arabia. Flow cytometric analysis at the small SWRO plant (maximal capacity 40,000 $m^3 day^{-1}$) located at the King Abdullah University of Science and Technology showed that microbial abundance changed as a function of distance away from the discharge. The changes were minor and may have resulted from normal dilution, At the same site, van der Merwe *et al.* (2014b) showed that the photophysiology of the algal symbiont of the coral *Fungia granulosa* was not influenced by rapid and prolonged changes in salinity (49.4 compared to 41.3 average ambient salinity), but varied with changes in light conditions. The coral's holobionts (natural and cultured symbionts of the genus *Symbiodinium*) were remarkably resilient to increased salinity

In their review of published research on brine impacts on the benthic biota Roberts *et al.* (2010) observed that effects range from plumes extending over tens or hundreds of meters to several kilometers in extreme cases. These authors are highly critical of the biological monitoring studies (Table 8.3, Roberts *et al.* 2010) which 'do not appear to be scientifically defensible assessments of impacts' and manifest 'a general lack of empirical evidence supporting conclusions regarding the effects of desalination brines' with the exception of those conducted in seagrass beds (Roberts *et al.* 2010: 5126). The published benthic monitoring programs had not complied with Before After Control-Impact (BACI) design which consists of multiple reference locations and repeated sampling before and after plant operation, nor with post-operational After, Control-Impact (ACI) design. Indeed, these shortcomings were evident in the results of post-operational studies conducted after a decade of operation (Riera *et al.* 2011; Riera *et al.* 2012) consisting of two sampling periods (May, 2008 and January, 2009) of the subtidal, soft-bottom meio- and macrofauna were marred by uncertainty due to shift in sediment particle size. Similar findings were reported for Perth's SWRO (Shute, 2009) emphasizing the need to long term monitoring.

Where properly planned and conducted, laboratory-based experiments, manipulative field experiments and monitoring studies establish the potential for desalination brines to cause negative biotic impacts. Hypersalinity may influence vital processes such as osmoregulation, rates of protein synthesis and oxygen uptake, scope for growth, and extracellular acid–base balance, the life history of species, and alter the structure and dynamics of local communities. Interactions among salinity changes and temperature, organic and inorganic contaminants, oxygen levels and other environmental stressors may confound and obscure its impacts. In recent decades concern of hypersalinity impact on marine biota has been intensified due to forecast of global salinification of shallower parts of the subtropical seas (Hosoda *et al.* 2009). Increasing temperatures and salinity may facilitate the establishment of non indigenous species, and brine plumes may supply the 'stepping stones' for their spread.

8.3.3 Effects other than salinity

While the impacts of hypersalinity resulting from brine have been studied (see above), impacts of chemicals other than natural marine salts are scarcely known. The co-occurrence of stressors such as co-discharged waste effluents or increased seawater temperature, confound the discussion of the results, preventing the establishment of a cause-response relationship. Chlorine is used in both desalination and power plants to prevent fouling, but whereas in RO plants the residual chlorine is reduced to prevent damage to the membranes by chlorine oxidation, in thermal desalination plants, as in power plants, residual chlorine is discharged with the brine. Residual chlorine reacts swiftly with seawater to form toxic complexes with the bromide and nitrogen-containing organic seawater constituents. Bromoform was the most important chlorination byproduct in coastal power stations effluents in the UK (Taylor, 2006). Laboratory experiments with the European seabass, *Dicentrarchus labrax* (L. 1758), showed it accumulates bromoform in the liver. It was impossible to separate the effect of the bromoform from that of the increased temperature in a study of the blue mussel, *Mytilus edulis* (L. 1758) that revealed high genoplasticity and the evolution of stress proteins.

Corrosion products from thermal desalination plants, in particular copper, a common material in heat exchangers, were shown to accumulate in the vicinity of outfalls. Many of the studies state that the presence of copper does not mean an adverse effect because copper is a natural compound found in nature (Lattemann & Hopner, 2008a). However, an early study found in laboratory studies that copper affected echinoderms, tunicates and Florida seagrass (Chesher, 1971). Copper is also used as a biocide in many marine applications such as antifouling paints and was found to affect micro-organisms (Brand *et al.* 1986). Recently, higher than natural concentrations of copper and zinc in sediments and bivalves was reported at the brine discharge of two SWRO in Taiwan (Lin *et al.* 2013).

Sodium metabisulphite ($Na_2S_2O_5$) is commonly used in cleaning reverse osmosis membranes. Short-term pulses may result in acidification and hypoxia. Portillo *et al.* (2013) studied the effects of this additive in the hypersaline discharged by the Maspalomas II desalination plant, Gran Canaria, Canary Islands, Spain, at a concentration of ca. 1,600 ppm. Toxicity bioassays performed on local marine species the lizard fish *Synodus synodus* (Linnaeus, 1758) revealed a high sensitivity to short-term exposure to low concentrations, with total mortality occurring at concentrations equal to or higher than 50 ppm. Observations of dead individuals of this species and other soft-bottom fish species (*Bothus podas* (Delaroche, 1809), *Microchirus azevia* (de Brito Capello, 1867), *Trachinus draco* Linnaeus, 1758) were not confined to the area delimited by critical DOsat values (10%), but occurred in a broader area.

Whole effluent toxicity testing (WET) were performed using locally relevant species as part of the EIA for the Olympic Dam SWRO plant, Australia (Hobbs *et al.* 2008). The observed toxicity was attributed to salinity for all test species except the diatom *Nitzschia closterium* (Ehrenberg) W. Smith 1853 where the high salinity explained 70% of the toxic effects while 30% was attributed to the polyphosphonate antiscalant.

An additional environmental concern in conjunction with brine discharge is the possible reduction of oxygen levels due to increase in water stratification. This concern was raised during the EIA of the Perth SWRO plant that discharges the brine into the Cockburn Sound, a semi enclosed embayment. Detailed modeling prior to disposal and monitoring after the start of the operations showed no influence of the SWRO discharge on the dissolved oxygen concentration (Holloway, 2009).

8.4 CASE STUDIES

The impacts of SWRO on the marine environment depend upon the choice of location, production volume, particular technology, and effluent disposal. To illustrate this three case studies of SWRO operating in regions differing in their geo- and hydrographical features, physical, chemical and biological characteristics are described: Mediterranean coast (Israel); Western Australia, and Southern California, USA.

8.4.1 Israel, South Eastern Mediterranean Sea

Environmental Setting: The waters of the southern Levantine basin are ultra-oligotrophic, with exceptionally low nutrient concentrations, chlorophyll-a, low primary and bacterial production, and pico and nano dominated phytoplankton community. Neritic waters, in particular in anthropogenically stressed regions such as ports, estuaries, effluent outfalls, power plants and desalination plants, are characterized by higher nutrient and chlorophyll-a concentrations than open sea waters, higher primary production and a greater abundance of larger size phytoplankton (Yacobi *et al.* 1995; Gitelson *et al.* 1996; Kress & Herut, 2001).

Off shore: surficial temperatures range from 17 to 29°C and salinity from 38.7 to 39.7. Nearshore, the natural seasonal variations are larger. The area is home to many non-indigenous species, introduced through the Suez Canal. Its recent expansion has also increased the inflow of the warmer and saltier Red Sea water into the Mediterranean, with an estimated flow of 100 km^3 y^{-1} (Rosen, 2008; Galil, 2009).

Four SWRO desalination plants operate along the 190 km long Israeli coastline, producing c.a. 490 Mm^3 y^{-1} – about 50% of the country's domestic and industrial use. Production is expected to rise to 750 Mm^3 y^{-1} by 2020 (Dreizin *et al.* 2008). Two large plants, in Ashkelon (operational since 2005, capacity 120 Mm^3 y^{-1}) and Hadera (operational since 2010, capacity 127 Mm^3 y^{-1}), are co-located with power stations and dispose of the effluents at the shoreline, next to, or mixed with the power stations' cooling waters. The warm effluents are positively buoyant and disperse mostly at the surface. The plant at Palmachim (operational since 2007, capacity 90 Mm^3 y^{-1}), and the largest plant at Soreq (operational since 2013, capacity 150 Mm^3 y^{-1}) discharge the effluents through two marine outfall,

1.5–2 km off shore, at water depth of 20 m, where it disperses near the bottom. By the mid of 2015, the fifth plant (100 $Mm^3\ y^{-1}$), in Ashdod, will commence full production (Fig. 8.1).

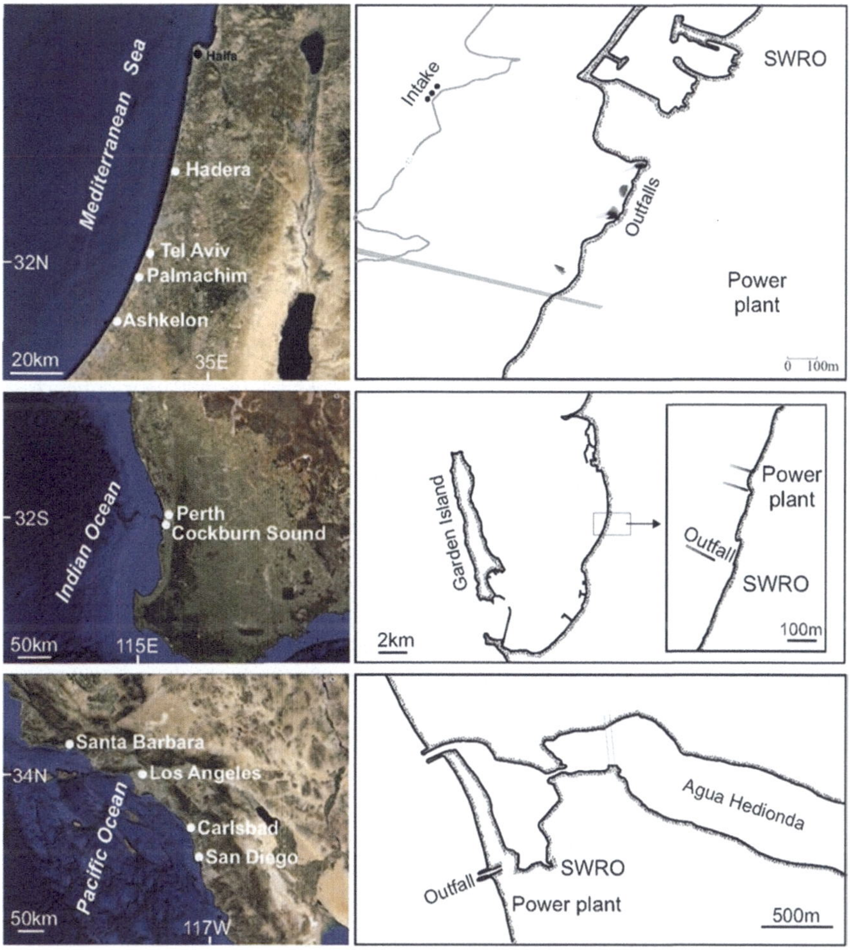

Figure 8.1 Map of SWRO locations in the Mediterranean coast of Israel, Western Australia and Southern California.

Israeli legislation and regulatory process requires an EIA as part of the planning of a desalination plant. The requirements include modeling the brine's dispersion, a curtailed survey of the local marine environment, before construction and operation of the plant, followed by regular environmental monitoring once operational. Plants are required to disclose the annual amount of the chemicals used, the commercial name but not the actual chemical structure of the additives, many of which are protected by international patents. No effluent toxicity tests are required. The requirements are not consistently enforced: the SWRO plant at Ashkelon went into operation with but a limited EIA (Einav *et al.* 2002; Einav & Lokiec, 2003), lacking a baseline survey. Since 2005 the plant has been discharging its effluents at the shoreline next to four cooling waters outlets of the adjacent power plant. Until 2010, backwash from the sand filters, containing iron hydroxide used as coagulant in the pre-treatment, was discharged in pulses, their frequency dependent on the intake seawater quality. The iron hydroxide formed a conspicuous 'red plume' (Safrai & Zask, 2008; UNEP, 2008) (Fig. 8.2). The results of a preliminary study revealed that the mixed effluent-cooling waters discharge were positively buoyant, dispersed at the surface to a distance of 1,340 m, and increased ambient salinity and temperature (Drami *et al.* 2011). Biotic impacts were severe: phytoplankton densities were lower at the outfall; chlorophyll-a and picophytoplankton cell numbers were negatively correlated with salinity, but significantly with temperature. The discharge of the pulsed backwash increased turbidity, suspended particulate matter and particulate iron and decreased phytoplankton growth efficiency at the outfall. These effects were noticed to a distance of 500 m from the outfall. The significantly reduced primary production could not be attributed to a specific component. Similar increase in salinity and temperature over ambient values were measured at the outfalls of the adjacent power plant operated by the Israel Electric Corporation (IEC). The IEC environmental monitoring reports failed to uncover effects other than temperature and salinity rise, with the exception of a slight decrease in numbers of benthic biota at the outfall (Glazer, 2011). The amount of discharged iron was reduced from 535 tons in 2007 to 100 tons in 2011, and since May 2010, the backwash effluent is mixed with the brine and discharged continuously, in an attempt to mitigate the visual effect ('red plume') at the outfall. The Soreq SWRO plant, which effluents are discharged through an outfall equipped with a diffuser, was required to treat and dispose the backwash iron on land.

At Palmachim, from 2007 to 2014, the negatively buoyant unmixed effluents discharged 1 km offshore through diffuser at 10 m water depth, raise the seawater salinity above bottom up to 50, 28% above ambient salinity of 39, increase the temperature by around 0.5°C as well as the turbidity. The effects are confined to waters 1 to 2 m above bottom and noted even at a distance of more than 1,000 m from the outfall, encompassing an area of ca.0.4 km^2. Concentrations of dissolved oxygen in seawater and chl-a have not been affected. In a few samples, particulate iron in seawater was higher than background, similar to the findings at Ashqelon. Infaunal assemblages at the vicinity of the outfall differed slightly compared to those further away, in particular in the fall (Kress *et al.* 2011; Kress & Galil, 2012). However, these results should be considered with caution for there are insufficient temporal data. Since 2014, the brine is disposed of through an outfall located at 20 m water depth.

Figure 8.2 Brine outfall of the SWRO plant in Ashkelon, Israel, April 2009 during the pulsed backwash discharge. B-Brine outfall. (Photos by Y. Gertner).

8.4.2 Cockburn Sound, Western Australia, Indian Ocean

Environmental Setting: Cockburn Sound is a shallow (maximum depth of 20 m), semi enclosed inlet in Western Australia, with an area of 100 km². Annual temperature and salinity ranges from 14 to 24°C and from 35 to 37.5, respectively. Hydrodynamics is driven by wind-stress and density gradients, with vertical mixing of the water column in winter-spring and strong vertical salinity stratification in the autumn. It explains the biological characteristics of the embayment, such as nuisance algal blooms and low oxygen occurrences. Industrial and port activities affected the environment, with a documented loss of seagrass meadows (D'Adamo, 2002; Kendrick *et al.* 2002).

The thrust to construct large scale desalination plants in Australia followed severe droughts in the past decade that heightened the urgent need to secure freshwater for the population. In 2008 the production of desalinated water was 107 $Mm^3\ y^{-1}$, 0.6% of the country's water consumption, and was expected to rise to 445 $Mm^3\ y^{-1}$ in 2013 (Hoang *et al.* 2009). Most of the production is concentrated in Western Australia (Perth, Kwinana, 45 $Mm^3\ y^{-1}$; Binningup, 100 $Mm^3\ y^{-1}$), with additional plants on the southern coast (Adelaide, 100 $Mm^3\ y^{-1}$), the north-east (Gold Coast, Queensland, 48 $Mm^3\ y^{-1}$), Sidney (2010, 91$Mm^3\ y^{-1}$) and Melbourne (2012, 150 $Mm^3\ y^{-1}$). However, as 2011–2012 were rainy, some plants are idle, on standby mode or not working at full capacity[3].

Australia's first major SWRO plant, Kwinana in Cockburn Sound, Western Australia, began operation in 2007 and supplies ca.17% of Perth's water. The briny effluent (salinity 65, ambient 33–37) is discharged through an outfall fitted with a 40 port diffuser, 500 m offshore in ~10 m of water. The ferric hydroxide coagulant is treated and discharged on land. An extensive EIA was required at the planning and construction stages, followed by marine environmental monitoring during plant operation. The main environmental concern at the planning stages was that the dense brine would stratify the water column, prevent mixing, and hence decrease the dissolved oxygen concentration in the bottom layer of seawater.

The baseline study found that environmental conditions (primarily wind strength and direction) play a dominant role in the mixing regime of coastal waters. Modeling of the brine discharge predicted a slight increase in the ambient salinity (about 1) in the shallow eastern shelf up to 2 km of the diffuser. Salinity, temperature, dissolved oxygen together with water and sediment quality, and the benthic assemblages were characterized at the baseline study (Holloway, 2009; Shute 2009). A monitoring and management program, dictated by the Australian EPA was put in place once operation began (Holloway, 2009; Shute, 2009). The results of the monitoring showed slight temperature stratification in bottom waters close to the diffuser, but within the Environment Quality Guidance (EQG) for the sound (AU, 2005), without appreciable change in seasonal trends. Salinity increased by up to 1 in bottom waters at impact sites, in agreement with the model predictions. The salinity complied with the EQG's requirements for salinity of ±1.2 of ambient within ~50 m of the discharge point and ±0.8 at 1 km from the outfall. Dissolved oxygen decreased at times but no anoxic or sub-oxic conditions were observed, and values did not exceed the EQG's.

Comparison of the pre and post operation survey results of the sediments and benthic assemblages revealed that sediment samples collected in 2008 had a higher proportion of silt than those collected in 2006, more evident in the northern part of the study area. The number of species, abundance, species richness and species diversity in 2008 were generally lower than those recorded in 2006, more

[3]Murray Griffin, 6 March 2013 in http://www.bloomberg.com/news/2013-03-06/drought-prompts-australia-to-turn-to-desalination-despite-cost.html

pronounced in the northern sites as with sediment composition. It was concluded that the differences did not result from brine discharge (Shute, 2009). Long term data series with proper controls are essential to normalize for natural temporal variability, but the only data available stems from the survey conducted after a single year of operation. Presently, only temperature, conductivity, wind speed and direction and dissolved oxygen are measured continuously and available at the water corporation of Australia website[4].

8.4.3 Southern California, USA

Environmental Setting: The Southern California Bight (SCB) consists of 700 km of shoreline from Point Conception, California, to Cabo Colnett, Mexico. The coastal region along the SCB is one of the most densely populated coastlines in the US. It includes diverse habitats for a broad range of marine life including more than 2000 species of invertebrates, 500 species of fish, and many marine mammals and birds. The high biological productivity in the SCB is derived from its unique circulation pattern: subtropical waters flow north close to the shore, while subarctic waters flow south offshore. Circulation is affected by El Nino Southern Oscillation. Temperatures range from 14 to 21°C and salinity from 31 to 34 (Hickey, 1993; Stein & Cadien, 2009; Jenkins *et al.* 2012).

Rapidly growing population and constrained water supplies have boosted plans for SWRO plants along the 1,350 km long coast of California. A few small scale SWRO plants are operational already, and proposals for over 20 plants are under consideration, the largest plants planned for the southern California coast (Carlsbad, 69 $Mm^3\ y^{-1}$ by 2016; Huntington beach, to supply 69 $Mm^3\ y^{-1}$ by 2018) (Cooley *et al.* 2006; Jenkins *et al.* 2012). The regulation of discharges and protection of water quality is based on federal and state laws.

After twelve years of planning and over six years in the state's permitting process, the Carlsbad SWRO plant received final approval in November 2012. Construction of the plant and pipeline is under way and expected to supply desalinated water by 2016[5]. The plant will be co-located with the Encina power plant (EPP) which uses seawater from the Agua Hedionda Lagoon for cooling. The SWRO plant will use the EPP cooling waters and the brine will be mixed with the cooling waters and discharged nearshore through the EPP discharge channel (Fig. 8.1).

An extensive EIA was performed, including brine discharge modeling, whole effluent toxicity testing, salinity tolerance tests (Le Page, 2005), description of

---

[4]http://www.watercorporation.com.au/water-supply-and-services/solutions-to-perths-water-supply/desalination/perth-seawater-desalination-plant (accessed 16 June 2013).
http://www.watercorporation.com.au/~/media/Files/Residential/Water%20supply%20and%20services/Desalination/PSDP/Annual_compliance_assessment_report_may_2011_april_2012-Accessed on June 16, 2013.

[5]http://carlsbaddesal.com/eir (accessed 2 February 2015).

the flora and fauna of the southern California bight, and intake effects assessment (Jenkins & Wasyl, 2001; Graham, 2005). The EIA agreed that the Carlsbad SWRO plant is not expected to have a significant impact on the biota in the vicinity of the co-disposal of brine and cooling waters from the EPP. This assessment was based on the findings described below.

Analysis of the marine habitats nearby revealed that the species diversity and abundances in the vicinity of the EPP discharge were comparable to sites located outside the discharge area, home to no endangered marine species and that the modeled zone of initial dilution (308 m) has no 'environmentally sensitive' habitats such as eel grass, surf grass, or kelp beds. Brine dispersion models revealed that under average conditions, the brine (salinity 67, ambient salinity 33.5) combined with the cooling waters will have an end-of pipe salinity of 36.5, decreasing to 34.4 at 300 m away. Under extreme conditions, the end-of pipe salinity would be 39. Salinity tolerance was experimentally assessed in 18 species commonly found at the discharge site, by exposure to a blend of desalination plant concentrate and power plant effluent with a salinity of 36. Organisms were monitored and evaluated for overall health based on qualitative parameters (appearance, willingness to feed, activity, and gonad production in sea urchins). No mortality was observed during the 51/2 months experiment. All organisms remained healthy and showed normal activity and feeding behavior at a salinity of 36. The purple sea urchin, *Strongylocentrotus purpuratus* (Stimpson, 1857), spawned successfully and no weight changes were detected. A toxicity study, in which species of concern (*S. purpuratus*, sand dollar *Dendraster excentricus* (Eschscholtz, 1829), and red abalone *Haliotis rufescens* Swainson, 1822) were kept at salinities of 37–40 over a 19 day period, revealed 100% survival in all test salinities. No acute or chronic toxicity was reveled in brine and cooling water blend toxicity tests on the giant Pacific kelp *Macrocystis pyrifera* (L.) C. Agardh 1820 (48 hours germination and growth test), topsmelt *Atherinops affinis* (Ayres, 1860) (7 day survival using 10-day old larva) and red abalone (48 hour post fertilization embryonic development test). The recommendations of the EIA were an absolute salinity limit of ≤40 at 1,000 ft from the discharge site. A recent report recommended an incremental salinity limit of not more than 5% from ambient at the mixing zone boundary (100 m from the discharge point) (Jenkins *et al.* 2012). Monitoring studies will be performed after the plant becomes operational and compared with the experimental findings.

8.5 OUTLOOK

SWRO desalination capacity has been increasing steeply in the last two decades and is slated to continue to expand. Yet, the necessary marine environmental research is in its infancy, lacking studies, a long term record and an integrative approach. Since SWRO effluents are growing in volume, our lack of full understanding of their long term impacts demands more, not less, caution.

Peer reviewed articles on SWRO mostly address the process itself, depicting quantitatively the improvements made in energy consumption, membrane efficiency, in increasing the resistance of membranes to biofouling, in the development of new membrane materials and novel processes. However, articles focused on SWRO's marine environmental impacts are in a large part descriptive and non-quantitative and resort to ambiguous and vague vocabulary such as: belief, potential adverse impacts, no effect with proper engineering design. Quantitative studies in the laboratory and in the field are few. Worse, important data, such as EIA's and monitoring reports, are sequestered in the grey literature, not peer reviewed, and mostly inaccessible. The whole field is fraught with sparse data and non-validated models. In an industrial context, we have essentially only 'pilot studies', which do not supply a scientifically-based rational for public policy. Brine discharge guidelines address mainly hypersalinity and consider the SWRO plant in detachment from the multiple anthropogenic stressors in the area, rather than a *delta* to the entire stress load. Monitoring at the brine's oufall is usually a compliance monitoring, measuring salinity, temperature and few other parameters, providing only a rudimentary and deficient view of the environment.

Understanding the effect of SWRO on the marine environment and its protection is dependent on cooperation among the desalination industry at both the operational and development level, research in the appropriate disciplines and regulators. Four complementary research and development courses should be coordinated and a mutual feedback mechanism established:

8.5.1 Desalination technology

- Optimization of intake strategies to reduce impingement and entrainment of biota.
- Optimization of brine discharge to increase dilution and reduce hypersalinity stress.
- Reduction of the use of chemicals in the desalination process, thus reducing their load into the marine environment.
- Reduction of seawater intake volume by improving process efficiency.
- Salts' extraction for re-use, thus reducing brine's salinity.
- Development of offshore plants, areas less environmentally sensitive than nearshore habitats.
- Development of new large scale, environmentally friendly technologies for seawater desalination.
- **Feedback: Use field and research results to focus and guide technological development to ameliorate the environmental impact.**

8.5.2 Research

- Compilation of an international database of the chemicals (or for proprietorial chemicals, their active compounds) used in the desalination process and their toxicological properties.

- Performance of toxicity studies of chronic, acute and sub-lethal effects of stressors, singly (such as salinity, temperature, coagulant, antiscalant, more) or synergetic, as appropriate. The studies should target species relevant to the SWRO site, encompass different taxa and different life stages.
- Integration of novel chemical and biological methodology, such as genomics and metagenomics.
- Expansion of laboratory scale studies to *in situ*, 'real life' experiments
- **Feedback: Use data from the detailed desalination process and from the environmental monitoring to fine tune the research.**

8.5.3 *In situ* monitoring

Pre-construction – baseline.

- Characterization and description of the marine environment: intake and discharge sites, hydrography, circulation, physical and chemical parameters. Special emphasis should be given to the biotic compartment and identification of sensitive species and life stages.
- Modeling brine dispersion at different hydrographic conditions and discharge strategies to optimize outfall design.
- **Feedback: Relay findings to researchers, environmental scientists, engineers and regulators to optimize intake and outfall design and to set up environmental guidelines and regulations.**

After plant construction and start of operations

- Regular monitoring of the marine environment. The monitoring should be a long term commitment, throughout the lifetime of the desalination plant. The monitoring data should be analyzed critically and monitoring design changed accordingly. The data should be published and disseminated to the community.
- Enforcement the permitting license's guidelines conditions by constant re-examination of the monitoring findings.
- Stay updated. Although the general SWRO process at a specific plant is well defined, changes that can influence the monitoring program do occur. Those include, among others, changes in chemicals' identities and quantities, in pre-treatment methodology, plant capacity, in intake and outfall position.
- Go beyond the standard monitoring and introduce novel basic research to environmental monitoring.
- **Feedback: Relay findings of the monitoring studies to the desalination industry, emphasizing problems that need technological solutions, to the regulator that can in turn analyze the guidelines and permits and when necessary request changes or further treatment at the plant, to researchers to fine tune laboratory experiments**

8.5.4 Integration

- Establish a global network for data and knowledge exchange and dissemination. The network should be easily accessible, open to all, and include individuals from diverse, complementary fields.
- Look outside the box. Relevant data exist at seemingly unrelated fields – such as climate change research, toxicology, economics, industrial engineering. In particular, increase in salinity and temperature in connection to desalination processes could be an indicator to the effects of climate change on the marine environment.

Paraphrasing John Donne's 'No man is an island', no SWRO plant is alone in the environment. Desalination is an ever changing industry, in the era of climate change and extensive anthropogenic use of the marine environment. Only a holistic approach, looking at seawater desalination as part of the larger picture within a multidisciplinary approach will be able to provide insight and understating to the workings of the marine environment in this age of rapid changes.

8.6 REFERENCES

Ahmed M., Shayya W. H., Hoey D. and Al-Handaly J. (2001). Brine disposal from reverse osmosis desalination plants in oman and the united arab emirates. *Desalination*, **133**, 135–147.

Ahmad N. and Baddour R. E. (2014). A review of sources, effects, disposal methods, and regulations of brine into marine environments. *Ocean & Coastal Management*, **87**, 1–7.

Alpert H., Borrowman C. and Haddad B. (2007). 'Evaluating Environmental Impacts of Desalination in California' Center for Integrated Water Research 27 July 2007. https://ciwr.ucsc.edu/sites/default/files/FirstWorkshopSummary.pdf (accessed 31 May 2015).

Argyrou M. (1999). Impact of desalination plant on marine macrobenthos in the coastal waters of Dhekelia Bay, Cyprus. Department of Fisheries, Ministry of Agriculture, Natural Resources and Environment, Cyprus.

AU. (2005). Environmental Quality Criteria Reference Document for Cockburn Sound (2003–2004). A supporting document to the state environmental (Cockburn Sound) policy. EPA AU Report 20.

Bar-Zeev E., Belkin N., Liberman B., Berman-Frank I. and Berman T. (2013). Bioflocculation: chemical free, pre-treatment technology for the desalination industry. *Water Research*, **47**, 3093–3102.

Barnthouse L. W. (2013). Impacts of entrainment and impingement on fish populations: a review of the scientific evidence. *Environmental Science & Policy*, **31**, 149–156.

Brand L. E., Sunda W. G. and Guillard R. R. L. (1986). Reduction of marine phytoplankton reproduction rates by copper and cadmium. *Journal of Experimental Marine Biology and Ecology*, **96**, 225–250.

Byrne M. (2011). Impact of ocean warming and ocean acidification on marine invertebrate life history stages: vulnerabilities and potential for persistence in a changing ocean. *Oceanography and Marine Biology An Annual Review*, **49**, 1–42.

Campbell R. L. and Jones A. T. (2005). Appropriate disposal of effluent from coastal desalination facilities. *Desalination*, **182**, 365–372.

Caron D. A., Garneau M.-È., Seubert E., Howard M. D. A., Darjany L., Schnetzer A., Cetinic I., Filteau G., Lauri P., Jones B. and Trussell S. (2010). Harmful algae and their potential impacts on desalination operations off southern California. *Water Research*, **44**, 385–416.

Chesher R. (1971). Biological impact of a large-scale desalination plant at Key West, Florida. *Elsevier Oceanography Series*, **2**, 99–164.

Cooley H., Gleick P. H. and Wolff G. A. (2006). Desalination, with a Grain of Salt. A California Perspective. Pacific Institute, Oakland, California.

D'Adamo N. (2002). Exchange and Mixing in Cockburn Sound, Western Australia: a Seasonally Stratified, Micro-tidal, Semi-enclosed Coastal Embayment. Ph.D. Dissertation, University of Canterbury, New Zeeland http://hdl.handle.net/10092/1243 (accessed 15 April 2013).

Darwish M., Hassabou A. H. and Shomar B. (2013). Using seawater reverse osmosis (SWRO) desalting system for less environmental impacts in qatar. *Desalination*, **309**, 113–124.

Del Pilar Ruso Y., la Ossa Carretero J. A. D., Casalduero F. G. and Lizaso J. L. S. (2007). Spatial and temporal changes in infaunal communities inhabiting soft-bottoms affected by brine discharge. *Marine Environmental Research*, **64**, 492–503.

Drami D., Yacobi Y. Z., Stambler N. and Kress N. (2011). Seawater quality and microbial communities at a desalination plant marine outfall. A field study at the israeli mediterranean coast. *Water Research*, **45**, 5449–5462.

Dreizin Y., Tenne A. and Hoffman D. (2008). Integrating large scale seawater desalination plants within israel's water supply system. *Desalination*, **220**, 132–149.

Dupavillon J. L. and Gillanders B. M. (2009). Impacts of seawater desalination on the giant australian cuttlefish *Sepia apama* in the upper spencer gulf, south australia. *Marine Environmental Research*, **67**(4–5), 207–218.

Einav R., Harussi K. and Perry D. (2002). The footprint of the desalination processes on the environment. *Desalination*, **152**, 141–154.

Einav R. and Lokiec F. (2003). Environmental aspects of a desalination plant in Ashkelon. *Desalination*, **156**, 79–85.

Elimelech M. and Phillip W. A. (2011). The future of seawater desalination: energy, technology, and the environment. *Science*, **333**, 712–717.

Fernandez-Torquemada Y., Gonzalez-Correa J. M., Loya A., Ferrero L. M., Diaz-Valdes M. and Sanchez-Lizaso J. L. (2009). Dispersion of brine discharge from seawater reverse osmosis desalination plants. *Desalination and Water Treatment*, **5**, 137–145.

Fernández-Torquemada Y., Sánchez-Lizaso J. L. and González-Correa J. M. (2005). Preliminary results of the monitoring of the brine discharge produced by the SWRO desalination plant of alicante (SE spain). *Desalination*, **182**, 395–402.

Fernández-Torquemada Y. and Sánchez-Lizaso J. (2011). Responses of two mediterranean seagrasses to experimental changes in salinity. *Hydrobiologia*, **669**, 21–33.

Fernández-Torquemada Y. and Sánchez-Lizaso J. L. (2013). Effects of salinity on seed germination and early seedling growth of the Mediterranean seagrass *posidonia oceanica* (L.) Delile. *Estuarine, Coastal and Shelf Science*, **119**, 64–70.

Fritzmann C., Löwenberg J., Wintgens T. and Melin T. (2007). State-of-the-art of reverse osmosis desalination. *Desalination*, **216**, 1–76.

Gacia E., Invers O., Manzanera M., Ballesteros E. and Romero J. (2007). Impact of the brine from a desalination plant on a shallow seagrass (Posidonia oceanica) meadow. *Estuarine, Coastal and Shelf Science*, **72**, 579–590.

Galil B. S. (2009). Taking stock: inventory of alien species in the Mediterranean sea. *Biological Invasions*, **11**, 359–372.

Ghaffour N., Missimer T. M. and Amy G. L. (2013). Technical review and evaluation of the economics of water desalination: current and future challenges for better water supply sustainability. *Desalination*, **309**, 197–207.

Gitelson A., Karnieli A., Goldman N., Yacobi Y. Z. and Mayo M. (1996). Chlorophyll estimation in the southeastern mediterranean using CZCS images: adaptation of an algorithm and its validation. *Journal of Marine Systems*, **9**, 283–290.

Glazer A. (2011). Monitoring the coastal and marine environment at the discharge site of the Rutenberg power plant, VID desalination plant and Mekorot's well amelioration plant. Results from 2010. Israel Electric Corp. RELP 13–2011. p. 319 (In Hebrew).

Graham J. B. (2005). Marine Biological Considerations Related to the Reverse Osmosis Desalination Project at the Encina Power Plan, Carlsbad, CA. Report submitted to Poseidon Resources.

Greenlee L. F., Lawler D. F., Freeman B. D., Marrot B. and Moulin P. (2009). Reverse osmosis desalination: water sources, technology, and today's challenges. *Water Research*, **43**, 2317–2348.

Hammond M., Blake N., Hallock-Muller P., Luther M., Tomasko D. and Vargo G. (1998). Effects of Disposal of Seawater Desalination Discharges on Near Shore Benthic Communities. Report of Southwest Florida Water Management District and University of South Florida.

Hickey B. M. (1993). Physical oceanography. In: Ecology of the Southern California Bight: A Synthesis and Interpretation. M. D. Dailey, D. J. Reish and J. W. Anderson (Eds.). Berkeley, CA: University of California Press, 19–70.

Hoang M., Bolto B., Haskard C., Barron O., Gray S. and Leslie G. (2009). Desalination in Australia. CSIRO Report.

Hobbs D., Stauber J., Kumar A. and Smith R. (2008). Ecotoxicity of Effluent from the Proposed Olympic Dam Desalination Plant. Final Report. Hydrobiology Pty Ltd. Aquatic Environmental Services.

Holloway K. (2009). Perth Seawater Desalination Plant Water Quality Monitoring Programme. Final Programme summary Report 2005–2008. Report No. 445_001/3. Prepared by Oceanica Consulting Pty LTD for the Water Corporation of Western Australia.

Hosoda S., Suga T., Shikama N. and Mizuno K. (2009). Global surface layer salinity change detected by argo and its implication for hydrological cycle intensification. *Journal of Oceanography*, **65**, 579–586.

Jenkins S. A. and Wasyl J. (2001). Hydrodynamic modeling of dispersion and dilution of concentrated seawater produced bu the ocean desalination project at the Encina power plant, Carlsbad CA. Environmental impact Report for Carlsbad Seawater Desalination Plant, City of Carlsbad.

Jenkins S., Paduan J., Roberts P., Schlenk D. and Weis J. (2012). Management of Brine Discharges to Coastal Waters. Recommentdations of a Science Advisory Panel. Technical Report 694, Southern California Coastal Water Research Project, Costa Mesa, CA.

Kendrick G. A., Aylward M. J., Hegge B. J., Cambridge M. L., Hillman K., Wyllie A. and Lord D. A. (2002). Changes in seagrass coverage in Cockburn Sound, Western Australia between 1967 and 1999. *Aquatic Botany*, **73**, 75–87.

Khan S. J., Murchland D., Rhodes M. and Waite T. D. (2009). Management of concentrated waste streams from high-pressure membrane water treatment systems. *Critical Reviews in Environmental Science and Technology*, **39**, 367–415.

Koch M. S., Schopmeyer S. A., Kyhn-Hansen C., Madden C. J. and Peters J. S. (2007). Tropical seagrass species tolerance to hypersalinity stress. *Aquatic Botany*, **86**, 14–24.

Kress N. and Herut B. (2001). Spatial and seasonal evolution of dissolved oxygen and nutrients in the southern levantine basin (eastern mediterranean sea): chemical characterization of the water masses and inferences on the N: P ratios. *Deep-Sea Research Part I-Oceanographic Research Papers*, **48**, 2347–2372.

Kress N., Galil B. and Shoham-Frider E. (2011). Marine Monitoring at the Brine Disposal Site of the Via Maris Desalination Plant off Palmahim. Results from May and September 2010. IOLR Report H25/2011. 89 p. (In Hebrew).

Kress N. and Galil B. S. (2012). Seawater desalination in israel and its environmental impact. *Desalination and Water Reuse* February-March 2012, 26–29.

Lattemann S. and Hopner T. (2008a). Environmental impact and impact assessment of seawater desalination. *Desalination*, **220**, 1–15.

Lattemann S. and Hopner T. (2008b). Impacts of seawater desalination plants on the marine environment of the gulf. In: Protecting the Gulf's Marine Ecosystems from Pollution. Eds. A. H. Abuzinada, H. J. Barth, F. Krupp, B. Böer and T. Z. Al Abdessalaam Birkhäuser, Verlag/Switzerland, 191–205.

Lattemann S. (2010). Development of an environmental impact assessment and decision support system for seawater desalination plants. Ph.D. Dissertation, Delft University of Technology, The Netherlands, 294 p.

Le Page S. (2005). Salinity Tolerance Investigations: A Supplemental Report for the Carlsbad, CA Desalination Project. Report presented to Poseidon Resources.

Lin Y.-C., Chang-Chien G.-P., Chiang P.-C., Chen W.-H. and Lin Y.-C. (2013). Potential impacts of discharges from seawater reverse osmosis on taiwan marine environment. *Desalination*, **322**, 84–93.

Macedonio F., Drioli E., Gusev A. A., Bardow A., Semiat R., and Kurihara M. (2012). Efficient technologies for worldwide clean water supply. *Chemical Engineering and Processing: Process Intensification*, **51**, 2–17.

Marín-Guirao L., Sandoval-Gil J. M., Bernardeau-Esteller J., Ruíz J. M. and Sánchez-Lizaso J. L. (2013). Responses of the mediterranean seagrass *posidonia oceanica* to hypersaline stress duration and recovery. *Marine Environmental Research*, **84**, 60–75.

Mayhew D. A., Jensen L. D., Hanson D. F. and Muessig P. H. (2000). A comparative review of entrainment survival studies at power plants in estuarine environments. *Environmental Science & Policy*, **3**, Supplement 1, 295–301.

McConnell R. (2009). Tampa Bay Seawater Desalination Facility – Environmental Impact Monitoring. Proceedings of 2009 Annual WateReuse Conference, Seattle.

Millero F. J. (1993). What is PSU? *Oceanography*, **6**, 67.

Missimer T. M., Ghaffour N., Dehwah A. H. A., Rachman R., Maliva R. G. and Amy G. (2013). Subsurface intakes for seawater reverse osmosis facilities: Capacity limitation, water quality improvement, and economics. *Desalination*, **322**, 37–51.

Monaco C. J. and Helmuth B. (2011). Tipping points, thresholds and the keystone role of physiology in marine climate change research. *Advances in Marine Biology*, **60**, 126–160.

Murray J. B. and Wingard G. L. (2006). Salinity and temperature tolerance experiments on selected florida bay mollusks. *U.S. Geological Survey Open-File Report*, **1026**, 59 p.

Nasrolahi A., Pansch C., Lenz M. and Wahl M. (2012). Being young in a changing world: how temperature and salinity changes interactively modify the performance of larval stages of the barnacle Amphibalanus improvisus. *Marine Biology*, **159**, 331–340.

NRC. (2008). Desalination, a national perspective National Research Council of the National Academies. The National Academies press, Washington, D.C.

Nunes-Vaz R. A. (2012). The salinity response of an inverse estuary to climate change & desalination. *Estuarine, Coastal and Shelf Science*, **98**, 49–59.

NWC. (2008). Emerging trends in desalination: A review. Waterlines Report Series No 9, Published by the National Water Commision (NWC), Australia, 96 p.

Perez-Gonzalez A., Urtiaga A. M., Ibanez R. and Ortiz I. (2012). State of the art and review on the treatment technologies of water reverse osmosis concentrates. *Water Research*, **46**, 267–283.

Portillo E., Ruiz de la Rosa M., Louzara G., Ruiz J. M., Marín-Guirao L., Quesada J., González J. C., Roque F., González, N., Mendoza H. (2013). Assessment of the abiotic and biotic effects of sodium metabisulphite pulses discharged from desalination plant chemical treatments on seagrass (*cymodocea nodosa*) habitats in the canary islands. *Marine Pollution Bulletin*, **80**(1–2), 222–233.

Purnama A., Al-Barwani H. H. and Smith R. (2005). Calculating the environmental cost of seawater desalination in the arabian marginal seas. *Desalination*, **185**, 79–86.

Raventos N., Macpherson E. and García-Rubiés A. (2006). Effect of brine discharge from a desalination plant on macrobenthic communities in the NW mediterranean. *Marine Environmental Research*, **62**, 1–14.

Richlen M. L., Morton S. L., Jamali E. A., Rajan A. and Anderson D. M. (2010). The catastrophic 2008–2009 red tide in the arabian gulf region, with observations on the identification and phylogeny of the fish-killing dinoflagellate Cochlodinium polykrikoides. *Harmful Algae*, **9**, 163–172.

Riera R., Tuya F., Sacramento A., Ramos E., Rodriguez M. and Monterroso O. (2011). The effects of brine disposal on a subtidal meiofauna community. *Estuarine, Coastal and Shelf Science*, **93**, 359–365.

Riera R., Tuya F., Ramos E., Rodríguez M. and Monterroso Ó. (2012). Variability of macrofaunal assemblages on the surroundings of a brine disposal. *Desalination*, **291**, 94–100.

Roberts D. A., Johnston E. L. and Knott N. A. (2010). Impacts of desalination plant discharges on the marine environment: a critical review of published studies. *Water Research*, **44**, 5117–5128.

Rosen D. S. (2008). Monitoring boundary conditions at mediterranean basin. Key element for reliable assessment of climate change, variability and impacts at mediterranean basin shores. In: Towarsd an Integrated System of Mediterranean Marine Observatories. CIESM workshop Monographs, no. 34, 107–111.

Ruiz J. M., Marin-Guirao L. and Sandoval-Gil J. M. (2009). Responses of the mediterranean seagrass posidonia oceanica to in situ simulated salinity increase. *Botanica Marina*, **52**, 459–470.

Safrai I. and Zask A. (2008). Reverse osmosis desalination plants – marine environmentalist regulator point of view. *Desalination*, **220**, 72–84.

Sale P. F., Feary D. A., Burt J. A., Bauman A. G., Cavalcante G. H., Drouillard K. G., Kjerfve B., Marquis E., Trick C. G., Usseglio P. and Van Lavieren H. (2011). The growing need for sustainable ecological management of marine communities of the persian gulf. *Ambio*, **40**, 4–17.

Sánchez-Lizaso J. L., Romero J., Ruiz J., Gacia E., Buceta J. L., Invers O., Fernández Torquemada Y., Mas J., Ruiz-Mateo A. and Manzanera M. (2008). Salinity tolerance of the mediterranean seagrass *posidonia oceanica*: recommendations to minimize the impact of brine discharges from desalination plants. *Desalination*, **221**, 602–607.

Sandoval-Gil J. M., Marin-Guirao L. and Ruiz J. M. (2012a). Tolerance of mediterranean seagrasses (*posidonia oceanica* and *cymodocea nodosa*) to hypersaline stress: water relations and osmolyte concentrations. *Marine Biology*, **159**, 1129–1141.

Sandoval-Gil J. M., Marín-Guirao L. and Ruiz J. M. (2012b). The effect of salinity increase on the photosynthesis, growth and survival of the Mediterranean seagrass cymodocea nodosa. *Estuarine, Coastal and Shelf Science*, **115**, 260–271.

Service R. F. (2006). Desalination Freshens Up. *Science*, **313**, 1088–1090.

Shannon M. A., Bohn P. W., Elimelech M., Georgiadis J. G., Marinas B. J. and Mayes A. M. (2008). Science and technology for water purification in the coming decades. *Nature*, **452**, 301–310.

Sheppard C., Al-Husiani M., Al-Jamali F., Al-Yamani F., Baldwin R., Bishop J., Benzoni F., Dutrieux E., Dulvy N. K., Durvasula S. R. V., Jones D. A., Loughland R., Medio D., Nithyanandan M., Pilling G. M., Polikarpov I., Price A. R. G., Purkis S., Riegl B., Saburova M., Namin K. S., Taylor O., Wilson S. and Zainal K. (2010). The gulf: a young sea in decline. *Marine Pollution Bulletin*, **60**, 13–38.

Shute S. (2009). Perth Desalination Plant- Cockburn Sound benthic macrofauna community and sediment habitaat, Repeat Macrobenthic survey. Oceanica Consulting. Report No. 604–011/1, 202pp.

Stein E. D. and Cadien D. B. (2009). Ecosystem response to regulatory and management actions: the southern california experience in long-term monitoring. *Marine Pollution Bulletin*, **59**, 91–100.

Tal A. (2006). Seeking Sustainability: israel's evolving water management strategy. *Science*, **313**, 1081–1084.

Tasker M. L. (2008). The effect of climate change on the distribution and abundance of marine species in the OSPAR maritime area. *ICES Cooperative Research Report*, **293**, 45 p.

Taylor C. J. L. (2006). The effects of biological fouling control at coastal and estuarine power stations. *Marine Pollution Bulletin*, **53**, 30–48.

Uddin S., Al Ghadban A. and Khabbaz A. (2011). Localized hyper saline waters in arabian gulf from desalination activity—an example from south kuwait. *Environmental Monitoring and Assessment*, **181**, 587–594.

UNEP (2008). Desalination Resource and Guidance Manual for Environmental Impact Assessments. United Nations Environment Programme, Regional Office for West Asia, Manama, and World Health Organization, Regional Office for the Eastern Mediterranean, Cairo Ed. S. Lattemann, 168 p.

UNEP/MAP/MEDPOL (2003). Sea Water Desalination in the Mediterranean: Assessment and Guidelines. MAP Technical Reports Series No. 139 UNEP/MAP, Athens.

Van der Merwe R., Hammes F., Lattemann S., Amy G. (2014a). Flow cytometric assessment of microbial abundance in the near-field area of seawater reverse osmosis concentrate discharge. *Desalination*, **343**, 208–216.

van der Merwe R., Röthig T., Voolstra C. R., Ochsenkühn M. A., Lattemann S. and Amy G. L. (2014b). High salinity tolerance of the red sea coral *fungia granulosa* under desalination concentrate discharge conditions: an in situ photophysiology experiment. *Frontiers in Marine Science*, **1**, 2014. doi: 10.3389/fmars.2014.00058

Vars S., Johnston M., Hayles J., Gascooke J., Brown M., Leterme S. and Ellis A. (2013). 29Si{1H} CP-MAS NMR comparison and ATR-FTIR spectroscopic analysis of the diatoms chaetoceros muelleri and thalassiosira pseudonana grown at different salinities. *Analytical and Bioanalytical Chemistry*, **405**, 3359–3365.

Voutchkov N. (2011). Overview of seawater concentrate disposal alternatives. *Desalination*, **273**, 205–219.

Walker D. I., Kendrick G. A. and McComb A. J. (1988). The distribution of seagrass species in shark bay, western australia, with notes on their ecology. *Aquatic Botany*, **30**, 305–317.

Walker D. I. and McComb A. J. (1990). Salinity response of the seagrass amphibolis antarctica (Labill.) sonder et aschers.: an experimental validation of field results. *Aquatic Botany*, **36**, 359–366.

Water-world. 22nd-GWI/IDA. International Desalination Association (IDA) and Global Water Intelligence (GWI). http://www.waterworld.com/index.html.

Wiltshire K., Kraberg A., Bartsch I., Boersma M., Franke H.-D., Freund J., Gebühr C., Gerdts G., Stockmann K. and Wichels A. (2010). Helgoland roads, north sea: 45 years of change. *Estuaries and Coasts*, **33**, 295–310.

WWAP (World Water Assessment Programme). (2012). The United Nations World Water Development Report 4: Managing Water under Uncertainty and Risk. Paris, UNESCO. http://unesdoc.unesco.org/images/0021/002156/215644e.pdf (accessed 31 May 2015).

Yacobi Y. Z., Zohary T., Kress N., Hecht A., Robarts R. D., Waiser M. and Wood A. M. (1995). Chlorophyll distribution throughout the southeastern mediterranean in relation to the physical structure of the water mass. *Journal of Marine Systems*, **6**, 179–190.

Yermiyahu U., Tal A., Ben-Gal A., Bar-Tal A., Tarchitzky J. and Lahav O. (2007). Rethinking desalinated water quality and agriculture. *Science*, **318**, 920–921.

Chapter 9

Charting the future course for reverse osmosis and nanofiltraton membranes – opportunities and challenges

Steven J. Duranceau

Associate Professor and Director of the Environmental Systems Engineering Institute in the Civil, Environmental, and Construction Engineering Department at the University of Central Florida in Orlando, Florida

Corresponding author: Steven.Duranceau@ucf.edu

9.1 EMERGING NEEDS

Available water supplies continue to be degraded due to the human experience. The availability of environmental resources on a global scale will diminish in the future, as population growth generates consumers that deteriorate Earth's air, land and water quality. New threats are emerging in the form of increased severity of natural disasters, species invasions, and the global transport of trace inorganic, organic and microbiolical pollutants. As world urbanization and industrial development grows, so does fresh water use. Along with this increase in water use, the availability of traditional water supplies is declining while the costs of these supplies are on the rise; hence future technologies will be needed to overcome these many challenges water and wastewater utilities will face in an effort to operate without interruption in the future.

9.1.1 Existing desalination technologies

Ocean water desalination is a vast water source that remains untapped in significant quantities. Because ninety-seven percent of the Earth's water takes the form of salt water in oceans, seas, and lakes, desalination technologies continue to gain credibility as a drought-proof water treatment means. Given a pressing worldwide water shortage, a need exists for alternative desalination technologies that would provide affordable reliable sources of water available for the world's growing population, as well as to meet its industrial and agricultural needs. Desalination technologies such as reverse osmosis (RO), nanofiltration (NF) and electrodialysis reversal (EDR) have proven to be effective in removing a number of dissolved contaminants from water and wastewater supplies but can require a significant electrical demand and often experience fouling problems uncommon with more conventional water and wastewater treatment methods (Duranceau & Taylor, 2010). Newer technologies will hence be needed to transcend current methods for the future, and innovative methods will be required to provide the power required to sustain these technologies. Potential green energy is available wherever one finds a river that flows into a sea, equivalent to the energy contained in a 900 ft high waterfall (Salter, 2005). Emerging membrane technologies will take advantage of these natural energy sources, providing opportunities to solve challenges faced in desalination markets.

9.1.2 Balancing acute and chronic water quality risks

There are alternatives for augmenting, extending, or increasing existing freshwater supplies. Conservation has been, and will continue to be, the simplest and more cost-effective means of managing water demands; however, there are limitations in the amount of water that can be conserved. Wastewater reclamation and reuse can replace existing freshwater supplies for non-potable and some limited indirect potable uses; however, salinity can still impact reuse options, requiring desalination to make up part of a direct potable reuse portfolio. Since people require drinking water to live, their exposure to any contaminant in a drinking water system is involuntary. Newer more stringent and comprehensive regulations will further motivate the water industry to develop innovative methods to ensure compliance. Continuous technological advancements that allow investigators to detect sub-nanogram contaminants will translate to more sophisticated regulations intended to protect public health from all known or perceived forms of acute and chronic toxicity. The added components of cost and risk assessment that includes stakeholder-negotiated rule making processes further add to the complexity of environmental regulations.

9.1.3 Emerging membrane technologies

There are emerging desalination technologies that may provide the needed solutions to resolve the water issues of the present and of the future. Forward osmosis (FO), pressure retarded osmosis (PRO), and capacitive deionization (CDI) are emerging desalination technologies that may provide effective solutions to the cost barriers associated with traditional desalination processes. The advent of nanocomposite membrane formulations and hollow fiber nanofiltration manufacturing will further aid in the advancement of the future application of membrane technologies for a variety of water treatment uses. Bioelectrical membrane desalination cells may provide a source of renewable energies to allow many technologies to advance unabated. Additionally, combining artificial intelligent systems with desalination technologies could provide an additional tool to aid is reducing the cost and providing enhanced solutions for complex operations currently faced. This chapter will explore these opportunities and identify some of the challenges that remain to be solved so that these emerging technologies can aid in overcoming the many ongoing societal water dilemmas.

9.2 FORWARD OSMOSIS AND PRESSURE RETARDED OSMOSIS

Forward osmosis (FO) and pressure retarded osmosis (PRO) are, as reported by McGinnis and Elimelech (2008), emerging membrane-based separation technologies that have the potential to be innovative, sustainable, and affordable alternatives (or enhancements) to RO, NF and EDR because of their ability to utilize the green energy available in natural systems. FO and PRO rely on osmotically-driven water flux across a semi-permeable dissolved solids rejecting membrane. The term 'osmosis' describes the natural diffusion of water through a semi-permeable membrane from a solution of a lower concentration to a solution with a higher concentration.

RO, FO and PRO are common in that they use semi-permeable membranes to separate dissolved solutes from water. The semi-permeable membrane acts as a barrier that allows small molecules such as water to pass through, while blocking larger molecules like salts, organics, and proteins, as well as viruses, bacteria, and other pathogenic material. FO and PRO exploit the osmotic pressure difference that develops when a semi-permeable membrane separates two solutions of different concentrations. Instead of employing hydraulic pressure as the driving force for separation in the RO process, FO and PRO use the osmotic pressure gradient across the membrane to induce a net flow of water through the membrane into what is referred to as the 'draw' solution, thus efficiently separating the freshwater from its solutes. Driven by an osmotic pressure gradient, FO and PRO do not require significant energy input as is required for RO.

In osmosis, the osmotic pressure itself is the driving force for mass transport. The osmotic potential depends on the molar concentration, not the weight of dissolved species. FO and PRO use the osmotic pressure differential $\Delta\pi$ across the membrane, rather than hydraulic pressure differential Δp (as in RO). The FO process results in dilution of a highly concentrated stream. The concentrated solution on the permeate side of the membrane is the source of driving force in the FO process. The flux direction of the permeating water in FO, PRO and RO is demonstrated in Figure 9.1 (McGinnis & Elimelech, 2008; Achilli, Cath & Childress, 2009; Duranceau & Taylor, 2010). The governing equation describing water transport through FO and PRO membranes and the volumetric flux of water into the draw solution compartment of osmotic pumps can be described by the relationship first reported by Kedem–Katchalsky (1958, 1963) and as shown below:

$$F_w = k_w(\Delta p - \Delta\pi)$$

Where F_w is the water flux, k_w is the water mass transfer coefficient, Δp is the applied pressure, and $\Delta\pi$ is the osmotic pressure. For FO, Δp is zero; for PRO $\Delta\pi > \Delta p$.

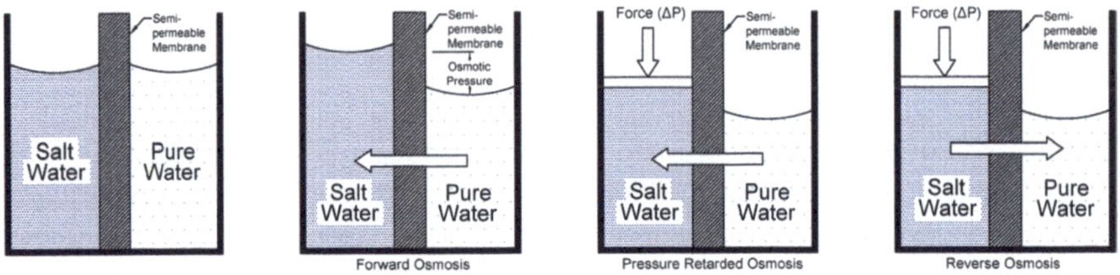

Figure 9.1 Solvent flows in FO, PEO, PRO and RO (*Source*: https://en.wikipedia.org/wiki/Forward_osmosis#/media/File:Osmotic_Processes_Diagram.jpg).

Figure 9.2 depicts the differences between PRO and FO. In the FO application for desalination, the active layer of the membrane faces the feed solution and the porous support layer faces the draw solution. However, in PRO application, the porous support layer faces the feed solution and the active layer of the membrane faces the draw solution. The chemical potential ($\Delta\mu_w$), which is determined by the osmotic pressure difference across the membrane, is the effective driving force in making energy and fresh water. This potential is usually lower than the potential generated by the osmotic pressure difference between the bulk feed and bulk draw solution due to external concentration polarization (McCutcheon, McGinnis & Elimelech, 2006).

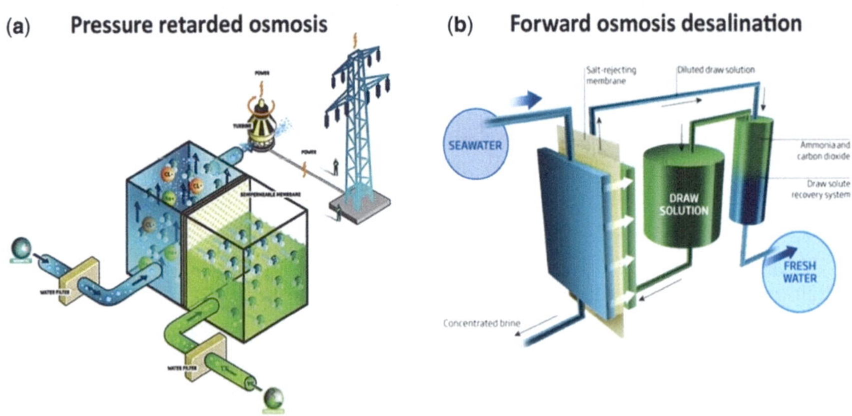

Figure 9.2 Schematic representation of (a) pressure retarded osmosis and (b) forward osmosis desalination (*Source*: Alsvik and Hagg, 2013).

The main advantages of using FO and PRO are that they can be operated at low or with no hydraulic pressure difference across the membrane, they possess a high rejection for a wide range of contaminants. Because the only pressure involved in the FO and PRO processes are due to flow resistance in the membrane module, it may have a lower membrane fouling tendency than pressure-driven membrane processes. The ideal material of membranes for FO and PRO would possess a high active-layer density and result in high solute rejection; a thin membrane film having minimum porosity of the support layer would minimize internal concentration polarization, and therefore, provide a higher water flux. In addition, it would be desired that the membrane be hydrophilic for enhanced flux and reduced membrane fouling; an important aspect of these alternative membranes would be high mechanical strength to sustain hydraulic pressure when used for PRO.

Different module configurations for membrane technologies historically have included plate-and-frame, spiral-wound, tubular, and bag configuration (Duranceau & Taylor, 2010). Plate-and-frame modules are the simplest device for packing flat sheet membranes and can be constructed in different sizes and shapes ranging from lab-scale to full-scale systems. The limitations of plate-and-frame elements are lack of adequate membrane support and low packing density. Spiral-wound configurations have been successfully modified for FO mode by placement of an additional glue line at the center of the membrane envelope that provides a path for the feed to flow inside the envelope. The feed flows into the first half of the perforated central pipe, is then forced to flow into the envelope, and then flows out through the second half of the perforated central pipe. According to Yip and colleagues (2011) the limitation of spiral-wound FO configuration is the difficulty to induce flow into the permeate channels for the purpose of cleaning or backwashing. In practical application, and the use of tubular membranes, tubes or hollow fibers for continuously operated FO and PRO processes may have some advantage as compared to the other configuration. This is because high hydraulic pressures can be supported without deformation, and can be packed easily in bundles directly inside a holding vessel. Also, hollow fiber membranes do not require thick support layers as is required for flat sheet RO membranes, thereby reducing internal concentration polarization and enhancing membrane performance.

With PRO, the water potential between fresh water and sea water corresponds to a pressure of 26 bars. This pressure is equivalent to a column of water (hydraulic head) of 885 feet (270 meters) high (Salter, 2005). However, the optimal working pressure is only half of this, ranging from 11 to 15 bar (U.S. House of Representatives, 2002). This method of generating power was invented in 1973 (Patent Application 1973) and first demonstrated in Norway with a 4 kW capacity as reported by Gartner (2009) and later confirmed by Achilli and Childress (2010). According to Pankratz (2010) it is estimated that 1600 TWh could be generated worldwide; however, no operation plant is currently known to exist at this time.

In another study, the use of a custom-made laboratory-scale membrane module enabled the collection of experimental PRO data. Results obtained with a flat-sheet cellulose triacetate (CTA) FO membrane and NaCl feed and draw solutions closely matched model predictions by Achilli and colleagues (2009). Maximum power densities of 2.7 and 5.1 W/m^2 were observed for 35 and 60 g/L NaCl draw solutions, respectively, at 970 kPa of hydraulic pressure. Power density was substantially reduced due to internal concentration polarization in the asymmetric CTA membranes, and to a lesser degree, to salt passage. External concentration polarization was found to exhibit a relatively small effect on reducing the osmotic pressure driving force. However Wang and coworkers (2010) demonstrated the development of a thin-film composite FO hollow fiber with an ultrathin RO-like skin layer on either the inner surface or outer surface of a porous hollow fiber substrate, achieved a 32.2 L/m^2-h water flux using a 0.5 M sodium chloride draw solution. Hence, this latest work demonstrates that the performance of FO can be increased.

9.2.1 Applications in desalination and water purification

FO has thus drawn much attention with applications developed in various fields, such as desalination, wastewater treatment, pharmaceutical and juice concentration, and even power generation and potable-water reuse in space. Recent studies of applications of FO in the field of water and wastewater purification include landfill leachate treatment, concentration of dilute industrial wastewater and potable reuse in advanced life support systems, and concentration of liquids from anaerobic sludge digestion at a domestic wastewater treatment facility.

One viable commercial drinking water application for FO has been developed by Hydration Technology Innovations, which has developed a small water purification bag (hydration bag) used mostly by the military and for emergency use (Butler *et al.* 2013). The bag is made of a semi-permeable FO membrane and contains a solid glucose draw solution. When immersed in contaminated water from puddles and ponds, purified water diffuses into the bag, slowly diluting the solid draw solution to produce a sweet drink. The sugar solution inside the pouch has a high osmotic pressure and, over time, clean water flows from the contaminated side to the sugar side to produce an energy drink, leaving the undesirable feed constituents behind. The diluted draw solution may then be ingested directly. A point of additional interest with such bags is that they may be readily used to recycle urine, greatly extending the ability of a backpacker or soldier to survive in arid environments. This process may also, in principle, be employed with highly concentrated saline feedwater sources such as seawater, as one of the first intended uses of FO with ingestible solutes was for survival in life rafts at sea.

One of the start-up companies in this new commercial arena is Oasys Water Inc., which uses FO technology developed at Yale University (McCutcheon, McGinnis & Elimelech, 2006; Oasys, 2013). The developers claim that its technology can produce desalinated water at less than half the cost and using 90 percent less energy than RO. This area of current research in FO involves the direct removal of draw solutes by thermal means. This process is typically referred to as the 'ammonia–carbon dioxide' FO process, as the draw solutes are salts formed from the mixing of ammonia and carbon dioxide gases in water.

As an example application, seawater is the feed and the accompanying draw solution is formed by using highly soluble ammonia and carbon dioxide gases, which produces a strong enough osmotic pressure to extract water from seawater. These salts can reach high concentrations, particularly as the ratio of ammonia to carbon dioxide is increased. An especially convenient property of these salts is that they readily dissociate into ammonia and carbon dioxide gases if a solution containing them is heated. The diluted draw solution is heated to about 60°C at 1 atm pressure to decompose the ammonia and carbon dioxide and the pure product water is then recovered after degassing via column or membrane distillation. Once the concentrated draw solution is used to effect separation of water from the FO feed solution, the diluted draw solution is directed to a reboiled stripper (distillation column) and the solutes are completely removed and recycled for reuse in the FO system (Lee, Baker & Lonsdale, 1981). An FO system of this type thereby effects membrane separation of water from the FO feed, using heat as its primary energy source. The quality of heat used by this process can be as low as 40°C. If FO of this type is used in a cogeneration environment (waste heat from a power plant, for example), its energy cost can be greatly reduced compared to RO (Nicoll, 2011).

A FO water plant developed by Modern Water producing 4.2 m^3/day of drinking water to the residents of a remote village in the Al Wusta region of the Sultanate of Oman as reported by Thompson and Nicoll (2011), has been operating at a 35 percent water recovery consuming 4.9 kWh/m^3 The membranes, manufactured by Modern Water (Bramley House, The Guildway, Old Portsmouth Rd, Guilford, GU3 lLR, United Kingdom) and used in the Oman plant are reportedly chlorine tolerant. Consequently, FO appears to be a viable process that should continue to be developed for desalination in the future.

9.2.2 Wastewater treatment

Another application of FO is treatment of impaired water and wastewater, with specific opportunities for direct potable reuse of treated sewage. Research carried out by the Colorado School of Mines, in Golden, Colo., has demonstrated the application that utilizes treated sewage or another impaired water source in a FO system to dilute seawater prior to desalination with RO, which is shown in Figure 9.3.

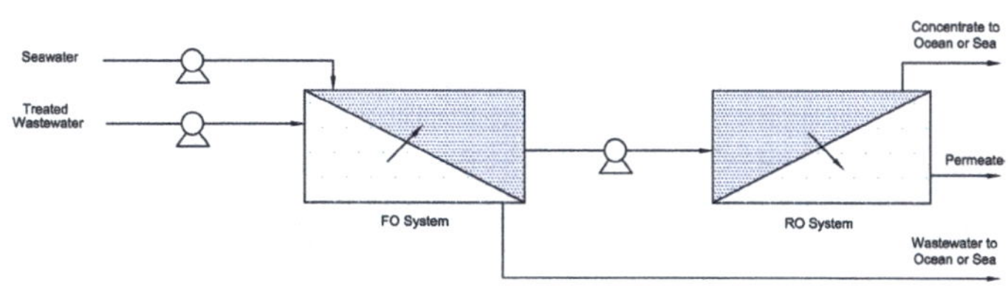

Figure 9.3 Schematic for an FO-RO for seawater and wastewater treatment.

With RO seawater desalination, most of the energy is used to overcome the osmotic pressure of the salt water, and this high osmotic pressure limits the amount of pure water that can be recovered. In coastal areas, treated wastewater is being discharged to the ocean, wasting a valuable resource. By using seawater as the draw solution and passing wastewater through the FO membrane prior to ocean discharge, purified wastewater can be used to dilute seawater prior to feeding to an RO system for desalination. The osmotic pressure of the seawater is then reduced which lowers the energy required for desalination and improves water recovery. Disposal of RO concentrate back to the ocean has always caused some environmental concerns, so a more dilute combination of FO and RO wastes may also overcome many of these concerns.

9.2.3 Moving forward with FO and PRO

The concept of utilizing the osmotic effect in many fields of science and engineering including water and wastewater treatment, seawater/brackish water desalination, and electric power production has received increasing attention over the last years. FO and PRO have the potential to convert the osmotic pressure difference between fresh water (i.e. river water) and seawater to electricity. Moreover, these technologies can recover energy from highly concentrated brine in seawater desalination. The increasing attention in FO and PRO is directed towards membranes with improved properties that will make it possible to commercialize these processes in the near future. For example, the development of nanofiltration-like FO membranes that have demonstrated high rejection of multivalent ions such as magnesium and sulfate with little to no rejection of monovalent ions such as sodium and chloride. Table 9.1 highlights some of the research accomplished in this area.

Table 9.1 Summary of research developing NF-like FO membranes using DI feed water.

Membrane Type*	Water Flux (L/m^2 h)	Salt Flux to Water Flux Ratio (g/L)	Draw Solution ($MgCl_2$)	Reference
6-layer LBL membrane	14.6	0.034	0.05 M	Liu *et al.* (2013)
	25.9	0.066	0.1 M	Liu *et al.* (2013)
	40.5	0.20	0.5 M	Liu *et al.* (2013)
	18.44	0.11	0.5 M	Liu *et al.* (2013)
	20.66	0.14	1.0 M	Liu *et al.* (2013)
CA double-skinned HF	15.7	0.06	0.5 M	Su *et al.* (2012)
PAI-PEI crosslinked HF	13.13	0.70	0.5 M	Su *et al.* (2012)
LBL flat sheet	25.1	0.57	1.0 M	Saren *et al.* (2011)
Dual-layer PBI HF	15.6	0.038	2.0 M	Yang *et al.* (2009)
PAI-PEI crosslinked HF	8.36	0.30	0.5 M	Setiawan *et al.* (2011)

Table 9.1 Summary of research developing NF-like FO membranes using DI feed water (*Continued*).

Membrane Type*	Water Flux (L/m^2 h)	Salt Flux to Water Flux Ratio (g/L)	Draw Solution ($MgCl_2$)	Reference
LBL flat sheet	22	0.30	1.0 M	Saren *et al.* (2011)
Dual-layer PBI HF	7.5	0.053	2.0 M	Yang *et al.* (2009)
PES HF	32.2	0.11	0.5 M	Wang *et al.* (2010)

**Notes:* LBL = layer-by-layer deposition; CA = cellulose-acetate; PAI = poly (amide-imide); PEI = polyethyleneime; PBI = polybenzimidazole; HF = hollow fiber; PES = polyethersulfone.

According to the Xu and coworkers (2010) as well as Wang and researchers (2010) hollow fiber membranes possess significant advantages over flat sheet membranes in the FO process due to their high packing density, favorable flow pattern and self-supported structure. Liu and colleagues have fabricated FO hollow fiber membranes in a layer by layer process polyelectrolyte technique that where the active layer faces the draw solution, water fluxes on the order of 25.9 L/m^2 h can be achieved using 0.1 M magnesium chloride draw solutions.

In contrast to FO that takes place when the hydraulic pressure difference is zero, PRO will occur where the applied pressure difference is between zero and the flux reversal point, as illustrated in Figure 9.4. At the point where the applied pressure difference is greater than the osmotic pressure difference, RO flow will occur. This osmotic pressure difference can be used to generate a flow of water which will turn a turbine such that FO and PRO could be viable sources of renewable energy in the future (Lee, 1981).

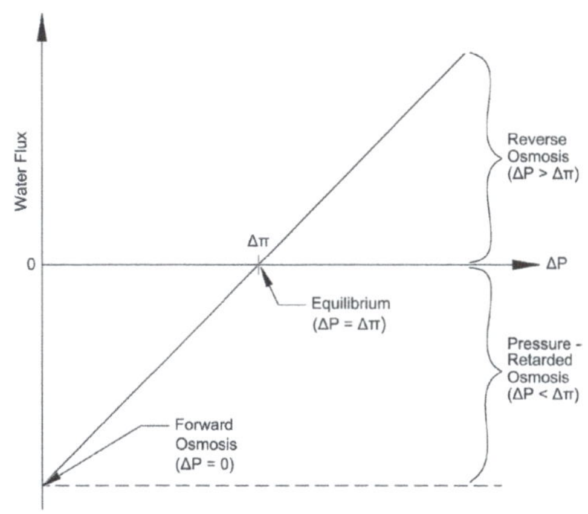

Figure 9.4 Graphical summary of the direction and magnitude of water flux as a function of applied pressure in FO, PRO, and RO (adapted from Lee, 1981, with permission from Elsevier).

9.3 NOVEL NANOCOMPOSITE MEMBRANES

It is well accepted that thin-film polyamide (PA) composite membranes are the widely accepted membrane material type providing accepted thermal, chemical, and hydrolytic stability in their application as reverse osmosis and nanofiltration membranes treating aqueous mediums. However its incompatibility with disinfectants and susceptibility to fouling limit this application. Although many researchers have taken advantage of the high reactivity of the end groups to modify the surface properties of commercial PA membranes through free radical grafting of monomers and plasma treatment their effect on significant membrane performance benefits remain limited (Mukherjee, Kulkarni & Gill, 1996; Wu *et al.* 1997; Belfer *et al.* 1998; Gilron *et al.* 2001; Ghosh, 2008; Kim, Shin & Lee, 2007; Fujimoro *et al.* 2010; Rana & Matsuura, 2010). Alternative membrane formulation and manufacturing techniques could provide our future with more affordable membrane materials of construction.

9.3.1 Mixed matrix membranes

Mixed matrix membranes (MMMs), also known as composite membranes, are produced by embedding inorganic fillers in a polymeric matrix. In MMMs a small filler material, such as a zeolite, is dispersed throughout a larger polymeric matrix has helped bring the development of advanced membrane materials with improved mechanical, chemical, and thermal stability, as well as enhanced separation capacity. Flux can also be increased by using porous nanomaterials in mixed-matrix membranes.

Research has shown that membrane water permeability, anti-fouling property, thermal and mechanical stability of pressure-driven membranes can be potentially enhanced by incorporating zeolite nanomaterials into the active polyamide layer of thin-film composite membranes (Jeong *et al.* 2007; Lind *et al.* 2009; Bae *et al.* 2009; Fathizadeh *et al.* 2011; Ma *et al.* 2012; Ma *et al.* 2013; Kiesow *et al.* 2013). Zeolites are crystalline aluminosilicate materials with uniform sub-nanometer structure is composed of 3–8 Å pores. Zeolite nanoparticle exhibits 0.4 nm pores and MFI-type has 0.56 nm pores which are lower than the size of the common ions presents in contaminated water but bigger than the size of a water molecule (0.278 nm).

Similar benefits for MMMs have been noted when using embedded silica nanoparticles (Jadav & Singh, 2009; Huang *et al.* 2012; Wu, Tang & Wu, 2013), embedded titanium oxide nanoparticles (Lee *et al.* 2008), and silver nanoparticles (Kim *et al.* 2012; Liu *et al.* 2013). Humplik and colleagues (2011) have shown that nanostructured materials are well-defined for used as filtration membranes. However, poor dispersion of the filler in the polymer matrix and incompatibility issue between the polymer and filler might result in poor selectivity of the MMMs (Shi, Yang & Chung, 2012; Zornoza, 2013).

9.3.2 Carbon nanotube membranes

Membranes comprising sub-nanometer diameter CNT can desalt water when used as RO membranes. The narrow pores reject ions effectively while conducting water up to a thousand times the rate of commercially available thin-film composite RO membranes. The primary causes of salt rejection and water transport in these studies is the narrow, smooth, nonpolar nature of the CNTs. As reported by Chavan and colleagues (2012), CNTs can be conceptually described as a one dimensional analog of zero-dimensional fullerene with a micrometer-scale graphene sheet being rolled into a cylinder of nanoscale diameter.

Ratto and coworkers (2010) compared membranes fabricated with and without embedded CNTs. Figure 9.5 presents a representation of a thin-film composite membranes embedded with carbon nanotubes. The results demonstrated enhanced flow generated by the CNT pathways, a near doubling of water flux (44l m^{-2} day^{-1} bar^{-1} as compared with 26l m^{-2} day^{-1} bar^{-1}), with a slightly higher salt rejection (97.69%) as compared to membranes without CNTs; however, there is a need for studies on much larger surface area membranes.

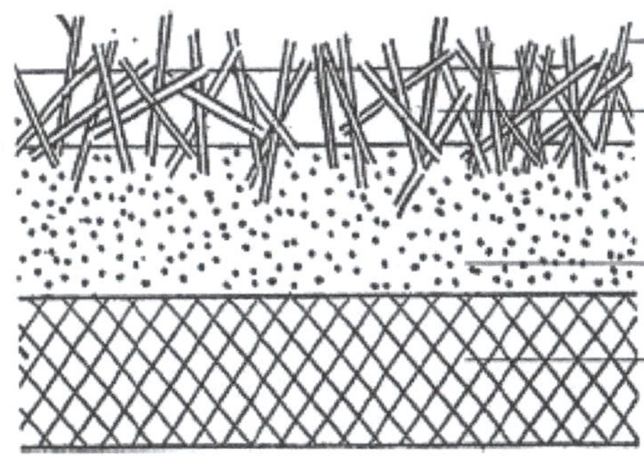

Figure 9.5 Representation of a thin-film composite membrane with embedded carbon nanotubes within the active polyamide layer (*Source*: Ratto, Holt, Szmodis 2010).

The use of graphene in the water filtration process had been demonstrated in the early 2010s. This involved atom-thick sheets of carbon, able to separate salt from water using much lower pressure, and hence, much lower energy. This was due to the extreme precision with which the perforations in each graphene membrane could be manufactured where each hole was the perfect size for a water molecule to fit through. An added benefit was the high durability of graphene, potentially making desalination plants more reliable and longer-lasting. Figure 9.6 reveals a field emission-scanning electron microscopy image of CNTs.

9.3.3 Application

With respect to MMMs, embedding the nanoparticles changes the properties of the membrane, making it hydrophilic, or water-attracting, so that water passes through more easily; in addition, the membrane retains its ability to filter out contaminants. As it turns out, zeolite and silica based thin-film nanocomposite membranes appear to be more readily adaptable to commercial use due to their similarity to current commercial RO membranes than carbon nanotube based membranes where alignment of CNT into the active thin-film layer remains a challenge.

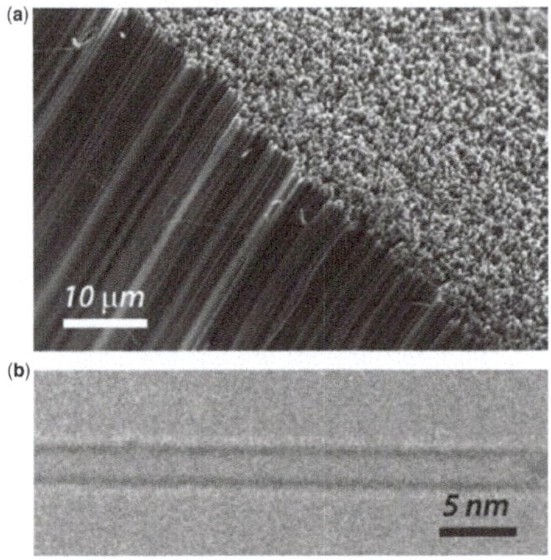

Figure 9.6 (a) Field emission-scanning electron microscopy image of a single walled carbon nanotube forest and (b) image of a single walled CNT (Reproduced from Miller *et al.* 2012 with permission of The Royal Society of Chemistry).

Incorporation of nano-based technologies into membrane formulation could offer an attractive alternative to prepare fouling-resistant membranes. Despite the many practical concerns related to MMMs, Jeong (2007) reported the development of thin-film nanocomposite membranes that incorporated nanoparticles into an interfacially formed polyamide thin film. This new membrane material that was initially tested in brackish water doubled membrane permeability while maintaining salt rejection at industry standards. Kurth and colleagues (2011) demonstrated the ability to increase the output of a small 84.5 m^3/d plant to 102 m^3/d while reducing the operating pressure from greater than approximately 55.5 bars to approximately 50.5 bars. In May 2012, according to LG NanoH2O (2012) a one million gallon per day RO seawater desalination plant on the island of St. Kitts, West Indies, operating at 46 bar (665 bar), 15.8 lmh (9.3 gfd) and 38 percent water recovery was placed into operation.

9.4 CAPACITIVE DEIONIZATION

9.4.1 Electrosorption

Capacitive deionization (CDI), also known as electrosorption, was first postulated by Claude and colleagues (1966) and effectively demonstrated by Johnson and Newman

(1971) as a promising desalination process (Oren, 2008). Figure 9.7 presents the basic concept behind CDI technologies. In CDI, feed water flows through spaces between oppositely charged porous electrodes. The opposing charges of the two electrodes create a low-voltage electromagnetic field that attracts the aqueous ions. The ions are adsorbed within the pores of the electrodes, and remain adsorbed until the current is shut-off when the ions can then be flushed from the process. Cycling the current on and off allows desalinated water to be the primary output of the process.

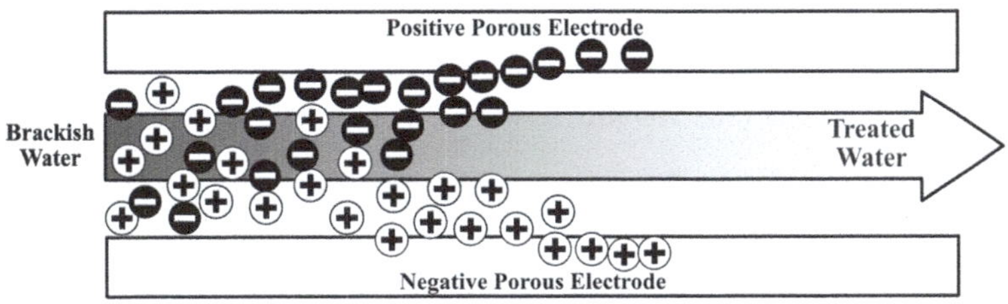

Figure 9.7 Schematic representation of capacitive deionization.

Capacitive deionization is based on an electrostatic process operating at low voltages and pressures (Zhao *et al.* 2012, Kuipers *et al.* 2013). Produced water is pumped through an electrode assembly. Ions in the water are attracted to the oppositely charged electrodes. The negative electrodes attract positively charged ions (cations) such as calcium (Ca^{2+}), magnesium (Mg^{2+}) and sodium (Na^{1+}), whereas the positively charged electrodes attract negative ions (anions) such as chloride (Cl^{1-}) and nitrate (NO_3^{1-}). When the electrodes' capacity is reached, the water flow is stopped and the polarity of the electrodes is reversed. This causes the ions to move away from the electrodes, where they had previously accumulated. The ions are thus released and a concentrated brine stream is flushed from the system. The cleaned water passes through the unit within cycling of the electrode voltage. In practice, for a gallon of brackish water fed to the a CDI process, more than 80% emerges as deionized potable water, and the remainder is discharged as a concentrated brine solution containing virtually all of the salts in the feed.

9.4.2 Considerations

Unlike reverse osmosis or nanofiltration membrane technologies, CDI does not require elevated pressures. Based on current industry trends, pairs of high surface area carbon electrodes that comprise the CDI technology are held at a potential difference of 1.3 V to 1.5 V (Zhao *et al.* 2012, Kuipers *et al.* 2013). However, as is the case with RO and NF technologies, scaling by precipitation of sparingly soluble salts, organic fouling and biofouling remain factors in CDI's applicability; consequently, pretreatment is thus required, particularly with respect for dissolved organic carbon. As is recognized in electrodialysis and electrodialysis reversal processes, and as described by Duranceau and Taylor (2010), charged foulant particles are also attracted to and held onto membranes of opposite charge; similarly, fouling by charged substances such as dissolved organic matter or the development of biofilms by attached extra-cellular polymers may impact CDI applicability (Oren, 2008).

The primary advantage claimed by proponents of CDI has been its low operating cost, which is about one third that of the main competitor, reverse osmosis (RO). However, as reported by Oren (2008), desalination with CDI is currently limited to water of lower salinities indicating that the application of CDI for seawater treatment that the energy consumption as measured by voltage for CDI is much larger than that for seawater RO processes.

9.4.3 Innovations and limitations

Additionally, as indicated in Zou and colleagues (2008), electrode material properties for CDI applications require further development. Improvements to CDI have been documented at the laboratory scale by Suss and coworkers (2012) where a flow-through aerogel electrode design enabled reduction in processing time desalinating higher salinity waters (up to 70 mM NaCl) per charge. However, it was recognized that additional research was required for optimization of the process, particularly with respect to fouling. For example, the impact of hydrogen sulfide on CDI performance has not been reported.

One innovative modification to CDI that has been recently recently reported by Kuipers and Porada (2013) may help advance CDI technologies. As was noted in the work of Zou *et al.* (2008) and Suss *et al.* (2012), the voltage applied to CDI is achieved by a direct physical connection between an external power source and the current collectors or electrodes. However, Kuipers and Porada (2013) have demonstrated as a proof-of-concept wireless desalination using porous carbon electrode cells. In this application, the electrodes are charged at constant current and an external transmitting coil that induces a magnetic field is received by a circuit which charges the electrodes one relative to the other. These aforementioned innovative methods, limited to laboratory-scale research, could enhance the opportunities for CDI technologies in the future. Although the inclusion of ion exchange membranes in front of the porous electrodes could improve the efficiency of the process, possibly upwards of 90% efficiency, questions remain as to the ultimate scalability of CDI technologies.

9.5 HOLLOW FIBER NANOFILTRATION

9.5.1 Spiral-wound membrane softening membranes

Softening membranes were first introduced in 1976 and marketed as RO softening membranes. The technology became more popular in the 1980s when FilmTec

Corporation introduced polyamide NF spiral-wound elements as reported by Cadotte and colleagues (1988). NF membranes are now used for a variety of applications including the removal of disinfection byproduct precursors and pesticides in drinking water. Nanofiltration membranes often have a molecular weight cut-off between 200 and 1000 Daltons, which enables NF membranes to target rejection of divalent ions. Additionally, NF membranes are generally able to remove 95% of naturally occurring color and DBP precursors (Van der Bruggen & Vandecasteele, 2003; Watson & Hornburg, 1989). Dissolved organic carbon with a molecular weight greater than 400, hardness ions, and limited monovalent ions can be removed by tight nanofiltration membranes (Hilal *et al.* 2004). Spiral-wound NF processes operate in cross-flow filtration where the feed water flows tangential to the membrane surface and produces two output streams. Concentrate refers to the membrane output stream that contains water that has not passed through the membrane and includes the concentrated constituents rejected by the membrane. Permeate refers to the portion of the feed stream that does pass through the membrane. Spiral-wound NF membranes are designed to operate at a constant flux by increasing the feed pressure of the system. The operating pressures and rejections are monitored over time to assess membrane performance.

9.5.2 Hollow-fiber ultrafiltration membranes

Ultrafiltration is a low pressure driven membrane filtration technique primarily used to remove particles and microorganisms by the mechanisms of straining or size exclusion. The most common membrane configuration for UF is hollow fiber (Howe *et al.* 2007). There are two different flow directions for hollow-fiber membranes: flow can be from inside out or outside-in. In the case of outside-in flow direction, there is more flexibility in the amount of feed to flow around the hollow fibers, whereas the inside-out flow direction has to consider the pressure drop through the inner volume of the hollow fibers. Inside-out flow, however, offers much more uniform flow distribution through the lumen of hollow fiber compared to the outside-in flow as found by Xu and coworkers (2008) but inside-out flow patterns are more susceptible to plugging (USEPA, 2005).

UF membranes can be operated in either cross-flow or dead-end filtration modes. In the case of cross-flow filtration, there are three streams, one inlet and two outlet streams referred to as the feed, filtrate, and retentate. In this process, the components in the water are separated by a semi-permeable membrane through application of pressure and flow parallel to the membrane surface. The cross-flow mode of filtration has a lower recovery rate compared to dead-end filtration (USEPA, 2005). In dead-end filtration, feed water is forced through the hollow fibers, which capture and retain particles. Dead-end filtration involves one inlet and outlet stream resulting in 100% of the feed water passing through the fibers without a recycle stream. For dead-end filtration, periodic backwashes with filtrate water and/or cleaning chemicals are required to recover membrane productivity and prevent irreversible fouling.

Fouling of UF membranes can be due to several issues including: (1) operation of membranes beyond a critical flux value (AWWA, 2005); (2) foulant compaction that reduces the effectiveness of backwashing (Smith *et al.* 2006); and (3) membrane material compatibility with feed water (Liu *et al.* 2001). A UF system can be considered to be fouling if the operating parameters such as feed pressure, temperature and flow rates are held constant but the flux rate or mass transfer coefficient (MTC) of water through the membrane is decreasing (Cheryan, 1998). The flux rate and mass transfer coefficient can be calculated using Equations 1–5. Once a membrane experiences fouling, it is common to clean the membrane by performing a chemically enhanced backwash (CEB) or a clean in place (CIP) to restore hydraulic performance.

A fair amount of research demonstrating the removal capabilities of UF membranes has been published in the literature. Hirata and Hashimoto (1998) demonstrated a > 7 log removal for *Cryptosporidium* oocyst using hollow-fiber UF membranes. Research conducted by Jacangelo, Trussell, and Watson (1997) found removals greater than 6.8 log for MS2 bacteriophage agreeing with earlier research conducted by Olivieri and coworkers (1991) and Simmons and colleagues (2001). Suspended solids, colloidal, and dissolved organic compounds can also be removed depending on their molecular mass and on the molecular mass cut-off of the membrane. UF membranes have pore size of approximately 0.01–0.05 μm in size and have a molecular weight cut-off of about 10,000–500,000 Daltons (AWWA, 2005). These characteristics enable UF membranes to remove particulate matter, bacteria, viruses, and parasites including Cryptosporidium and Giardia (Jacangelo, Adham, & Laîné, 1995). In well-designed and operated systems, ultrafiltration membranes can consistently produce filtered water with turbidity values below 0.05 NTU and can result in SDI < 1 (Duranceau & Taylor, 2010). UF membranes often operate at transmembrane pressures between 3–15 psi (AWWA, 2005). However, UF membranes have historically not been designed to remove dissolved solids such as divalent ions, small organics, and hardness but rather are implemented as a pretreatment step to NF or RO or as a replacement to media filtration in conventional treatment. Advancing technology has led to more developed hollow fiber membranes, with the potential to remove dissolved organics contributing to color and disinfection byproduct precursors.

9.5.3 Emerging hollow-fiber nanofiltration membrane technology

The application of NF membranes has typically been limited to high-pressure spiral wound configurations. This configuration typically has a lower capital cost relative to hollow fiber membranes. However, compared to hollow fiber membranes, spiral wound nanofiltration membranes are more prone to fouling, more difficult to clean, and often require advanced pretreatment when treating water containing colloidal and particulate matter (Sethi & Wiesner, 2000). Sethi and Wiesner (2000) conducted modeling to investigate the different conditions under which the HF-NF configuration or HF-UF and SW-NF configuration would be the optimal selection for cost-effectiveness. Their study was conducted by modeling the operation of both a HF-NF and a HFUF-SWNF application. The model parameters were estimated using operational data collected from a HF-NF pilot conducted in previous research by Jacangelo and Laine (1994). The cost difference between a HF-NF application and a HFUF-SWNF application made the HFUF-SWNF more economical for large scale implementation. However, additional research and development of HF-NF has continued over the past decade and the parameters used in this modeling effort may no longer be valid.

Kiso and colleagues (2002) tested a HF-NF for rejection of pesticides. The HFNF used in their study (referred to as HFNF-1) was a composite membrane with polyamide skin layer and polysulfone (PSF) support layer. The HF-NF-1 demonstrated a salt rejection of thirty-five percent and a magnesium sulfate rejection of ninety-three percent at 0.5 MPa. Their work investigated the rejection properties of HF-NF using hydrophilic compounds that do not adsorb to the membrane material and are often used to evaluate if a membrane rejects solutes through size exclusion. The results of their study indicated that rejection was highly correlated to molecular weight and molecular width. Adsorption of pesticides on the membrane surface was controlled by the hydrophobicity (log P) of the solute. In their work the effects of concentration polarization were not considered.

Darvishmanesh and coworkers (2011) prepared and characterized a polyphenylsulfone (PPSU) HF-NF for organic solvent removal. PPSU has better solvent resistance than polyethersulfone (PES) or PSF. He and colleagues (2008) developed a new type of HF-NF membrane by dip coating a PES support membrane with a sulfonated poly ether ether keytone (SPEEK) layer. Rejection was characterized with dextran mixtures and various salt solutions.

9.5.4 Pilot testing demonstrates partial divalent removals

Duranceau and colleagues (2013) conducted a pilot demonstration of hollow-fiber nanofiltration using Pentair's hollow-fiber HFW 1000 membrane. The HFW HF-NF lumens have diameters of 0.8 mm and a module area of 430 ft^2. The study investigated the treatment of an aerated groundwater that contained elevated levels of sulfate (and natural organic matter). The transmembrane pressure varied over the course of the study from approximately 8 psi to 17 psi. The results from the hollow-fiber pilot testing show an average sulfate removal of 7% at a flux of 13 gal/

ft^2-day, although sulfate removals as high as 9% were observed. The hollow-fiber membrane achieved an average TOC removal of approximately twenty-five percent while producing water with an average turbidity of 0.1 NTU. Additionally, the membrane did not experience fouling under the conditions tested and therefore the use of cleaning chemicals and scale inhibitors were not necessary to maintain production during the pilot operation.

9.5.5 Emergence of HF-NF as a viable technology

A technological 'gap' currently exists in membrane technology in that the removal of dissolved multi-valent ions is currently limited to high-pressure spiral-wound NF modules. A low-pressure hollow fiber membrane that is capable of softening could prove useful to the drinking water industry. Van der Bruggen and colleagues (2003) showed that capillary nanofilters were capable of good performance, having superior water flux for most applications of permeable commercial flat sheet membranes. This proves that the concept of using capillary nanofiltration membranes without an extensive feed pretreatment works well in the case of surface water filtration Technological breakthroughs are needed to enhance the water permeability of hollow fiber membranes while still retaining their separation efficiency.

9.6 MICROBIAL DESALINATION AND REVERSE ELECTRODIALYSIS

The concept of microbial desalination cells (MDC) evolved from microbial fuel cell (MFC) technology. Both MDC and MFC employ bio-electrochemical reactions where oxidation of organic compounds on the anode supplies the electrons and protons that reduce oxygen on the cathode. This reaction produces electricity as electrons flow between the two electrodes. MDCs differ from MFC in that they employ an additional chamber, or chambers, between the anode and the cathode that contains saline water. An anion exchange membrane (AEM) is the boundary of the saline water chamber on the side of the anode and a cation exchange membrane (CEM) lies between the saline chamber and the cathode. Naturally occurring bacteria are used on the anode. These bacteria survive and grow as a result of the addition of organic matter (commonly acetate). Through a bio-electochemical reaction, the bacteria release protons into the water. Protons cannot move to the cathode because they cannot diffuse through the AEM. Therefore, in order to maintain change balance in the anode chamber, anions flow from the middle desalination chamber to the anode. At the cathode, protons are removed from water, so cations move from the middle chamber to the cathode chamber to maintain the charge balance. An illustration of a basic schematic of a microbial electrolysis cell as presented by Nguyen (2014), can be found at http://www.smartplanet.com/blog/thinking-tech/breakthrough-may-lead-to-limitless-supply-of-hydrogen-fuel/.

Reverse electrodialysis ion-exchange membrane stacks in microbial reverse-electrodialysis cells was demonstrated by Cusick, Kim and Logan (2012) to produce salinity-gradient energy from ammonium bicarbonate salt solutions continuously regenerated with waste heat (≥40°C). The maximum power density using acetate was found to approximate 5.6 watts per square meter of cathode surface area. Additionally, Vermaas and colleagues (2013) showed that 95 percent of the available energy when mixing salt and fresh water can be obtained in reverse electrodialysis.

Table 9.2 presents a summary of MDC studies, as adapted from Kim and Logan (2013). The first study published in the literature regarding MDCs was in 2009 by Cao *et al.* who conducted pioneering experiments demonstrating a three chamber, batch operated, MDC (Cao, Huang *et al.* 2009). The first MDCs proposed for water desalination consisted of three chambers: the anode, a middle desalination chamber, and the cathode. In three-chamber MDCs, wastewater in the anode chamber has a much smaller ion concentration than the desalination chamber. Domestic wastewater has an ion content of ~0.5 g/L and rarely exceeds 0.9 g/L of total dissolved solids. Typical seawater in the middle chamber, with a total dissolved solid concentration of 35 g/L, creates a significant concentration gradient across an ion-exchange membrane in contact with the wastewater. Due to this concentration gradient, salt ions in the middle chamber of a MDC are driven to the adjacent anode and cathode chambers.

Since that time a number of studies have been conducted studying MDCs. The highest current density observed in MDCs and as reported by Chen and colleagues (2012) is 8.4 A/m^2. However, the efficiency of the MDC process can be improved by adding additional saline water chambers to the cell and through the use of stacking (USBR, 2014; Kim & Logan, 2013).

MDCs could find their best use as a pretreatment method for RO and NF, as reducing the salinity will proportionally decrease the energy requirement for the downstream RO process. And if fouling can be controlled in MDCs, then the pretreatment may provide an enhancement for the control of RO and NF biofouling. As noted by Saeed and colleagues (2015), MDCs produce their own electricity while treating wastewater and desalinating water simultaneously by exposing the organic matter in waste water to exoelectrogenic bacteria. Certainly partial slip-stream desalination using conventional RO and NF could be integrated into an overall strategy to treat brackish wastewater supplies, possibly for consideration in direct potable reuse.

The MDC systems of the future will be used as a pretreatment system for desalination processes, as well as a stand-alone process for decentralized treatment of wastewater. Although there appear to be advantages of incorporating MDC's into wastewater treatment processes, such a system has not been implemented. This is because this new technology must overcome many technical problems as MDCs are not well understood, and further investigations are required for scale-up.

Table 9.2 Summary of MDC Studies (3C: three-chamber MDC; Stack: sacked MDC; Bipolar: Bipolar MDC; Osmotic: Osmotic MDC).

Reactor Types	Anolyte				Catholyte			Salt Water			Reference
	Sodium Acetate (g/L)	PBS (mM)	Vol. (mL)	COD Removal (%)	Reaction	Buffer	Vol. (mL)	NaCl (g/L)	Vol. (mL)	Salinity Removal (%)	
3C	1.6	50	300	–	$Fe(CN)_6^{3-}$	PBS	100	35	3	93	Cao *et al.* (2009)
3C	4	10	1.0L/d[a]	–	ORR	H_2SO^4	4.3 L/d[a]	30	0.1 L/d[a]	100	Jacobson *et al.* (2011a)
3C	4	10	1.0L/d[a]	–	ORR	H_2SO^4	4.3 L/d[a]	30	0.4 L/d[a]	11	Jacobson *et al.* (2011a)
3C	3	10	5.7L/d[a]	02	ORR	H_2SO^4	5.8 L/d[a]	35	0.2 L/d[a]	94	Jacobson *et al.* (2011b)
3C	3	10	5.7L/d[a]	92	ORR	H_2SO^4	5.8 L/d[a]	35[g]	0.2 L/d[a]	74	Jacobson *et al.* (2011b)
3C	1–2	50	14	77–82	ORR	PBS	14	20	14	63	Mehanna *et al.* (2010a)
3C	2	50	14	38–54	HER[d]	PBS	27	20	14	37	Mehanna *et al.* (2010b)
3C	1	50	100[c]	–	HER[e]	PBS	144	10	17	98.8	Luo and Jenkins (2011)
Stack	1.64	47	1000	–	ORR	PBS	300	20	21	72.1	Chen *et al.* (2011)
Stack	1	100	120	91	ORR	None	72	35	60[a]	44	Kim and Logan (2011)

Stack	1	100	3800	91	ORR	None	2300	35	300[a]	98	Kim and Logan (2011)
3C	–[b]	None	140	53.8	$Fe(CN)_6^{3-}$	PBS	140	14.2[h]	60	66	Luo *et al.* (2012)
3C	1[c]	25	28	74–79	ORR	PBS	14	20	14	34–55	Qu *et al.* (2012)
Bipolar	1	50	30	–	ORR[f]	None	30	10	10	86	Chen *et al.* (2012)
3C	4	10	735	–	$Fe(CN)_6^{3-}$	PBS	735	20	75	41	Zhang and He (2012)
Osmotic	4	10	735	–	$Fe(CN)_6^{3-}$	PBS	735	20	75	60	Zhang and He (2012)
Osmotic	4	10	735	–	$Fe(CN)_6^{3-}$	PBS	735	35[g]	75	63	Zhang and He (2012)

[a] Fed by continuous flow.
[b] Wastewater.
[c] Xylose.
[d] Eap 0.55 V.
[e] Eap 0.8 V.
[f] Eap 1.0 V.
[g] Sea salt.
[h] Mixture of NaCl and $NaHCO_3$.

9.7 ADAPTIVE CONTROL

Artificial intelligence (AI) is the intelligence exhibited by machines or software; these techniques include rule-based expert systems and object-oriented systems (Clive & Raymond, 1991). The emphasis is declarative representation: separating the description (the knowledge) of a process, from the subsequent analysis of that knowledge by an inference machine. Current online industrial applications are generally built within shells, which package a combination of tools (Zilouchian & Jafar, 2001). For desalination plants that are installed in remote locations where no operator is routinely on site, AI can be used to diagnose the process itself when the equipment breaks down. Consequently the facility often requires more control than the typical reverse osmosis plant, hence there is a significant amount of sequential control, and sometimes continuous control. The input sensors include pressure, flow, pH, salinity, conductivity, switch position, and various equipment statuses. Outputs include valves, pump power, circuit breakers and similar equipment. Such processes could advance the operating paradigm of desalination technologies.

The central issues (or goals) of AI research include reasoning, knowledge, planning, learning, natural language processing (communication), perception and the ability to move and manipulate objects. AI has been used demonstrate the effectiveness of a method developed by Biswas and colleagues in building a detailed model of the reverse osmosis (RO) system for the water recovery system (WRS) of the advanced life support system. Desalination systems could be operated by AI where operational specifications for the controller are represented by utility functions, and the corresponding resource management problem is formulated as a safety control problem. Abdelwahab (2005) and colleagues used a controller designed as a limited-horizon online supervisory controller that performed utility functions to decide on the best operating action. The feasibility and accuracy of the online algorithm were used to demonstrate the effectiveness of the system by running a set of experiments on a RO process. Badreddin (2004) developed a laboratory set-up for education and research on automation of reverse osmosis processes employing a sustainable energy source. A control system design of reverse osmosis plants by using advanced optimization techniques has also been reported (Gambier, 2009).

Artificial intelligence, which can be applied to the design, operation, control, and performance optimization of desalination plants, can be used to improve process efficiency and enhance operational safety. Desalination processes have been effectively modeled using neural networks due to their complex process design and operation as was outlined by Voros and colleagues (1990). The desalination processes can be evaluated based on physical properties such as flow, temperature, pressure, and salinity. AI techniques, including expert system, neural networks, and fuzzy logic could be used for optimization of plant performance, manage control and operational decision making procedures, and diagnose the desalination process using alarms and fault logic, similar to current human methods.

Unfortunately, the use of AI neural networking efforts has not been pursued for use in RO desalination, with prior emphasis has been placed on the use of AI to manage RO processes in efforts to explore space (Bonasso, 2001). Future efforts that integrate quantum computing into AI are recommended, and when integrated into replicating machines and other automated methods, then desalination may take another leap forward relative to optimal production and cost effectiveness applications.

9.8 PROVIDING OPPORTUNITIES AND OVERCOMING CHALLENGES

Opportunities for using reverse osmosis for desalination have increased since the inception of the technology in the 1950s. As reported by Ghaffour and colleagues (2013), water recovery in reverse osmosis seawater desalination facilities treating 35,000 mg/L of salinity has increased from twenty-five percent in the 1980s to thirty-five percent in the 1990s; currently, water recoveries average about forty-five percent. Many new technologies are just beginning to be developed. Many challenges for designing cost-effective systems can be expected for FO, PRO, MDC, CDI and HF-NF in the future. The increase of desalination capacity has not only been due to increases in water demand due to population growth and depletion of natural resources but also by the significant reduction in desalination cost as a result of technological advances in reverse osmosis membrane manufacturing and energy recovery appurtenances (Reddy & Ghaffour, 2007; Greenlee *et al.* 2009; Khawaji *et al.* 2008). Improvements to the membranes typically include better resistance to compression, longer life, higher possible recovery, improved flux, and reduced salt passage.

Although FO and PRO offer great promise for future application, the technologies must overcome challenges to their advancement. For example, reverse permeation of solutes from the draw solution into the feed solution decreases the osmotic driving force in FO and PRO processes. Consequently this reduces the membrane efficiency in both PRO and FO. Also, a complete understanding of the extent of draw solutions, as well as fouling of FO and PRO membranes remains limited in scope. In addition, few FO and PRO membranes are commercially available at this time.

Yip and Elimelech (2014) compared reverse electrodialysis with PRO, demonstrating that PRO can achieve both greater efficiencies (54–56 percent) and higher power densities (2.4–38 W/m^2) than reverse electrodialysis (18–38 percent efficiency and 0.77–1.2 W/m^2 power densities). PRO's efficiency is attributed to the ability of the membranes to effectively utilize the salinity difference to drive water permeation and suppress the detrimental leakage of salts, unlike the low conductivity of currently available ion exchange membranes that impedes reverse electrodialysis ion flux and, thus, constrains the power density.

In some specific areas, desalination has now been able to successfully compete with water resources that rely on dams or water transfers for potable water supply. Energy usage in seawater RO remains higher than 1.0 kWh/m^3, the theoretical minimal energy level for desalination of typical seawater (35 g/L of total dissolved solids), assuming a thermodynamically reversible process at fifty percent water recovery (Semiat, 2008). The most energy efficient systems for seawater desalination have recently achieved an energy requirement between 1.8 and 2.2 kWh/m^3; hence, there remains an opportunity for improvement in this arena.

The emergence of new low-energy desalination technologies, such as forward osmosis, capacitive deionization, and microbial desalination cells, will provide opportunities to lower the overall future desalting costs. Additionally, nanocomposite support membranes undergo far less physical compaction than pure polysulfone supports, which suggest that these membranes may have multiple applications and uses. A combination of systems may be developed depending on the intended use of these emerging technologies.

9.9 REFERENCES

Achilli A., Cath T. Y. and Childress A. E. (2009). Power generation with pressure retarded osmosis: an experimental and theoretical investigation. *Journal of Membrane Science*, **343**(1–2), 42–52.

Achilli A. and Childress A. E. (2010). Pressure retarded osmosis: from the vision of Sidney Loeb to the first prototype installation – review. *Desalination*, **261**(3), 205–211.

Alsvik I. L. and Hagg M. B. (2013) Pressure retarded osmosis and forward osmosis membranes: materials and methods. *Polymers*, **5**, 303–327.

Badreddin, Gambier, Aboul-Foouh and Cadotte J. (2004). Laboratory set-up for education and research on automation of reverse osmosis plants employing a sustainable energy source. *Desalination*, **166**(1–3), 307–314.

Bae T.-H. et al. (2009). Facile high-yield solvothermal deposition of inorganic nanostructures on zeolite crystals for mixed matrix membrane fabrication. *J. Am. Chem. Soc.* **131**, 14662–14663.

Beaudry E. G. and Herron J. R. (1997). Direct Osmosis for Concentrating Wastewater, in Proceedings of the 27th International Conference on Environmental Systems. Lake Tahoe, NV, July 14–17.

Belfer S., Purinson Y., Fainshtein R., Radchenko Y. and Kedem O. (1998). Surface modification of commercial composite polyamide reverse osmosis membranes, *J. Membr. Sci.* **139**(2), 175–181.

Biswas G., Manders E. J., Ramirez J., Mahadevan N. and Abdelwahab S. (2004). Online model-based diagnosis to support autonomous operation of an advanced life support system. *Habitation*, **10**(1), 21–37.

Bonasso B. (2001). Intelligent control of a NASA advanced water recovery system, *Proc. Int'l Symp.* Artificial Intelligence Robotics and Automation in Space, Canadian Space Agency, Montreal.

Butler E., Silva A., Horton K., Rom Z., Chwatko M., Havasov A. and McCutcheon J. R. (2013) Point of use water treatment with forward osmosis for emergy relief. *Desalination*, **312**, 23–30.

Cao X., Huang X., Liang P., Xiao K., Zhou Y., Zhang X. and Logan B. E. (2009). A new method for water desalination using microbial desalination cells. *Environmental Science & Technology*, **43**(18), 7148–7152.

Caudle D. D., Tucker J. H., Cooper J. L., Arnold B. B. and Papastamataki A. (1966). Electrochemical demineralization of water with carbon electrodes. Research Report, Oklahoma University Research Institute.

Chavan R., Desai U., Mhatre P. and Chinchole R. (2012). A review: carbon nanotubes, *Int. J. Pharm. Sci. Rev. Res.*, **13**(1), 124–134.

Chen X., Xia X., Liang P., Cao X., Sun H. and Huang X. (2011). Stacked microbial desalination cells to enhance water desalination efficiency. *Environmental Science & Technology*, **45**(6), 2465–2470.

Chen S., Liu G., Zhang R., Qin B. and Luo Y. (2012). Development of themicrobial electrolysis desalination and chemical-production cell for desalination as well as acid and alkali productions, *Environ. Sci. Technol.*, **46**, 2467–2472.

Cheryan M. (1998). Ultrafiltration and Microfiltration Handbook. Lancaster, PA, Technomic Publishing Comany, Inc.

Clive L. D. and Raymond L. E. (1991). Knowledge-Based Systems in Engineering, McGraw-Hill, Singapore.

Cusick R. D., Kim Y. and Logan B. E. (2012). Energy capture from thermolytic solutions in microbial reverse-electrodialysis cells. *Science*, **335**(6075), 1474–1477.

Darvishmanesh S., Tasselli F., Jansen J. C., Tocci E., Bazzarelli F., Bernardo P. and Van der Bruggen B. (2011). Preparation of solvent stable polyphenylsulfone hollow fiber nanofiltration membranes. *Journal of Membrane Sci.*, **384**(1–2), 89–96.

Duranceau S. J. (2009). 'Modeling the permeate transient response to perturbations from steady-state in a nanofiltration process.' *Desalination and Water Treatment*, **1**, 7–16.

Duranceau S. J. and Taylor J. S. (2010). Chapter 11 Membrane Processes' in Water Quality and Treatment, 6th Edition. In: Water Quality and Treatment, A Handbook on Drinking Water.

J. K. Edzwald. (ed.) New York, McGraw-Hill, 11–1 to 11–106.

Duranceau S. J. and Yonge D. T. (2013). Volume 2: Verna Treatment Assessment: Reverse Osmosis Process Assessment Study and Verna Nanofiltration Process Evaluation. Final Report; UCF Office of Research and Commercialization Research Funding Agreement 16208081. City of Sarasota Utilities, 1750–12th Street, Sarasota, FL 34236.

Fathizadeh M., Aroujalian A. and Raisi A. (2011). Effect of added NaX nano-zeolite into polyamide as a top thin layer of membrane on water flux and salt rejection in a reverse osmosis process. *J. Membr. Sci.*, **375**, 88–95.

Forester R., Kim M., Petersen R. and Stocker T. (1988). Nanofiltration membranes broaden the use of membrane separation technology. *Desalination*, **70**(1–3), 77–88.

Forrestal C., Xu P. et al. (2012). Microbial desalination cell with capacitive adsorption for ion migration control. *Bioresource Technology*, **120**(0), 332–336.

Fujumori A., Chiba S., Sato N., Abe Y. and Shibasaki Y. (2010). Surface morphological changes in monolayers of aromatic polyamides containing various N-alkyl side chains, *J. Phys. Chem., B*. **114**, 1822–2835.

Gambier A., Wellenreuther A. and Badreddin E. (2009). Control system design of reverse osmosis plants by using advanced optimization techniques. *Desalination and Water Treatment*, **9**(1–3), (2009).

Gartner J. World's First Osmotic Power Plant Open. Reuters (11-24-2009).

Ghaffour N., Missimer T. M. and Amy G. L. (2013). Technical review and evaluation of the economics of water desalination: current and future challenges for better water supply sustainability. *Desalination*, **309**, 197–207.

Ghosh A. K., Jeong B.-H., Huang X. and Hoek E. M. V. (2008). Impacts of reaction and curing conditions on polyamide composite reverse osmosis membrane properties, *J. Membr. Sci.*, **311**(1–2), 34–45.

Gilron J., Belfer S., Vaisanen P. and Nystrom M. (2001). Effects of surface modification on antifouling and performance properties of reverse osmosis membranes, *Desalination*, **140**(2), 167–179.

Greenlee L. F., Lawler D. F., Freeman B. D., Marrot B. and Moulin P. (2009). Reverse osmosis desalination: water source, technology, and today's challenges. *Water Research*, **43**, 2317–2348.

He T., Frank M., Mulder M. H. V. and Wessling M. (2008). Preparation and characterization of nanofiltration membranes by coating polyethersulfone hollow fibers with sulfonated poly(ether ether ketone). *J. Membr. Sci.*, **307**(1), 62–72.

Hilal N., Al-Zoubi H., Darwish N. A., Mohamma A. W. and Abu Arabi M. (2004). A comprehensive review of nanofiltration membranes: treatment, pretreatment, modelling, and atomic force microscopy. *Desalination*, **170**(3), 281–308.

Hirata T. and Hashimoto A. (1998). Experimental assessment of the efficacy of microfiltration and ultrafiltration for Cryptosporidium removal. *Water Science and Technology*, **38**(12), 103–107.

Howe K. J., Marwah A., Chiu K.-P. and Adham S. S. (2007). Effect of membrane configuration on bench-scale MF and UF fouling experiments. *Water research*, **41**(17), 3842–3849.

HTI (2013). Hydration Technology Innovations, 2484 Ferry Street SW, Albany, OR 97322. Hydration Technologies Inc., Hydration bags–technology overview.

Huang J. et al. (2012). Fabrication of polyethersulfone-mesoporous silica nanocomposite ultrafiltration membranes with antifouling properties. *J. Membr. Sci.*, **423**, 362–370.

Humlik T., Lee J., O'Hern S. C., Fellman B. A. and Baig M. A. Nanostructured materials for water desalination. *Nanotechnology*, **22**, 1–19.

Jacobson K. S., Drew D. M. and He Z. (2011a). Efficient salt removal in a continuously operated upflow microbial desalination cell with an air cathode. *Bioresource Technology*, **102**(1), 376–380.

Jacobson K. S., Drew D. M. and He Z. (2011b). Use of a liter-scale microbial desalination cell as a platform to study bioelectrochemical desalination with salt solution or artificial seawater. *Environmental Science & Technology*, **45**(10), 4652–4657.

Jacangelo J. G. and Laine J.-M. (1994). Evaluation of Ultrafiltration Membrane Pretreatment and Nanofiltration of Surface Waters. Denver, CO, AWWA Research Foundation.

Jacangelo J. G., Adham S. S. and Laîné J. M. (1995). Mechanism of cryptosporidium, giardia, and MS2 virus removal by MF and UF. *Journal AWWA*, **87**(9), 107–121.

Jacangelo J. G., Rhodes Trussell R. and Watson M. (1997). Role of membrane technology in drinking water treatment in the United States. *Desalination*, **113**(2), 119–127.

Jadav G. L. and Singh P. S. (2009). Synthesis of novel silica-polyamide nanocomposite membrane with enhanced properties. *J. Membr. Sci.*, **328**, 257–267.

Jeong B. H. et al. (2007). Interfacial polymerization of thin film nanocomposites: a new concept for reverse osmosis membranes. *J. Membr. Sci.*, **294**, 1–7.

Johnson A. M. and Newman J. (1971). Desalting by means of porous carbon electrodes. *J. Electrochem. Soc.*, **118**(3), 510–517.

Kedem O. and Katchalsky A. (1958). Thermodynamic analysis of the permeability of biological membranes to non-electrolytes. *Biochemica et Biophysica Acta.*, **27**, 123–147.

Kedem O. and Katchalsky A. (1963). Permeability of composite membranes, part I: electric current, volume flow and flow of solute through membranes. *Trans. Faraday Soc.*, **59**, 1918–1930.

Kiesow I., Marczewski D., Reinhardt L., Mühlmann M., Possiwan M. and Goedel W. A. (2013). Bicontinuous. zeolite polymer composite membranes prepared via float casting. *J. Am. Chem. Soc.*, **135**, 4380–4388.

Kim H., Shin D. H. and Lee Y. T. (2007). Effect of silane coupling agents on the performance of RO membranes, *J. Membr. Sci.*, **300**, 224–231.

Kim Y. and Logan B. (2011). Series assembly of microbial desalination cells containing stacked electrodialysis cells for partial or complete seawater desalination, *Environ. Sci. Technol.*, **45**, 5840–5845.

Kim E. S., Hwang G., El-Din M. G. and Liu Y. (2012). Development of nanosilver and multi-walled carbon nanotubes thin-film nanocomposite membrane for enhanced water treatment. *J. Membr. Sci.*, **394**, 37–48.

Kim Y. and Logan B. E. (2013). Microbial desalination cells for energy production and desalination. *Desalination*, **308**, 122–130.

Kiso Y., Mizuno A., Othman R. A. A. b., Jung Y.-J., Kumano A. and Ariji A. (2002). Rejection properties of pesticides with a hollow fiber NF membrane (HNF-1). *Desalination*, **143**(2), 147–157.

Khawaji A. D., Kutubkhanah I. K. and Wie J. (2008). Advances in seawater desalination technologies. *Desalination*, **221**, 47–69.

Kurth C. J., Burk R. and Green J. (2011). Utilizing nanotechnology to enhance RO membrane performance for seawater desalination. Proceedings of the International Desalination Association World Congress. Perth Convention and Exhibition Centre, Perth, Western Australia. Paper IDA-WC/PER11–323, September 4–9.

Lee K. (1981). Membranes for power-generation by pressure-retarded osmosis. *Journal of Membrane Science*, **8**(2), 141–171.

Lee K. L., Baker R. W. and Lonsdale H. K. (1981). Membranes for power generation by pressure-retarded osmosis, *J. Membrane Science*, **8**(2), 141–171.

Lee H. S. et al. (2008). Polyamide thin-film nanofiltration membranes containing TiO2 nanoparticles. *Desalination*, **219**, 48–56.

Lee K. P., Arnot T. C. and Mattia D. (2011). A review of reverse osmosis membrane materials for desalination-development to date and future potential. *J. Mem. Sci.*, **370**(1–2), 1–22.

Liu C., Caothien S., Hayes J. C. and Otoyo T. O. (2001). Membrane Chemical Cleaning: From Art of Science. AWWA 2000 Water Quality Technology Conference. Denver, CO.

Liu X. et al. (2013). Synthesis and characterization of novel antibacterial silver nanocomposite nanofiltration and forward osmosis membranes based on layer-by-layer assembly. *Water Res.*, **47**, 3081–3092.

Liu C., Fang W. X., Chou S. R., Shi L., Fane A. G. and Wang R. (2013). Fabrication of layer-by-layer assembled FO hollow fiber membranes and their performances using low concentration draw solutions. *Desalination*, **308**, 147–153.

Lind M. L. et al. (2009). Influence of zeolite crystal size on zeolite-polyamide thin film nanocomposite membranes. *Langmuir*, **25**, 10139–10145.

Luo H., Xu P., Roane T. M., Jenkins P. E. and Ren Z. (2012) Microbial desalination cells for improved performance in wastewater treatment, electricity production, and desalination. *Bioresource Technology*, **105**, 60–66.

Ma N., Wei J., Liao R. and Tang C. Y. (2012). Zeolite-polyamide thin film nanocomposite membranes: towards enhanced performance for forward osmosis. *J. Membr. Sci.*, **405–406**, 149–157.

Ma N. et al. (2013). Nanocomposite substrates for controlling internal concentration polarization in forward osmosis membranes. *J. Membr. Sci.*, **441**, 54–62.

McCutcheon J. R., McGinnis R. L. and Elimelech M. (2006). Desalination by a novel ammonia–carbon dioxide forward osmosis process: influence of draw and feed solution concentrations on process performance, *J. Membr. Sci.*, **278**, 114–123.

McGinnis R. and Elimelech M. (2008). Global challenges in energy and water supply: the promise of engineered osmosis. environmental. *Science & Technology*, **42**(23), 5230–5239.

Mehanna M., Saito T., Yan J., Hickner M., Cao X., Huang X. and Logan B. E. (2010a). Using microbial desalination cells to reduce water salinity prior to reverse osmosis. *Energy & Environmental Science*, **3**(8), 1114–1120.

Mehanna M., Kiely P. D., Call D. F. and Logan B. E. (2010b). Microbial electrodialysis cell for simultaneous water desalination and hydrogen gas production. *Environmental Science & Technology*, **44**(24), 9578–9583.

Miller T. S., Ebejer N., Guell A. G., Macpherson J. V. and Unwin P. R. (2012). Electrochemistry at carbon nanotube forests: sidewalls and closed ends allow fast electron transfer. *Chem. Commun.*, 7435–7437.

Mukherjee D., Kulkarni A. and Gill W. N. (1996). Chemical treatment for improved performance of reverse osmosis membranes, *Desalination*, **104**(3), 239–249.

Nicoll P. (2011). Manipulated osmosis – an alternative to reverse osmosis? *Climate Control Middle East, April 2011*, 46–49.

Nguyen T. C. (2014). Breakthrough may lead to 'limitless' supply of hydrogen fuel. *Smart Planet.* http://www.smartplanet.com/blog/thinking-tech/breakthrough-may-lead-to-limitless-supply-of-hydrogen-fuel/ (accessed 31 August 2014).

Oasys Water Inc. (2013). (21 Drydock Avenue, 7th Floor, Boston, MA 02210.

Olivieri V., Parker D. Jr., Willinghan G. and Vickers J. (1991). Continuous microfiltration of surface water. AWWA Membrane Technology Conference Proceedings, Orlando, FL.

Pankratz T. 2010. FO Plant Completes One Year of Operation. Water Desalination Report: 2–3. 15 Nov.

Patent Application. Israel Patent 42658. (July 3, 1973). (U.S. patent 3,906,250 granted September 16, 1975).

Qu Y., Feng X., Wang J., Liu J., Lv W. and He B. E. (2012). Simultaneous water desalination and electricity generation in a microbial desalination cell with electrolyte recirculation for pH control, *Bioresour. Technol.*, **106**, 89–94.

Rana D. and Matsuura T. (2010). Surface modifications for antifouling membranes, *Chem. Rev.*, **110**, 2448–2471.

Ramon G. Z., Feinberg B. J. and Hoek E. M. V. (2011). Membrane-based production of salinity-gradient power. *Energy Environ. Science*, **4**(11), 4423–4434.

Ratto T. V., Holt J. K. and Szmodis A. W. (2010). Membranes with embedded nanotubes for selective permeability, Patent Application No. 20100025330. United States Patent and Trademark Office.

Reddy K. V. and Ghaffour N. (2007). Overview of the cost of desalinated water and costing methodologies. *Desalination*, **205**(1–3), 340–353.

Saeed H. M., Husseini G. A., Yousef S., Saif J., Al-Asheh S., Fara A. A., Azzam S., Khawaga R. and Aidan. (2015). Microbial desalination cell technology: a review and a case study. *Desalination*, **359**, 1–13.

Salter R. J. (2005) Forward osmosis. *Water Conditioning &Purification*, **48**(4), 36–38.

Saren Q., Qiu C. Q. and Tang C. Y. (2011). Synthesis and characterization of novel forward osmosis membranes based on layer-by-layer assembly. *Environ. Sci. Technol.,* **45**, 5201–5208.

Semiat R. (2008). Energy issues in desalination processes. *Environ. Sci. Technol.,* **42**, 8193–8202.

Sethi S. and Wiesner M. R. (2000). Simulated cost comparisons of hollow-fiber and integrated nanofiltration configurations. *Water research*, **34**(9), 2589–2597.

Setiawan L., Wang R., Li K. and Fane A. G. (2011). Fabrication of novel poly(amide–imide) forward osmosis hollow fiber membranes with a positively charged nanofiltration-like selective layer. *J. Membr. Sci.,* **369**(1–2), 196–205.

Shi G. M., Yang T. and Chung T. S. (2012). Polybenzimidazole (PBI)/zeolitic imidazolate frameworks (ZIF-8) mixed matrix membranes for pervaporation dehydration of alcohols. *J. Membr. Sci.,* **415**, 577–586.

Simmons O. D., Sobsey M. D., Heaney C. D., Schaefer F. W. and Francy D. S. (2001). Concentration and detection of cryptosporidium oocysts in surface water samples by method 1622 using ultrafiltration and capsule filtration. *Applied and Environmental Microbiology*, **67**(3), 1123–1127.

Smith P. J., Vigneswaran S., Ngo H. H., Ben-Aim R. and Nguyen H. (2006). A new approach to backwash initiation in membrane systems. *Journal of Membrane Sci.*, **278**(1), 381–389.

Su J., Chung T. S., Helmer B. J. and de Wit J. S. (2012). Enhanced double-skinned FO membranes with inner dense layer for wastewater treatment and macromolecule recycle using Sucrose as draw solute. *J. Membr. Sci.* **396**, 92–100.

Suss M. E., Baumann T. F., Bourcier W. L., Spadaccini C. M., Rose K. A., Santiago J. G. and Stadermann M. (2012). Capacitive desalination with flow-through electrodes. *Energy and Environmental Science*, **5**, 9511–9519.

USBR (2014). Microbial Desalination Fuel Cells: Assessment of technology status and potential benefit for Reclamation. White Paper (K. Guerra). Bureau of Reclamation, Denver, CO 80225.

USEPA (2005). Membrane Filtration Guidance Manual. (815-R-06–009). USEPA Office of Water.

U.S. House of Representatives 2002. Energy From the Ocean: A report prepared for the Subcommittee on Advanced Energy Technologies and Energy Conservation Research, Development, and Demonstration of the Committee on Science and Technology, Ninety-fifth Congress, Second Session, Prepared by the Congressional Research Service, Science Policy Research Division.

Van der Bruggen B. and Vandecasteele C. (2003). Removal of pollutants from surface water and groundwater by nanofiltration: overview of possible applications in the drinking water industry. *Environmental Pollution*, **122**(3), 435–445.

Van der Bruggen B., Hawrijk I., Cornelissen E. and Vandecasteele C. (2003). Direct nanofiltration of surface water using capillary membranes: comparison with flat sheet membranes. *Separation and Purification Technology*, **31**, 193–201.

Vermaas D. A., Veerman J., Yip N. Y., Elimelech M., Saakes M. and Nijmeijer K. (2013). High effficiency in energy generation from salinity gradients with reverse electrodialysis. *ACS Sustainable Chem. Eng.,* **1**(10), 1295–1302.

Voros N., Maroulis Z. B. and Marinos-Kouris D. (1996). Optimization of reverse osmosis networks for seawater desalination, *Computers chem. Engng.*, **20**, Suppl, S345–S350.

Wang R., Shi L., Tang C. Y., Chou S., Qiu C. and Fane A. G. (2010). Characterization of novel forward osmosis hollow fiber membranes. *J. Membr. Sci.,* **355**(1–2), 158–167.

Wu S., Xing J., Zheng C., Xu G., Zheng G. and Xu J. (1997). Plasma modification of aromatic polyamide reverse osmosis composite membrane surface, *J. Appl Polym Sci.,* **64**, 1923–1926.

Wu H. Q., Tang B. B. and Wu P. Y. (2013). Optimizing polyamide thin film composite membrane covalently bonded with modified mesoporous silica nanoparticles. *J. Membr. Sci.,* **428**, 341–348.

Xu J., Ruan G., Gao X., Pan X., Su B. and Gao C. (2008). Pilot study of inside-out and outside-in hollow fiber UF modules as direct pretreatment of seawater at low temperature for reverse osmosis. *Desalination*, **219**(1), 179–189.

Xu Y., Peng X. Y., Tang C. Y. Y., Fu Q. S. A. and Nie S. Z. (2010). Effect of draw solution concentration and operating conditions on forward osmosis and pressure retarded osmosis performance in a spiral wound module. *J. Membr. Sci.,* **348**(1–2), 298–309.

Yang Q., Wang K. Y. and Chung T. S. (2009). Dual-layer hollow fibers with enhanced flux as novel forward osmosis membranes for water production. *Environ. Sci. Technol.,* **43**, 2800–2805.

Yip N. Y. and Elimelech E. M. (2014). Comparison of energy efficiency and power density in pressure retarded osmosis and reverse electrodialysis. *Environ. Sci. Technol.,* **48**, 11002–11012.

Yip N. Y., Tiraferri A., Phillip W. A., Schiffman J. D., Hoover L. A., Kim Y. C and Elimelech M. (2011). Thin-film composite pressure retarded osmosis membranes for sustainable power generation from salinity gradients. *Environ. Sci. Technol.,* **45**, 4360–4369.

Zhang B. and He Z. (2012). Integrated salinity reduction and water recovery in an osmotic microbial desalination cell, *RSC Adv.,* **2**, 3265–3269.

Zhao R., van Soestbergen M., Rijnaarts H. H. M., van der Wal A., Bazant M. Z. and Biesheuvel P. M. (2012). Time-dependent ion selectivity in capacitive charging of porous electrodes. *Journal of Colloid and Interface Science*, **384**, 38–44.

Zilouchian A. and Jafar M. (2001). Automation and process control of reverse osmosis plants using soft computing methodologies. *Desalination*, **135**(1–3), 51–59.

Zornoza B., Tellez C., Coronas J., Gascon J. and Kapteijn F. (2013). Metal organic framework based mixed matrix membranes: an increasingly important field of research with a large application potential. *Microporous Mesoporous Matter.*, **166**, 67–78.

Zou L., Li L., Song H. and Morris G. (2008). Using mesoporous carbon electrodes for brackish water desalination, *Water Research,* **42**, 2340–2348.

Chapter 10

Desalination – reverse osmosis *versus* the rest

Winson C. L. Lay, Aik Num Puah and Harry Seah*

Technology Department, PUB

**Corresponding author: winson_lay@pub.gov.sg*

10.1 INTRODUCTION

With population growth, rapid urbanisation, and climate change, there is an escalating demand on the global fresh water resources (Daigger, 2012). Where fresh water from natural sources is limited, alternative options for water supply will be necessary to meet the demand for water. Given the abundance of seawater, desalination often presents a key option for water supply. The water supply made available by desalination, however, comes at a cost. The major concern is energy, as this constitutes a significant portion of the desalination costs. Desalination is an energy-intensive process.

It can be, therefore, expected that the increasing use of seawater desalination to augment water supply – while necessary – will pose a challenge in view of the considerable energy usage that this process needs. Among the current desalination technologies, reverse osmosis (RO) is regarded as the most energy-efficient technology for seawater desalination, and serves as the benchmark for comparison of new desalination technologies (Elimelech & Phillip, 2011).

Conventional seawater reverse osmosis (SWRO) is well established and will likely continue to be the dominant technology for desalination (Semiat, 2008). However, conventional SWRO technology requires an energy footprint of around 3–4 kWh/m^3 of electrical energy (Jacangelo *et al.* 2011; Veerapaneni *et al.* 2011). To put things in perspective, it has been shown that a 'business as usual' policy for SWRO and assuming a 15% annual growth in SWRO production could result in the energy requirement of SWRO desalination to exceed that of all other separations industries by 2050 (Fane, 2011). A critical goal for desalination is, therefore, to improve its energy efficiency, and concomitantly address the environmental concerns related to it (Subramani *et al.* 2011).

This chapter aims to provide a brief review of emerging alternative desalination technologies and concepts, and to provide a discussion on how these could lead to greater efficiency for seawater desalination vis-a-vis conventional SWRO.

10.2 CONVENTIONAL SEAWATER REVERSE OSMOSIS (SWRO) DESALINATION TECHNOLOGY

Since the invention of the first technically viable RO membrane by Loeb and Sourirajan in the early 1960s, the technology has progressed tremendously. A distinct advantage of the RO technology is that it is more energy efficient than thermal desalination technologies. The global trend is moving towards the increasing use of RO membrane desalination worldwide.

As illustrated in Figure 10.1, the breakdown of the total installed desalination capacity in 2011 shows that nearly 60% employed RO, approximately 26% employed multi-stage flash distillation (MSF) and the rest used multiple effect distillation (MED) and the marginal technologies. The total installed desalination capacity was around 66.4 million m^3/d.

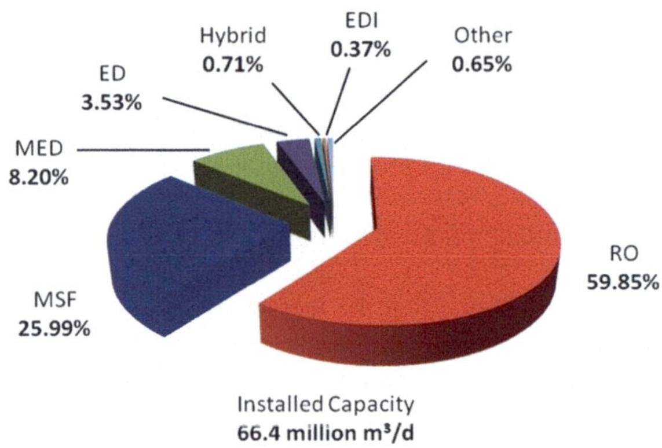

Figure 10.1 Total installed desalination capacity by technology (Adapted from IDA, 2012).

The key component of conventional SWRO desalination plant is the RO membrane separation process. The RO applies hydraulic pressure to force water through a semi-permeable membrane. RO is capable of rejecting nearly all colloidal or dissolved matter from an aqueous solution, producing a product stream which consists of almost pure water, known as permeate, and a concentrated waste stream, known as brine. RO, however, is also the most energy-consuming component in the entire plant, and may amount to more than 80% of the total energy consumption of a desalination system (Figure 10.2).

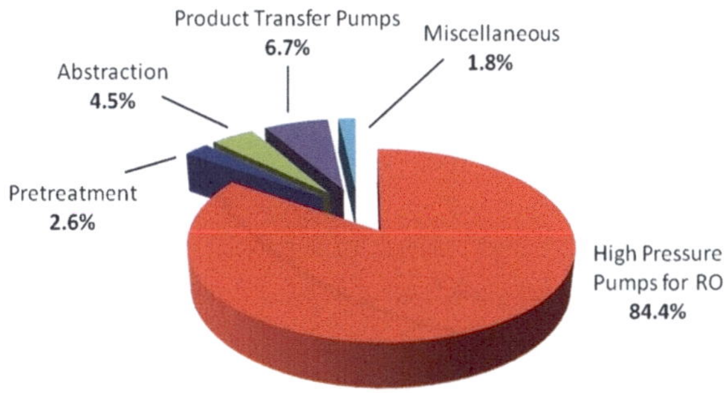

Figure 10.2 Breakdown of the power consumption of a SWRO plant. (Adapted from Wilf & Bartels, 2005.)

10.2.1 Minimum desalination energy

Theoretically, the minimum energy requirement for a RO desalination process occurs when the applied transmembrane hydraulic pressure (Δp) equals the osmotic pressure difference across the membrane ($\Delta\pi$), which is dependent on the osmotic pressure of the feed water (π_F), recovery (φ) and rejection (R). The following relationship can be derived by applying a mass balance around a RO desalination process.

$$\Delta p_{\min} = \Delta\pi(\pi_F, \phi, R) = \pi_F \cdot \left[1 + \frac{\phi \cdot R}{(1-\phi)}\right] \quad (10.1)$$

The osmotic pressure of a solution water is related to its total dissolved solids (TDS) concentration and may be estimated by assuming 6.9 kPa for each 100 mg/L TDS (AWWA, 2004). For seawater, the TDS concentration may range between 15,000 mg/L and 50,000 mg/L (UNESCO, 2008). The representative TDS value for seawater desalination typically assumed is 35,000 ppm (Elimelech & Phillip, 2011), and typical recovery for a SWRO plant is between 30% and 50% (AWWA, 2004).

Based on equation (10.1), a graph with the minimum energy requirement for a RO desalination process can be plotted against φ for different feed TDS concentrations. Illustrations are provided in Figure 10.3. Here, it is assumed that $R = 99.5\%$, as typical SWRO membrane has R between 99.4% and 99.8% (Lee *et al.* 2011). As shaded in the figure, it can be seen that for a representative concentration of 35 g/L TDS, the minimum desalination energy is around 1 kWh/m^3 for the interested range of recovery.

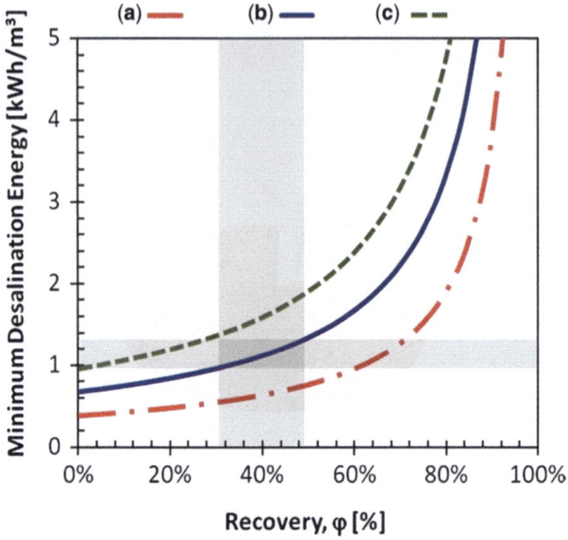

Figure 10.3 Minimum desalination energy as function of recovery and different seawater TDS concentration derived for a desalination process based on equation (10.1). The plotted TDS concentrations are (a) 20 g/L TDS, (b) 35 g/L TDS, and (c) 50 g/L TDS.

10.2.2 Improving energy efficiency of conventional SWRO

Conventional SWRO based on polymeric membrane is an established technology for desalination. The current power consumption of SWRO is around 3–4 kWh/m^3, and is significantly more efficient compared to that in the 1970s, when SWRO had power consumption >10 kWh/m^3 (Fritzmann *et al.* 2007).

Recent efforts on further improving the energy efficiency of conventional SWRO technology include enhancing system design, staging of elements of different permeability, developing novel feed spacer to minimise pressure drop across membrane module and recovering hydraulic energy from the pressurised brine stream with an energy recovery device (Subramani *et al.* 2011). However, there is reservation on the extent of improvement that such efforts may attain.

For instance, it is known that there is a water/salt selectivity trade-off for conventional SWRO polymeric membrane for desalination. Conventional SWRO polymeric membrane may have reached its upper bound performance limit, and it would be difficult to further increase water permeability of a SWRO polymeric membrane without impairing its salt rejection property (Geise *et al.* 2011). Another example is the recovery of hydraulic energy from pressurised brine stream using an energy recovery device to improve the overall energy efficiency of SWRO desalination. With state-of-the-art energy recovery devices already attaining 97% efficiency (Stover & Andrews, 2011), it is difficult to enhance this efficiency further.

The foregoing discussion clarifies that although there is still room to optimise conventional SWRO technology, the impact of such improvement may be incremental. Pursuing incremental improvements could be one strategy for technology development, and it has also been shown that the compounding effect of continuous incremental improvements in several fronts could result in a substantial figure of merit in the long-term (Birkett & Truby, 2007).

However, new technology development will be needed to achieve a breakthrough in desalination with substantially reduced power consumption and enhanced energy efficiency. Such new developments may include advanced membrane materials and innovative processes and technologies (Subramani *et al.* 2011).

10.3 ALTERNATIVE DESALINATION TECHNOLOGIES

The following provides a brief review of selected alternative desalination technologies. Current research and development interests include biomimetic techniques, electrochemical desalination, engineered osmosis and membrane distillation. A shortlist of other emerging alternative desalination technologies will also be discussed at the end of this section. However, the list is not meant to be exhaustive.

10.3.1 Biomimetic techniques

The potential of biomimetic techniques to lead to a breakthrough in desalination has been confirmed by various research groups (Bowen, 2006; Lee *et al.* 2011; Pendergast & Hoek, 2011; Tang *et al.* 2012). One identified strategy is to fabricate biomimetic membrane that has the concomitant advantages of ultrahigh water permeability and high selectivity (or rejection) towards ions (Tang *et al.* 2012). Such an attribute is not found in conventional polymeric RO membranes, whereby there is an inverse relationship between water permeability and salt rejection, and is a limitation of conventional SWRO desalination technology (Geise *et al.* 2011). Biomimetic membrane has the potential to transcend this limitation.

The separation mechanism of biomimetic membrane is provided by pore-forming proteins known as aquaporins. This family of proteins with molecular weight of 24–30 kDa, was a discovery that was awarded a Nobel Prize in 2003, and explains the remarkably fast rate of water transport in living cells (Agre, 2003). Aquaporins, which may be referred as water channels, have specific structural and functional properties that can selectively allow water molecules to pass through, while retaining small solutes that may be charged or uncharged.

Current studies indicate that biomimetic membrane, based on a mechanism of channel-mediated water transport through the Aquaporins, could achieve water permeability that is at least an order of magnitude larger than that of conventional

SWRO membranes with rejection of solutes such as glucose, glycerol, salt and urea (Kumar *et al.* 2007).

The water permeability of a desalination membrane, typically expressed in $ms^{-1}Pa^{-1}$ or $Lm^{-2}h^{-1}bar^{-1}$, is a measure of its productivity and energy efficiency. Current SWRO membranes may have water permeability around 1–2 $Lm^{-2}h^{-1}bar^{-1}$ with salt rejection between 99.4% and 99.8% (Lee *et al.* 2011). Assuming complete rejection of solutes, and disregarding effects due to fouling and concentration polarisation, the following theoretical expression may apply (Belfort, 1984).

$$\frac{Water\ flux}{Water\ Permeability} = (\Delta p - \Delta\pi) \qquad (10.2)$$

The significance of water permeability can be observed from the above expression. At the same productivity (constant water flux), a two-fold increase of water permeability would mean that the term (Δp–$\Delta\pi$) required for the process could be halved correspondingly. Although the desalination energy requirement is determined by the pumping pressure (Δp) and not (Δp–$\Delta\pi$), what this analysis shows is that if, theoretically, a membrane of infinite water permeability could be invented without sacrificing its salt rejection property, the total energy required for this process would approach the minimum energy of desalination ($\Delta p = \Delta\pi$, see Figure 10.4). Higher water permeability could also result in smaller membrane area being required for desalination, and hence smaller plant footprint, lower capital cost and reduced use of cleaning chemicals (Lee *et al.* 2011). These are underlying motivations for developing biomimetic membrane.

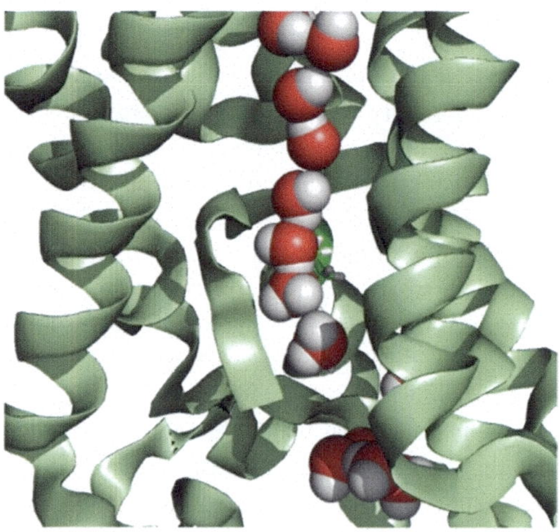

Figure 10.4 Illustration of water molecules transporting through Aquaporins (PUB, 2012b).

The development of Aquaporin membrane for desalination has made progresses in recent years (Tang *et al.* 2012), but these may have only scratched the surface of biomimetic techniques. The biomimetic techniques have a range of potentialities that can be further developed. Beside Aquaporin membrane, another area is research on highly specialised, ion-selective proteins in natural biological systems. These proteins, collectively known as ion channels, have unique and complex transport properties that allow gated and catalysed transport of selective ions (Nielsen, 2009). The concept is to be able mimic natural processes in selectively pumping ions. Thermodynamically, it does not matter how the separation is achieved. However, for practical application where the overall energy consumption is dependent on non-idealities due to process and equipment's efficiency, it can be expected that pumping salt ions will be more energetically favourable than water molecules as in the conventional SWRO desalination process due to the relative low concentration of ions in relation to water molecules in seawater. (UNESCO, 2008). The mechanism of selectively pumping ions may also have a better energy requirement if the final water quality is of higher TDS concentration due to the smaller osmotic pressure gradient ($\Delta\pi$) that needs to be overcome. Highly efficient ion channels mechanism found in nature include euryhaline fish and mangrove plants and are currently being studied (PUB, 2012b).

While the biomimetic techniques have shown promise, translation of the technology from laboratory scale research into commercial application will need to address a number of critical issues including scale-up, cost, stability, durability, and membrane fouling under operating condition. Overcoming these issues will be a key task in the next stage of development for biomimetics techniques and may need a long-term time frame (UNESCO, 2008; Pendergast & Hoek, 2011).

10.3.2 Electrochemical desalination

Electrochemical desalination technologies are voltage-driven processes, where a voltage gradient is applied to achieve desalination by moving ions from a bulk solution (typically through ion-exchange membrane) into a concentrate stream. The electrochemical technologies may have several advantages over conventional pressure-driven SWRO desalination processes. One advantage is that electrochemical technologies do not require high pressure for desalination. Therefore, mechanically and structurally less demanding equipment and materials may be used for the process. Another advantage that had been reported is that separation is achieved by means of moving smaller number of ions in feed water, which could be more energy efficient as compared to moving significantly larger number of water molecules as in RO (UNESCO, 2008; Shaw *et al.* 2011). Although thermodynamically it does not matter how the separation is achieved, the overall energy consumption for practical application is dependent on non-idealities due to process and equipment's efficiency as discussed above.

With regard to conventional Electrodialysis (ED), this electrochemical technology has made progress over the years. One development is the continuous electrochemical deionisation (CEDI) technology. The CEDI technology has the unique attribute that the cells which contain the diluate stream are filled with ion-exchange resins (Figure 10.5). As a result, the electrical resistance in these cells can be substantially reduced, which improves the efficiency of the system. Recent studies reported that a 50 m^3/d demonstration-scale desalination system based on the CEDI technology was able to produce product water that met WHO guideline while achieving an average power consumption of around 1.8 kWh/m^3 for feed water of approximately 32,000 ppm TDS and at recovery of approximately 30% (Shaw *et al.* 2011).

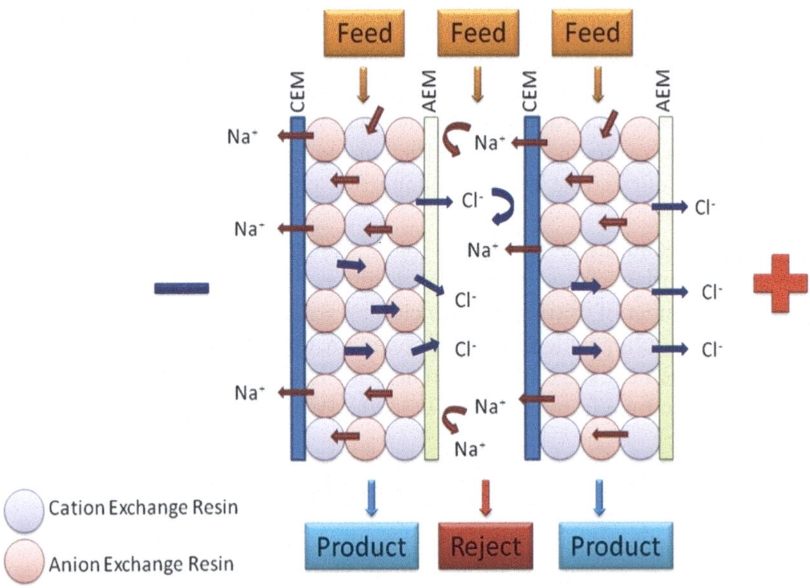

Figure 10.5 Working principles of the CEDI technology. (adapted from Shaw *et al.* 2011).

Another development is the capacitive deionisation (CDI) technology. CDI technology is an electro-sorption process that makes use of basic capacitor concept to remove dissolved ions from an electrolyte (water with ions) stream. Electrodes provide electrochemical surfaces where ions are adsorbed in accordance to the physical principle of electrical double layer. Counter ions are stored in the electrical double layers which form at the solution interface inside the porous electrodes, with cations stored in the negatively charged electrode (cathode), and anions stored in the positively charged electrode (anode). Ideally, no redox process occurs, and as a consequence, the process would be reversible. When an energy recovery system can be incorporated to tap into the store energy that will be released during regeneration, a low energy requirement can be achieved for the CDI process. The CDI process could be particularly attractive with suitable high surface area electrodes (e.g. carbon aerogel) that would enhance the electro-sorption process (Oren, 2008; Anderson *et al.* 2010).

The electrochemical technologies provide a valid alternative to conventional SWRO. The next stage of development for the electrochemical technologies is to address issues including development of high performance and low cost electrodes and ion-exchange membranes, improved module design, fouling control and process optimisation. For desalination, ongoing efforts for electrochemical technologies target at reducing power consumption for desalination to below 1.5 kWh/m^3 over a short-term time frame (Anderson *et al.* 2010; Shaw *et al.* 2011; PUB 2012b). Compared to optimised RO process with specific power consumption around 2 kWh/m^3 for similar TDS values (Peñate B. & García-Rodríguez L., 2011), the electrochemical technologies show promise.

Besides application in desalination, an electrochemical technology could also be applied to produce power from salinity gradient. One such technology is the reverse electrodialysis (RED) (Post *et al.* 2010). RED is one of the two potential salinity gradient power technologies that generate power by exploiting a salinity gradient between a high salinity water stream and a low salinity water stream. The other potential salinity gradient power technology is pressure-retarded osmosis. The salinity gradient power technologies are currently being researched and may present an option to improve the energy efficiency of desalination (Ramon *et al.* 2011). Further discussion is provided in the subsequent section on engineered osmosis.

10.3.3 Engineered osmosis

Engineered osmosis may generally be referred to the technology of exploiting the natural phenomenon of osmosis for an engineering purpose (Cath *et al.* 2006; McGinnis & Elimelech, 2008). For the purpose of water supply, this technology is known as Forward Osmosis (FO). For the purpose of power generation, the technology is known as Pressure-Retarded Osmosis (PRO). Alongside biomimetics and carbon nanotubes, FO has been identified as one of the three promising technologies to reduce energy consumption of desalination by the National Geographic magazine (Lange, 2010).

The attractiveness of FO as a desalination technology is that it does not require hydraulic pressure and hence may have low electrical energy requirement when compared to RO. Illustration of a FO desalination system is provided in Figure 10.6.

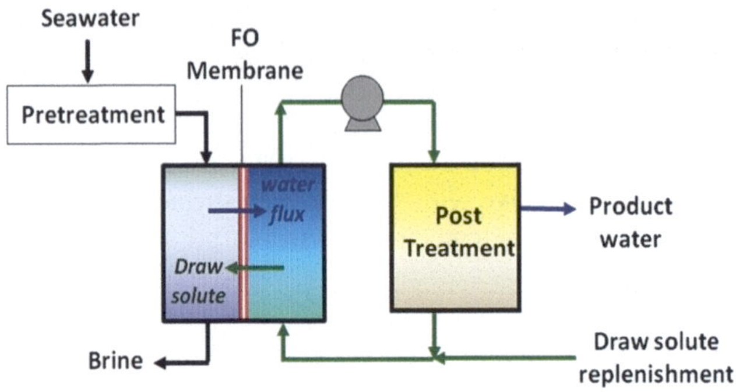

Figure 10.6 A FO desalination system.

It was reported that a FO system using thermolytic ammonia–carbon dioxide draw solute could consume less than 0.25 kWh/m^3 of electrical energy, while achieving recovery of up to 85% for seawater desalination (Elimelech, 2007). This concept is theoretically elegant, but there are practical challenges that need to be overcome including product contamination due to the thermolytic draw solute and costs associated with replenishment of substantial amount of draw solute lost through the FO membrane into the feed (Qin *et al.* 2012).

Small- to mid-scale desalination systems that are based on FO are commercially available. An example is the 200 m^3/d FO-based Desalination Plant built by Modern Water at Al Najdah in Oman (Nicoll, 2013). However, a FO desalination system is unlikely to be competitive with conventional SWRO technology or implemented for large scale (≥100,000 m^3/d) application in the short-term, unless several critical issues can be adequately addressed. These issues include the need to develop FO membrane with high water flux with concomitant low solute transmission and high fouling resistance, development of appropriate draw solution that is safe, low cost and has high osmotic efficiency, and integration with a suitable post-treatment that can economically regenerate draw solute and produce product water of drinking quality (Qin *et al.* 2012).

Engineered osmosis, however, may possess significant potential for other niche applications, where product water quality is not critical. Examples may be applying FO to evaporative cooling tower make-up water (Nicoll *et al.* 2011) or for fertiliser drawn agricultural application (Phuntsho *et al.* 2012). Another promising area for application of engineered osmosis is in applying PRO as a technology for exploiting salinity gradient power. PRO presents a clean and green way to tap into a renewable energy resource. A notable proponent for this technology was Sidney Loeb, who was also founder of the RO membrane. Loeb's vision was to use PRO to harvest the potential energy released from the mixing of low salinity water (e.g. river water) and high salinity water (e.g. seawater) (Loeb, 2002). Theoretically, 1/2 of the osmotic energy of a saline stream can be extracted by operating a PRO at hydraulic pressure equal to midpoint of the osmotic pressure gradient to achieve maximum power density for the membrane (Lee *et al.* 1981; Achilli & Childress, 2010). Practically, however, only around 1/3 of the osmotic energy can be harvested in PRO due to operational constraints and losses due to inefficiencies and pretreatment. For a seawater-river water PRO system, the extractable osmotic energy is estimated to be around 0.2–0.3 kWh/m^3 (Elimelech, 2012; Yip & Elimelech, 2012).

An option to apply PRO is to exploit the salinity gradient between brine streams from a SWRO desalination facility and a water reclamation facility (Chung *et al.* 2012). The power generated by the PRO could partially offset the energy requirement for desalination, in addition to the current practice of recovering hydraulic energy from SWRO brine stream by means of energy recovery device. This approach of turning brine streams into a resource offers advantages that include greater osmotic power generated (0.4–0.6 kWh/m^3) due to the greater salinity gradient between the two streams as compared to seawater-river water, complimentary pretreatment as this is incorporated as part of the desalination and water reclamation processes, and reduced environmental impact by co-mingling the brine streams. An illustration is provided in Figure 10.7.

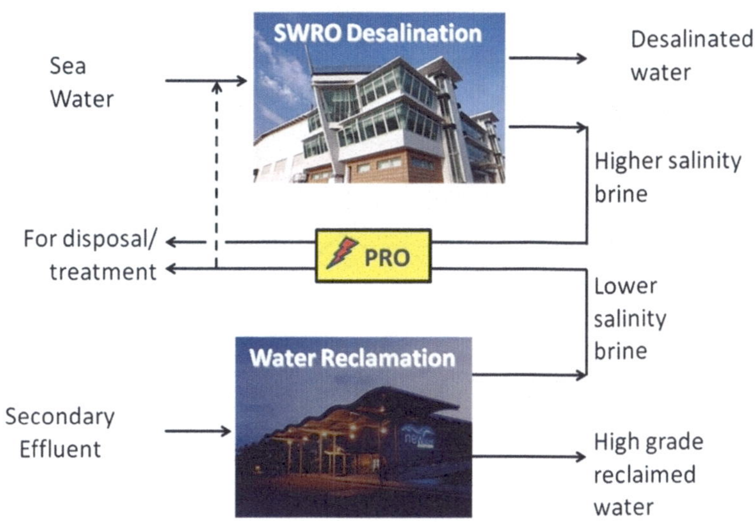

Figure 10.7 Application of PRO to exploit the salinity gradient between brine streams from a SWRO desalination facility and a water reclamation facility.

PRO shows promise as a salinity gradient power technology that may be implementable in the short-term. However, recent experiences from a PRO prototype plant revealed that intensive research efforts are still needed in areas that include improved module design, optimised PRO membrane, and effective fouling control (Brekke, 2012).

10.3.4 Membrane distillation

Membrane distillation (MD) process is a combination of both membrane and distillation processes that involve transport of water vapour through the pores of hydrophobic membranes due to a vapour pressure driving force provided by a temperature gradient across the membrane. For desalination, seawater is warmed to enhance water vapour production. While non-volatile dissolved solutes are retained, water vapour passes through the MD membrane and is condensed on a cooler surface to produce fresh water. This condensed permeate cannot pass back through the membrane, so it is trapped and collected as the product of the MD process. An illustration of the working principle of MD is provided in Figure 10.8.

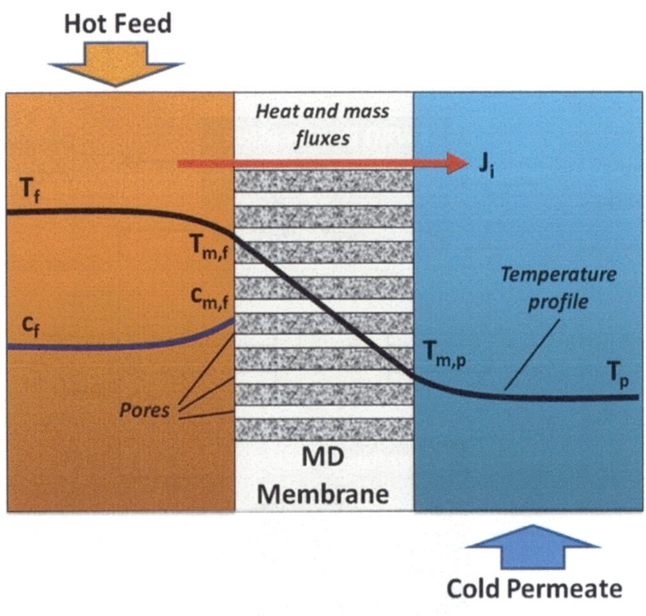

Figure 10.8 Working principle of MD. (Adapted from Khayet, 2011).

The MD is a membrane-based thermal process. Its distinct advantage is that it allows for process intensification such that the MD process can produce distillate water by operating at temperatures lower than those applied for conventional thermal desalination processes (e.g. MSF) and with more compact footprint due to the membrane modules (Drioli & Fontananova, 2012). This attribute allows MD to utilise low grade heat with temperature typically between 60°C and 90°C, and tap into waste heat from industrial processes and renewable energy sources such as solar (Lawson & Lloyd, 1997).

As the driving force for the separation is thermal, the electrical energy that is required for a MD process may be comparatively small, making MD a technology that can potentially bring down the energy cost of seawater desalination significantly (UNESCO, 2008). Pilot studies indicated that the electrical energy input required for a MD process could be below 1 kWh/m^3 with available supply of waste heat (Dotremont *et al.* 2010). Another potential advantage of MD is that it could achieve greater water recovery for desalination than conventional SWRO, as the process performance is less limited by the osmotic pressure of the retentate (Tun *et al.* 2005).

Despite the potential of MD, there are practical challenges that need to be overcome for the successful application of this technology. These challenges include fabrication of high performance and cost-effective MD membrane and module, as well as scale-up and operational issues such as membrane wetting and fouling (Khayet, 2011).

10.3.5 Other emerging desalination technologies

10.3.5.1 Nanotechnology-based membrane technologies

Advancement in nanotechnology provides an additional avenue to enhance membrane performance for desalination. Targeted areas for improvement are water permeability, solute selectivity and operational robustness (Pendergast & Hoek, 2011). Current research on nanotechnology-based membrane technologies includes mixed matrix membranes, nanocomposite membranes and nanotube membranes. Although aquaporin membrane has also been reviewed as a nanotechnology-enabled desalination, it may be more appropriately considered as part of the biomimetic techniques and is discussed in section 6.3.1.

In the near term, thin-film nanocomposite (TFN) membrane, for example zeolite nanoparticle-based membrane, may have the greatest potential to be commercialised. These membranes could provide enhancement in terms of compaction resistance and anti-fouling properties. While these membranes may only provide marginal enhancement of productivity (water permeability), its advantage is in enhancement of membrane robustness (Pendergast & Hoek, 2011).

In the longer term, carbon-nanotechnology-based membrane is an area for research. An example is carbon nanotube (CNT) membranes (Majumder & Ajayan, 2010). Membranes based on CNT have the potential to overcome shortcomings of polymeric membranes due to its unique structure. The underlying mechanism is thought to be due to the molecular smoothness of nanotube walls, the structure that water molecules take within nanotubes, and the smooth energetic landscape inside nanotubes (Elimelech & Phillip, 2011).

Another recent development in carbon-nanotechnology-based membranes is nanoporous graphene. The concept is to introduce nanopores into one or several atomic layers of graphene with unsaturated carbon atoms at the pore edge passivated by chemical functional groups (Figure 10.9). Making use of graphene's inherent mechanical strength, this ultrathin membrane could greatly increase water permeability while rejecting salt and achieving desalination with low energy requirement (Cohen-Tanugi & Grossman, 2012).

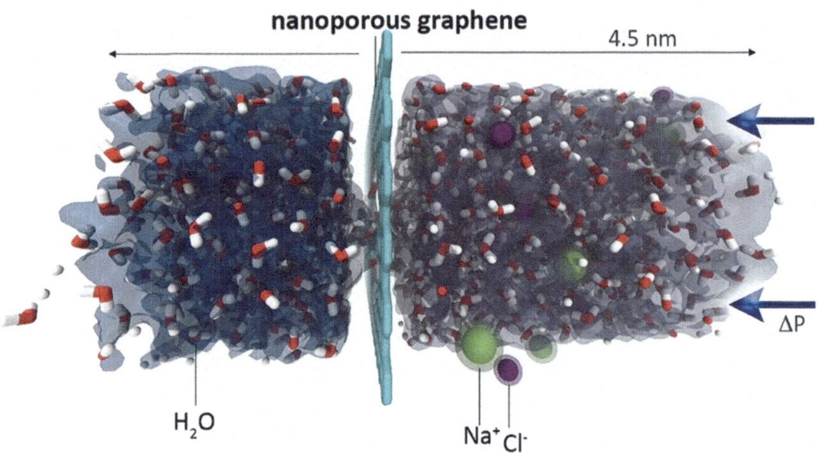

Figure 10.9 Water desalination across nanoporous membrane (courtesy of David Cohen-Tanugi).

However, in spite of the potential of nanotechnology-based membranes, reliable and cost-effective production of such membranes could be a challenge and there is still uncertainty in both the fate and stability of nanoparticles in the environment, and their possible toxicity effects on human beings (Hofman, 2011).

10.3.5.2 Closed circuit desalination

Closed circuit desalination (CCD) technology is a novel development of RO processes. The process works by re-circulating pressurised feed water until the desired recovery level is reached. Permeated water is replaced with fresh feed water without stopping operation. The low initial pressure of each CCD sequence means a lower average operating pressure, higher recovery and less pumping energy required compared to conventional RO systems. The potential of CCD technology for seawater desalination is that it may achieve an energy requirement that is lower than 1.5 kWh/m^3 at typical flux and recovery rates (Stover & Efraty, 2011).

10.3.5.3 Combined heating, cooling and desalination

The Combined Heating, Cooling and Desalination (CHCD) technology is a multi-purpose concept that seeks to desalinate salt water with little additional energy other than the amount used by a facility to provide for heating and cooling purposes. The method of desalination is freezing, such that fresh water can be mechanically separated in the form of ice from residual brine. The merit of the technology is not to reduce desalination energy requirement on the municipal scale, but rather it is to concomitantly fulfil the three functions of heating, cooling and water supply in a cost-effective manner, and improve system performance in localised smaller scale applications (PUB, 2012a).

10.3.5.4 Adsorption desalination

Adsorption desalination (AD) makes use of adsorbents such as silica gel (SiO2·nH2O) packed in an evaporator. When seawater is introduced to the system, water vapour is adsorbed onto the adsorbent, which leads to lower temperature in the liquid. At the same time, the temperature of the adsorbent bed increases; the heat is then channelled to the liquid so that its temperature can be maintained. The adsorbent after vapour adsorption is saturated with water. Waste heat from a nearby industrial source or renewable energy sources will then be needed to drive out water from the saturated adsorbent. The outputs of the process are chilling service and purified water. The technology has some potential and its electrical energy consumption for seawater desalination is reported to be around 1.38 kWh/m^3 (Ng *et al.* 2012). However, the system is technically complex, and as a heat-based technology, it has yet been proven advantageous compared to membrane distillation.

10.3.5.5 Gas hydrate desalination

Gas hydrate is a crystalline solid consisting of gas molecules, usually methane, each surrounded by a cage of water molecules. The gas hydrates are formed at seawater depths > 100 m and under low temperature and high pressure conditions. Gas hydrate may also form during offshore gas drilling, when water is condensed in the presence of methane at a high pressure and low temperature condition. The interest in gas hydrate is due to it being a potential resource for both energy and water. This technology is currently being researched and studies show that water could be extracted from gas hydrate with 72%–80% removal of the dissolved minerals (Park *et al.* 2011). However, the technology would be geographically limited to regions where gas hydrate can be found.

10.3.5.6 Deep sea desalination

The concept of deep sea desalination has been introduced since the 1960s (Drude, 1967). Recently, there has been an increasing interest in this concept as a low energy desalination option. The concept is to harness the natural hydrostatic pressure of deep seawater to perform desalination that is not limited by recovery, because feed water osmotic pressure remains constant throughout operation. The technology would also offer advantages in terms of the non-requirement of the following: intake facilities, pretreatment, pressure vessel, process chemical and issues related to brine disposal (Melin, 2011). The technology, however, could be limited by relatively low membrane flux, capital costs associated with piping of product water and maintenance issues. This technology would be geographically limited to regions where deep sea with depth > 350 m can be found.

10.4 OTHER ISSUES OF RELEVANCE

While the focus of this study is on improving desalination energy efficiency, there are other issues of relevance that should also be considered.

The type and quality of feed water has a significant impact on the type of desalination technology to be applied (UNESCO, 2008). Fouling is a central issue for all membrane processes (Belfort, 1984). In general, lower salinity seawater with lower fouling potential may be more suitable for membrane-based processes. For higher salinity seawater with higher fouling potential, thermal-based processes may be more suitable. The fouling propensity of the feed water will also determine the operating costs associated with chemicals and the extent of pretreatment required. Over prediction of the fouling propensity will result in unnecessary use of chemicals and as such increasing operating costs, or unnecessary pre-treatment steps and as such increasing the capital costs. On the other hand, under-prediction may result in operational problems with unexpected plant shut downs and cleaning, and as such, higher operating costs and loss of production. Measures to mitigate fouling as well as characterisation techniques for fouling are also significant areas that need consideration (Heitmann, 1990).

Although the separation step is the central component of a desalination process, the other components could also have significant impact on the overall desalination performance (Elimelech & Phillip, 2011). A systems engineering approach will be needed for optimisation of the entire desalination system comprising components such as intake, pre-treatment and post-treatment. For instance, the types of intake can be differentiated between open seawater intake, below-ground beach wells, infiltration galleries and cooling water discharge conduits of power plants. The different types of intake will have different impacts on a desalination process in terms of plant capital cost and pretreatment method, as well as impacts on marine environment and ecology (UNESCO, 2008).

The role of the pretreatment is to minimise fouling/ scaling potential of the incoming feed water, and its effectiveness is a prerequisite for sustainable operation of an entire desalination process. Conventional pretreatment methods typically involve flocculation, sedimentation, sand filtration and cartridge filtration.

Recently, there is increasing focus on applying porous membrane processes such as Microfiltration (MF) or Ultrafiltration (UF) to improve the effectiveness of the pretreatment step. On the other end of a desalination process is the post-treatment step. The post-treatment is required to deal with issues relevant to water distribution to the consumers. These issues include re-carbonation and re-mineralisation, disinfection, and boron removal (Fritzmann *et al.* 2007).

Environmental impact of desalination is another area that needs attention. The effect in the form of greenhouse gas (GHG) emissions is due to use of fossil fuels for power supply for desalination. Another effect is in the form of brine discharge from a desalination facility that needs to be managed. The brine, or concentrate waste stream, from a desalination facility consists of salts plus the chemicals added during the process. For typical seawater feed of 35,000 mg/L TDS, the TDS concentration of the brine from a desalination facility can be up to around 70,000 mg/L at 50% recovery. The concentrated brine has the potential to harm ecosystems, especially in regions with high evaporation rate that would already have elevated salinity. As the brine is denser than the surrounding sea water due to the higher TDS concentration, the brine could sink to the bed of the water body and damage the local ecosystems there. Brine management can be particularly important for inland desalination facilities, where the discharged brine can pose a threat to nearby water sources such as aquifers (AWWA, 2004).

10.5 LEVERAGING WATER-ENERGY-WASTE NEXUS FOR THE FUTURE OF DESALINATION

The foregoing discussion has alluded to the linkages between water, energy and waste in desalination. This could be understood in the sense that desalination requires energy to produce fresh water, and concomitantly, waste in the form of a brine stream. The brine stream, however, may also present as a resource, where minerals and energy could be tapped. The salinity gradient power (SGP) technologies, such as RED and PRO as discussed in section 6.3.2 and 6.3.3 respectively, provide a case in point of leveraging the water-energy-waste nexus for the future of desalination.

This strategy lies in the co-location of facilities. This concept is not new, and has been applied in the co-location of power plant and desalination facility (Heitmann, 1990). Such a system allows for capital savings in terms of sharing common seawater intake and outfall, and an increase in water production of the desalination facility due to the supply of warmed feed water from the cooling system of the power plant (Fritzmann *et al.* 2007).

The new thrust of the concept is that the water-energy-waste nexus could be expanded to exploit a range of synergies between water supply, power generation and waste incineration facilities (Seah, 2012). An illustration is provided in Figure 10.10.

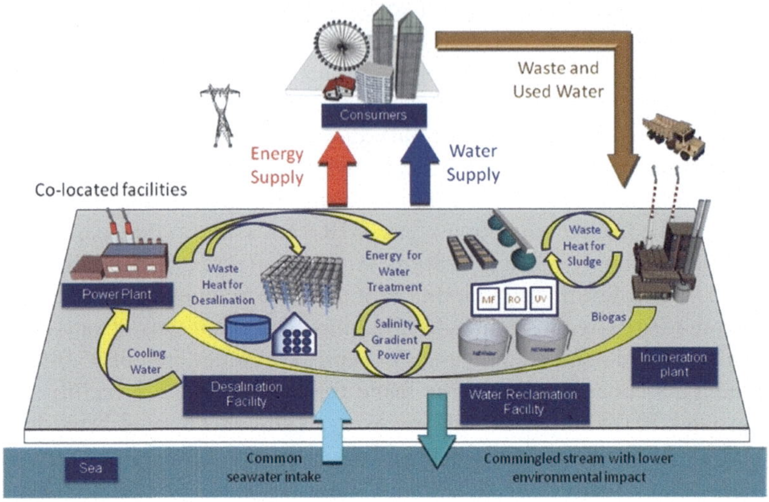

Figure 10.10 Co-location of facilities to tap into water-waste-energy nexus.

For this illustrated concept, the power plant provides energy for water supply. Waste heat from the power plant can also be used for desalination by using MD technology (Section 6.3.4). There can be a common seawater intake to provide cooling water for the power plant and feed water for the desalination facility. This would supply warmer feed water to the desalination facility and increase its productivity. A nearby water reclamation facility can treat collected used water to high-grade reclaimed water to augment water supply. Low salinity brine stream from the water reclamation facility could be synergistically coupled with high salinity brine stream from the desalination facility to harness osmotic energy to produce power by means of a SGP technology. The power generated by the SGP technology could then be used to partially offset the energy required to produce desalinated water with increased energy efficiency. Through this process, the concentration of the high salinity brine stream from the desalination facility gets diluted and the volume of the low salinity brine stream from the water reclamation facility gets reduced. Consequently, the environmental impact of the brine streams could be reduced when the co-mingled streams are discharged back into the sea.

From the treatment of used water, biogas produced from digestion of sludge possesses heat value that could be utilised to produce power. The digested sludge could be dried by waste heat from the incineration plant. Where feasible, CHCD technology could also be applied to serve a combined heating, cooling and desalination purpose by utilising waste heat (Section 6.3.5).

The co-location of facilities allows for a systems strategy to tap into the water-energy-waste nexus. Beside achieving higher energy efficiency and turning waste into resource, this strategy could also result in better land utilisation, more streamlined logistics, reduced manpower requirement and more favourable economics (Seah, 2012).

10.6 CONCLUDING REMARKS

The conventional SWRO is an established technology for desalination, and will likely continue to be the dominant technology for the near future. The current power consumption of SWRO is around 3–4 kWh/m^3, and there is still room to optimise the process and improve its energy efficiency incrementally. However, research and development of alternative desalination technologies will be needed to achieve breakthrough in desalination with substantially reduced power consumption and enhanced energy efficiency.

Biomimetic techniques show potential to transcend the limitation of conventional SWRO membrane and attain higher energy efficiency for desalination with targeted power consumption of around 1 kWh/m^3. Electrochemical desalination technologies also show potential for substantially reducing desalination energy consumption to below 1.5 kWh/m^3. Engineered osmosis is interesting, but its application potential for desalination is more relevant to exploiting salinity gradient power to enhance energy efficiency, and not treatment per se. Membrane distillation could potentially reduce electrical energy consumption to below 1 kWh/m^3, on the provision that a low-grade heat source is freely available. Other emerging desalination technologies have varying degrees of potential, and may have targeted applications.

Other issues of relevance for desalination that should also be considered include type and quality of feed water, fouling, intake, pretreatment, post-treatment, environmental issues such as greenhouse gas emission and brine management. For the future of desalination, the water-energy-waste nexus provides an opportunity to apply a systems strategy by co-location of facilities to meet multiple objectives that include achieving higher energy efficiency, turning waste into resource, better land utilisation, more streamlined logistics, reduced manpower requirement and more favourable economics.

10.7 ACKNOWLEDGEMENTS

The authors wish to thank various colleagues for their input to this study. They include Jodie Chin, Kelvin Koh, Kiran Kekre, Muhd Firdaus Hamzah, Qin Jianjun, Tao Guihe and Wang Yingjie. Daniel Ng's assistance in the technical drawings as part of his internship at PUB is also acknowledged and thanked.

10.8 REFERENCES

Achilli A. and Childress A. E. (2010). Pressure retarded osmosis: from the vision of Sidney Loeb to the first prototype installation – review. *Desalination*, **261**(3), 205–211.

Agre P. (2003). Aquaporin water channels. Nobel Lecture.

Anderson M. A., Cudero A. L. and Plamab J. (2010). Capacitive deionization as an electrochemical means of saving energy and delivering clean water. comparison to present desalination practices: will it compete? *Electrochimica Acta*, **55**, 3845–3856.

AWWA (2004). Water Desalting: Planning Guide for Water Utilities, John Wiley & Sons.

Belfort G. (1984). Synthetic membrane processes: fundamentals and water applications, Academic Press.

Birkett J. and Truby R. (2007). A figure of merit for appreciating improvements in RO membrane performance. IDA Newsletter March 2007.

Bowen W. R. (2006). Biomimetic separations – learning from the early development of biological membranes. *Desalination*, **199**(1–3), 225–227.

Brekke G. (2012). Experiences with the tofte prototype plant. The 3rd Osmosis Membrane Summit, Barcelona.

Cath T. Y., Childress A. E. and Elimelech M. (2006). Forward osmosis: principles, applications, and recent developments. *Journal of Membrane Science*, **281**(1–2), 70–87.

Chung T.-S., Li X., Ong R. C., Ge Q., Wang H. and Han G. (2012). Emerging forward osmosis (FO) technologies and challenges ahead for clean water and clean energy applications. *Current Opinion in Chemical Engineering*, **1**(3), 246–257.

Cohen-Tanugi D. and Grossman J. C. (2012). Water desalination across nanoporous graphene. *Nano Letters*, **12**(7), 3602–3608.

Daigger G. T. (2012). Technologies and practices to support the cities of the future. Keynote Presentation at the Water Convention, Singapore International Water Week, Singapore.

Dotremont C., K. B., Sih R., Lai K. C., Koh K. and Seah H. (2010). Seawater desalination with memstill technology – a sustainable solution for the industry. *Water Practice & Technology*, **5**(2). wpt.2010.026.

Drioli E. and Fontananova E. (2012). Membrane materials for addressing energy and environmental challenges. *Annual Review of Chemical and Biomolecular Engineering*, **3**(1), 395–420.

Drude B. C. (1967). Submarine units for reverse osmosis. *Desalination*, **2**(3), 325–328.

Elimelech M. (2007). Yale constructs forward osmosis desalination pilot plant. *Membrane Technology*, **2007**(1), 7–8.

Elimelech M. (2012). High performance thin-film composite membranes for pressure retarded osmosis. The 3rd Osmosis Membrane Summit, Barcelona.

Elimelech M. and Phillip W. A. (2011). The future of seawater desalination: energy, technology, and the environment. *Science*, **333**, 712–717.

Fane A. G. (2011). Membranes and the water cycle: challenges and opportunities. *Appl Water Sci*. **1**, 3–9.

Fritzmann C., Löwenberg J., Wintgens T. and Melin T. (2007). State-of-the-art of reverse osmosis desalination. *Desalination*, **216**(1–3), 1–76.

Geise G. M., Park H. B., Sagle A. C., Freeman B. D. and McGrath J. E. (2011). Water permeability and water/salt selectivity tradeoff in polymers for desalination. *Journal of Membrane Science*, **369**(1–2), 130–138.

Heitmann H.-G. (1990). Saline water processing. desalination and treatment of seawater, brackish water, and industrial waste water, VCH Publishers.

Hofman J. (2011). Membranes: the next generation … nanotechnology. 6th IWA Specialist Conference on Membrane Technology for Water & Wastewater Treatment (4–7 Oct 2011), Aachen.

IDA (2012). IDA Desalination Yearbook 2011–2012.

Jacangelo J. G., Oppenheimer J. A., Subramani A. and Badruzzaman M. (2011). Desalination of water supplies: strategies for energy optimization and renewable energy use. IDA World Congress, Perth.

Khayet M. (2011). Membranes and theoretical modeling of membrane distillation: a review. *Advances in Colloid and Interface Science*, **164**(1–2), 56–88.

Kumar M., Grzelakowski M., Zilles J., Clark M. and Meier W. (2007). Highly permeable polymeric membranes based on the incorporation of the functional water channel protein aquaporin Z. *Proc. Natl. Acad. Sci. USA*, **104**(52), 20719–20724.

Lange K. E. (2010). Get the salt out. *National Geographic Magazine.*

Lawson K. W. and Lloyd D. R. (1997). Membrane distillation. *Journal of Membrane Science*, **124**(1), 1–25.

Lee K. L., Baker R. W. and Lonsdale H. K. (1981). Membranes for power generation by pressure-retarded osmosis. *Journal of Membrane Science*, **8**(2), 141–171.

Lee K. P., Arnot T. C. and Mattia D. (2011). A review of reverse osmosis membrane materials for desalination—development to date and future potential. *Journal of Membrane Science*, **370**(1–2), 1–22.

Loeb S. (2002). Large-scale power production by pressure-retarded osmosis, using river water and sea water passing through spiral modules. *Desalination*, **143**(2), 115–122.

Majumder M. and Ajayan P. M. (2010). 1.14 – Carbon nanotube membranes: a new frontier in membrane science. Comprehensive Membrane Science and Engineering. D. Editor-in-Chief: Enrico and G. Lidietta. Oxford, Elsevier, 291–310.

McGinnis R. L. and Elimelech M. (2008). Global challenges in energy and water supply: the promise of engineered osmosis. *Environmental Science & Technology*, **42**(23), 8625–8629.

Melin T. (2011). Energy consumption, a critical success factor in membrane filtration. 6th IWA Specialist Conference on Membrane Technology for Water & Wastewater Treatment (4–7 Oct 2011), Aachen.

Ng K. C., Thu K., Kim Y., Chakraborty A. and Amy G. (2013). Adsorption desalination: an emerging low-cost thermal desalination method. *Desalination*, **308**: 161–179

Nicoll P. G., Thompson N. A. and Bedford M. R. (2011). Manipulated osmosis applied to evaporative cooling make-up water – revolutionary technology. IDA World Congress, Perth.

Nicoll P. G. (2013).Forward Osmosis as a Pre-Treatment to Reverse Osmosis. IDA World Congress, Tianjin.

Nielsen C. (2009). Biomimetic membranes for sensor and separation applications. *Analytical and Bioanalytical Chemistry*, **395**(3), 697–718.

Oren Y. (2008). Capacitive deionization (CDI) for desalination and water treatment — past, present and future (a review). *Desalination*, **228**(1–3), 10–29.

Park K.-n., Hong S. Y., Lee J. W., Kang K. C., Lee Y. C., Ha M.-G. and Lee J. D. (2011). A new apparatus for seawater desalination by gas hydrate process and removal characteristics of dissolved minerals (Na+ , Mg2+ , Ca2+ , K+ , B3+). *Desalination*, **274**(1–3), 91–96.

Peñate B. and García-Rodríguez L. (2011). Energy optimisation of existing SWRO (seawater reverse osmosis) plants with ERT (energy recovery turbines), technical and thermoeconomic assessment. *Energy*, **36**: 613–626.

Pendergast M. M. and Hoek E. M. V. (2011). A review of water treatment membrane nanotechnologies. *Energy & Environmental Science*, **4**(6), 1946–1971.

Phuntsho S., Shon H. K., Majeed T., El Saliby I., Vigneswaran S., Kandasamy J., Hong S. and Lee S. (2012). Blended fertilizers as draw solutions for fertilizer-drawn forward osmosis desalination. *Environmental Science & Technology*, **46**(8), 4567–4575.

Post J. W., Goeting C. H., Valk J., Goinga S., Veerman J., Hamelers H. V. M. and Hack P. J. F. M. (2010). Towards implementation of reverse electrodialysis for power generation from salinity gradients. *Desalination and Water Treatment*, **16**(1–3), 182–193.

PUB (2012a). Innovation in Water I Singapore (2nd Issue).

PUB (2012b). Innovation in Water I Singapore (3rd Issue).

Qin J.-J., Lay W. C. L. and Kekre K. A. (2012). Recent developments and future challenges of forward osmosis for desalination: a review. *Desalination and Water Treatment*, **39**: 123–136.

Ramon G. Z., Feinberg B. J. and Hoek E. M. V. (2011). Membrane-based production of salinity-gradient power. *Energy & Environmental Science*, **4**(11), 4423–4434.

Seah H. (2012). Sustainable Water Supply – Water, Energy and Waste Nexus. Keynote Presentation at the UPW-Asia conference, Singapore International Water Week.

Semiat R. (2008). Energy issues in desalination processes. *Environmental Science & Technology*, **42**(22), 8193–8201.

Shaw M., LeTournea T., Liang L.-S. and Ng K. H. (2011). Low energy desalination: siemens' solution to a global challenge. IDA World Congress, Perth.

Stover R. L. and Andrews B. (2011). Isobaric energy recovery devices – past, present and future. IDA World Congress, Perth.

Stover R. L. and Efraty N. (2011). Record low energy consumption with closed circuit desalination. IDA Congress (4–9 Sep 2011), Perth.

Subramani A., Badruzzaman M., Oppenheimer J. and Jacangelo J. G. (2011). Energy minimization strategies and renewable energy utilization for desalination: a review. *Water Research*, **45**(5), 1907–1920.

Tang C. Y., Zhao Y., Wang R., Hélix-Nielsen C. and Fane A. G. (2012). Desalination by biomimetic aquaporin membranes: review of status and prospects. *Desalination*.

Tun C. M., Fane A. G., Matheickal J. T. and Sheikholeslami R. (2005). Membrane distillation crystallization of concentrated salts—flux and crystal formation. *Journal of Membrane Science*, **257**(1–2), 144–155.

UNESCO (2008). UNESCO Centre for Membrane Science and Technology: Emerging Trends in Desalination: A Review. Waterlines Report Series, Australian Government, National Water Commission.

Veerapaneni S., Wang S. and Bond R. (2011). Emerging desalination technologies – an overview. IDA World Congress, Perth.

Wilf M. and Bartels C. (2005). Optimization of seawater RO systems design. *Desalination*, **173**(1), 1–12.

Yip N. Y. and Elimelech M. (2012). Thermodynamic and energy efficiency analysis of power generation from natural salinity gradients by pressure retarded osmosis. *Environmental Science & Technology*, **46**(9), 5230–5239.

IWA Publishing's authorised EU representative for General Product Safety Regulations is Diane D'Arras, 15 rue Duret, 75116 Paris, France, e-mail: safety@iwap.co.uk.

Printed and bound by CPI Group (UK) Ltd, Croydon, CR0 4YY

12/05/2026

02108869-0003